Lecture Notes in Computer Science   14574

Advanced Research in Computing and Software Science
Subline of Lecture Notes in Computer Science

More information about this series at https://link.springer.com/bookseries/558

Naoki Kobayashi · James Worrell
Editors

Foundations
of Software Science and
Computation Structures

27th International Conference, FoSSaCS 2024
Held as Part of the European Joint Conferences
on Theory and Practice of Software, ETAPS 2024
Luxembourg City, Luxembourg, April 6–11, 2024
Proceedings, Part I

 Springer

Editors
Naoki Kobayashi
The University of Tokyo
Tokyo, Japan

James Worrell
University of Oxford
Oxford, UK

ISSN 0302-9743   ISSN 1611-3349 (electronic)
Lecture Notes in Computer Science
ISBN 978-3-031-57227-2   ISBN 978-3-031-57228-9 (eBook)
https://doi.org/10.1007/978-3-031-57228-9

This Springer imprint is published by the registered company Springer Nature Switzerland AG
The registered company address is: Gewerbestrasse 11, 6330 Cham, Switzerland

Paper in this product is recyclable.

ETAPS Foreword

Welcome to the 27th ETAPS! ETAPS 2024 took place in Luxembourg City, the beautiful capital of Luxembourg.

ETAPS 2024 is the 27th instance of the European Joint Conferences on Theory and Practice of Software. ETAPS is an annual federated conference established in 1998, and consists of four conferences: ESOP, FASE, FoSSaCS, and TACAS. Each conference has its own Program Committee (PC) and its own Steering Committee (SC). The conferences cover various aspects of software systems, ranging from theoretical computer science to foundations of programming languages, analysis tools, and formal approaches to software engineering. Organising these conferences in a coherent, highly synchronized conference programme enables researchers to participate in an exciting event, having the possibility to meet many colleagues working in different directions in the field, and to easily attend talks of different conferences. On the weekend before the main conference, numerous satellite workshops took place that attracted many researchers from all over the globe.

ETAPS 2024 received 352 submissions in total, 117 of which were accepted, yielding an overall acceptance rate of 33%. I thank all the authors for their interest in ETAPS, all the reviewers for their reviewing efforts, the PC members for their contributions, and in particular the PC (co-)chairs for their hard work in running this entire intensive process. Last but not least, my congratulations to all authors of the accepted papers!

ETAPS 2024 featured the unifying invited speakers Sandrine Blazy (University of Rennes, France) and Lars Birkedal (Aarhus University, Denmark), and the invited speakers Ruzica Piskac (Yale University, USA) for TACAS and Jérôme Leroux (Laboratoire Bordelais de Recherche en Informatique, France) for FoSSaCS. Invited tutorials were provided by Tamar Sharon (Radboud University, the Netherlands) on computer ethics and David Monniaux (Verimag, France) on abstract interpretation.

As part of the programme we had the first ETAPS industry day. The goal of this day was to bring industrial practitioners into the heart of the research community and to catalyze the interaction between industry and academia. The day was organized by Nikolai Kosmatov (Thales Research and Technology, France) and Andrzej Wąsowski (IT University of Copenhagen, Denmark).

ETAPS 2024 was organized by the SnT - Interdisciplinary Centre for Security, Reliability and Trust, University of Luxembourg. The University of Luxembourg was founded in 2003. The university is one of the best and most international young universities with 6,000 students from 130 countries and 1,500 academics from all over the globe. The local organisation team consisted of Peter Y.A. Ryan (general chair), Peter B. Roenne (organisation chair), Maxime Cordy and Renzo Gaston Degiovanni (workshop chairs), Magali Martin and Isana Nascimento (event manager), Marjan Skrobot (publicity chair), and Afonso Arriaga (local proceedings chair). This team also organised the online edition of ETAPS 2021, and now we are happy that they agreed to also organise a physical edition of ETAPS.

ETAPS 2024 is further supported by the following associations and societies: ETAPS e.V., EATCS (European Association for Theoretical Computer Science), EAPLS (European Association for Programming Languages and Systems), and EASST (European Association of Software Science and Technology).

The ETAPS Steering Committee consists of an Executive Board, and representatives of the individual ETAPS conferences, as well as representatives of EATCS, EAPLS, and EASST. The Executive Board consists of Marieke Huisman (Twente, chair), Andrzej Wąsowski (Copenhagen), Thomas Noll (Aachen), Jan Kofroň (Prague), Barbara König (Duisburg), Arnd Hartmanns (Twente), Caterina Urban (Inria), Jan Křetínský (Munich), Elizabeth Polgreen (Edinburgh), and Lenore Zuck (Chicago).

Other members of the steering committee are: Maurice ter Beek (Pisa), Dirk Beyer (Munich), Artur Boronat (Leicester), Luís Caires (Lisboa), Ana Cavalcanti (York), Ferruccio Damiani (Torino), Bernd Finkbeiner (Saarland), Gordon Fraser (Passau), Arie Gurfinkel (Waterloo), Reiner Hähnle (Darmstadt), Reiko Heckel (Leicester), Marijn Heule (Pittsburgh), Joost-Pieter Katoen (Aachen and Twente), Delia Kesner (Paris), Naoki Kobayashi (Tokyo), Fabrice Kordon (Paris), Laura Kovács (Vienna), Mark Lawford (Hamilton), Tiziana Margaria (Limerick), Claudio Menghi (Hamilton and Bergamo), Andrzej Murawski (Oxford), Laure Petrucci (Paris), Peter Y.A. Ryan (Luxembourg), Don Sannella (Edinburgh), Viktor Vafeiadis (Kaiserslautern), Stephanie Weirich (Pennsylvania), Anton Wijs (Eindhoven), and James Worrell (Oxford).

I would like to take this opportunity to thank all authors, keynote speakers, attendees, organizers of the satellite workshops, and Springer Nature for their support. ETAPS 2024 was also generously supported by a RESCOM grant from the Luxembourg National Research Foundation (project 18015543). I hope you all enjoyed ETAPS 2024.

Finally, a big thanks to both Peters, Magali and Isana and their local organization team for all their enormous efforts to make ETAPS a fantastic event.

April 2024   Marieke Huisman
                                                    ETAPS SC Chair
                                              ETAPS e.V. President

Preface

This volume contains the papers presented at the 27th International Conference on Foundations of Software Science and Computation Structures (FoSSaCS 2024), which was held during April 8–11, 2024 in Luxembourg City, Luxembourg. The conference is dedicated to foundational research with a clear significance for software science and brings together research on theories and methods to support the analysis, integration, synthesis, transformation, and verification of programs and software systems.

In addition to an invited talk by Jérôme Leroux (Laboratoire Bordelais de Recherche en Informatique, France) on "Ackermannian Completion of Separators", the program consisted of 24 talks on contributed papers, selected from 79 submissions. Each submission was assessed by three or more Program Committee members, with the help of external reviewers. The conference management system EasyChair was used to handle the submissions, to conduct the electronic Program Committee discussions, and to assist with the assembly of the proceedings.

We wish to thank all the authors who submitted papers for consideration, the members of the Program Committee for their conscientious work, and all additional reviewers who assisted the Program Committee in the evaluation process. We would also like to thank Andrzej Murawski, the FoSSaCS Steering Committee Chair for various pieces of advice, and the members of the ESOP/FASE/FoSSaCS joint Artifact Evaluation Committee for the artifact evaluation. Finally, we would like to thank the ETAPS organization for providing an excellent environment for FoSSaCS, the other conferences and the workshops.

February 2024

Naoki Kobayashi
James Worrell

Organization

Program Committee Chairs

Naoki Kobayashi   The University of Tokyo, Japan
James Worrell   University of Oxford, UK

Program Committee

Sandra Alves	University of Porto, Portugal
Mauricio Ayala-Rincón	Universidade de Brasília, Brazil
Stephanie Balzer	CMU, USA
Udi Boker	Reichman University, Israel
James Brotherston	University College London, UK
Corina Cirstea	University of Southampton, UK
Yuxin Deng	East China Normal University, China
Claudia Faggian	CNRS - Université Paris Cité, France
Pierre Ganty	IMDEA Software Institute, Spain
Ichiro Hasuo	National Institute of Informatics, Japan
Naoki Kobayashi	The University of Tokyo, Japan
Robbert Krebbers	Radboud University, the Netherlands
Antonin Kucera	Masaryk University, the Czech Republic
Karoliina Lehtinen	CNRS - Université Aix-Marseille, France
Bas Luttik	Eindhoven University of Technology, the Netherlands
Rasmus Ejlers Møgelberg	IT University of Copenhagen, Denmark
Luca Padovani	Università di Camerino, Italy
Catuscia Palamidessi	Inria, France
Paritosh Pandya	IIT Bombay, India
Elaine Pimentel	University College London, UK
Damien Pous	CNRS - ENS Lyon, France
Ana Sokolova	University of Salzburg, Austria
Lidia Tendera	University of Opole, Poland
Nikos Tzevelekos	Queen Mary University of London, UK
Tarmo Uustalu	Reykjavik University, Iceland
Franck van Breugel	York University, Canada
James Worrell	University of Oxford, UK

ESOP/FASE/FoSSaCS Joint Artifact Evaluation Committee

AEC Co-chairs

Tobias Kappé	Open Universiteit and ILLC, University of Amsterdam, The Netherlands
Ryosuke Sato	University of Tokyo, Japan
Stefan Winter	LMU Munich, Germany

AEC Members

Arwa Hameed Alsubhi	University of Glasgow, UK
Levente Bajczi	Budapest University of Technology and Economics, Hungary
James Baxter	University of York, UK
Matthew Alan Le Brun	University of Glasgow, UK
Laura Bussi	University of Pisa, Italy
Gustavo Carvalho	Universidade Federal de Pernambuco, Brazil
Chanhee Cho	Carnegie Mellon University, USA
Ryan Doenges	Northeastern University, USA
Zainab Fatmi	University of Oxford, UK
Luke Geeson	University College London, UK
Hans-Dieter Hiep	Leiden University, Belgium
Philipp Joram	Tallinn University of Technology, Estonia
Ulf Kargén	Linköping University, Sweden
Hiroyuki Katsura	University of Tokyo, Japan
Calvin Santiago Lee	Reykjavík University, Iceland
Livia Lestingi	Politecnico di Milano, Italy
Nuno Macedo	University of Porto and INESC TEC, Portugal
Kristóf Marussy	Budapest University of Technology and Economics, Hungary
Ivan Nikitin	University of Glasgow, UK
Hugo Pacheco	University of Porto, Portugal
Lucas Sakizloglou	Brandenburgische Technische Universität Cottbus-Senftenberg, Germany
Michael Schröder	TU Wien, Austria
Michael Schwarz	TU Munich, Germany
Wenjia Ye	University of Hong Kong, China

Additional Reviewers

Abraham, Erika
Ajdarow, Michal
An, Jie
Asada, Kazuyuki
Avanzini, Martin
Balasubramanian, A. R.
Barbosa, João
Basold, Henning
Batz, Kevin
Beohar, Harsh
Bertrand, Nathalie
Beyersdorff, Olaf
Bohn, León
Bonelli, Eduardo
Bonsangue, Marcello
Breuvart, Flavien
Bruyère, Véronique
Carette, Titouan
Chadha, Rohit
Clemente, Lorenzo
Cockett, Robin
Czerwiński, Wojciech
D'Osualdo, Emanuele
Dagnino, Francesco
De Moura, Flavio L. C.
De, Abhishek
Di Stasio, Antonio
Espírito Santo, José
Fahrenberg, Uli
Feng, Yuan
Fijalkow, Nathanaël
Filiot, Emmanuel
Fokkink, Wan
Frumin, Daniil
Galal, Zeinab
Geatti, Luca
Geuvers, Herman
van Glabbeek, Rob
van Gool, Sam
Goy, Alexandre
Guha, Shibashis
Guttenberg, Roland
Hague, Matthew

Hainry, Emmanuel
Harper, Robert
Hausmann, Daniel
Hedges, Jules
Hinrichsen, Jonas Kastberg
Ho, Hsi-Ming
Jaber, Guilhem
Jafarrahmani, Farzad
Jakl, Tomas
Jancar, Petr
Kanazawa, Makoto
Kaposi, Ambrus
Katsumata, Shin-Ya
Kavvos, Alex
Keiren, Jeroen J. A.
Kelmendi, Edon
Klaška, David
Klock Ii, Felix S.
Knight, Sophia
Koutavas, Vasileios
Krivine, Jean
König, Barbara
Laurent, Olivier
Leroux, Jérôme
Lhote, Nathan
Li, Yong
Long, Huan
Lopez, Aliaume
Loreti, Michele
Maarand, Hendrik
Madnani, Khushraj
Mallik, Kaushik
Martens, Jan
Marti, Johannes
Mascle, Corto
Mazzocchi, Nicolas
McDermott, Dylan
Melliès, Paul-André
Mery, Daniel
Michaliszyn, Jakub
Michielini, Vincent
Miculan, Marino
Moot, Richard

Morawska, Barbara
Mulder, Ike
Nguyễn, Lê Thành Dũng
Novotný, Petr
Paquet, Hugo
Piedeleu, Robin
Pinto, Luís
Proença, José
Pérez, Jorge A.
Rehak, Vojtech
Riba, Colin
Rivas, Exequiel
Rogalewicz, Adam
Rot, Jurriaan
Rowe, Reuben
Sakayori, Ken
Sarkis, Ralph
Schmid, Todd
Schmitz, Sylvain
Schröder, Lutz
Sin'Ya, Ryoma

Skrzypczak, Michał
Sobociński, Paweł
Staton, Sam
Stein, Dario
Takagi, Tsubasa
Tini, Simone
Totzke, Patrick
Urbat, Henning
Valencia, Frank
Vandenhove, Pierre
Varacca, Daniele
Veltri, Niccolò
Ventura, Daniel
Waga, Masaki
Wagemaker, Jana
Wan, Cheng-Syuan
Weil-Kennedy, Chana
Winskel, Glynn
Witkowski, Piotr
Wißmann, Thorsten
Wolter, Frank

Contents – Part I

Invited Talk

Ackermannian Completion of Separators 3
  Jérôme Leroux

Infinite Games

Fair ω-Regular Games. 13
  Daniel Hausmann, Nir Piterman, Irmak Sağlam,
  and Anne-Kathrin Schmuck

Stochastic Window Mean-Payoff Games 34
  Laurent Doyen, Pranshu Gaba, and Shibashis Guha

Symbolic Solution of Emerson-Lei Games for Reactive Synthesis 55
  Daniel Hausmann, Mathieu Lehaut, and Nir Piterman

Parity Games on Temporal Graphs 79
  Pete Austin, Sougata Bose, and Patrick Totzke

Categorical Semantics

Drawing from an Urn is Isometric. 101
  Bart Jacobs

Enriching Diagrams with Algebraic Operations. 121
  Alejandro Villoria, Henning Basold, and Alfons Laarman

Monoidal Extended Stone Duality. 144
  Fabian Birkmann, Henning Urbat, and Stefan Milius

Towards a Compositional Framework for Convex Analysis
(with Applications to Probability Theory). 166
  Dario Stein and Richard Samuelson

Automata and Synthesis

Determinization of Integral Discounted-Sum Automata is Decidable 191
  Shaull Almagor and Neta Dafni

Checking History-Determinism is NP-hard for Parity Automata 212
    Aditya Prakash

Tighter Construction of Tight Büchi Automata 234
    Marek Jankola and Jan Strejček

Synthesis with Privacy Against an Observer 256
    Orna Kupferman, Ofer Leshkowitz, and Naama Shamash Halevy

Author Index 279

Contents – Part II

Types and Programming Languages

From Rewrite Rules to Axioms in the $\lambda\Pi$-Calculus Modulo Theory   3
  Valentin Blot, Gilles Dowek, Thomas Traversié,
  and Théo Winterhalter

Light Genericity 24
  Beniamino Accattoli and Adrienne Lancelot

Logical Predicates in Higher-Order Mathematical Operational Semantics 47
  Sergey Goncharov, Alessio Santamaria, Lutz Schröder, Stelios Tsampas,
  and Henning Urbat

On Basic Feasible Functionals and the Interpretation Method 70
  Patrick Baillot, Ugo Dal Lago, Cynthia Kop, and Deivid Vale

Logic and Proofs

Succinctness of Cosafety Fragments of LTL via Combinatorial
Proof Systems 95
  Luca Geatti, Alessio Mansutti, and Angelo Montanari

A Resolution-Based Interactive Proof System for UNSAT 116
  Philipp Czerner, Javier Esparza, and Valentin Krasotin

Craig Interpolation for Decidable First-Order Fragments 137
  Balder ten Cate and Jesse Comer

Clones, closed categories, and combinatory logic 160
  Philip Saville

Infinite-State Systems

Reachability in Fixed VASS: Expressiveness and Lower Bounds 185
  Andrei Draghici, Christoph Haase, and Andrew Ryzhikov

From Innermost to Full Almost-Sure Termination of Probabilistic
Term Rewriting 206
  Jan-Christoph Kassing, Florian Frohn, and Jürgen Giesl

Dimension-Minimality and Primality of Counter Nets 229
    Shaull Almagor, Guy Avni, Henry Sinclair-Banks, and Asaf Yeshurun

Parameterized Broadcast Networks with Registers: from NP to the Frontiers
of Decidability 250
    Lucie Guillou, Corto Mascle, and Nicolas Waldburger

Author Index 271

Invited Talk

Ackermannian Completion of Separators

Jérôme Leroux[(✉)] [ID]

Univ. Bordeaux, CNRS, Bordeaux INP, LaBRI, UMR 5800, F-33400 Talence, France
jerome.leroux@labri.fr

Abstract. Vector addition systems (VAS for short), or equivalently vector addition systems with states, or Petri nets are a long established model of concurrency with extensive applications in modeling and analysis of hardware, software and database systems, as well as chemical, biological and business processes. The central algorithmic problem is reachability: whether from a given initial configuration there exists a sequence of valid execution steps that reaches a given final configuration. The complexity of the problem has remained unsettled since the 1960s, and was recently proved to be Ackermannian-complete.

In 2009, we proved that the reachability problem can be decided with a simple algorithm by observing that negative instances of the reachability problem can be witnessed by partitioning the set configurations into semilinear sets called *complete separators*. Since we can decide in elementary time if a pair of semilinear sets denotes a complete separator, the size of such a witness is Ackermannian in the worst case.

In this paper, we show how recent results about the reachability problem can be combined to derive a matching upper-bound, i.e. for every negative instance of the reachability problem, we can effectively compute in Ackermannian time a complete separator witnessing that property.

1 Introduction

Vector addition systems [8] (VAS for short), or equivalently vector addition systems with states [7], or Petri nets are one of the most popular formal methods for the representation and the analysis of parallel processes [3]. The central algorithmic problem is reachability: whether from a given initial configuration there exists a sequence of valid execution steps that reaches a given final configuration. Many important computational problems in logic and complexity reduce or are even equivalent to this problem [22,6].

After an incomplete proof by Sacerdote and Tenney [20], decidability of the problem was established by Mayr [17,19], whose proof was then simplified by Kosaraju [9]. Building on the further refinements made by Lambert in the 1990s [10], in 2015, a first complexity upper-bound of the reachability problem was provided [12] more than thirty years after the presentation of the algorithm introduced by Mayr [9,10]. The upper-bound given in that paper is "cubic Ackermannian", i.e. in F_{ω^3} (see [21]). This complexity bound was obtained by analyzing the Mayr algorithm. With a refined algorithm and a new ranking

N. Kobayashi and J. Worrell (Eds.): FoSSaCS 2024, LNCS 14574, pp. 3–10, 2024.
https://doi.org/10.1007/978-3-031-57228-9_1

function for proving termination, an Ackermannian complexity upper-bound was obtained in [15]. This means that the reachability problem can be solved in time bounded by $F_\omega(p(n))$ where p is a primitive recursive function and where F_ω is an Ackermann function. Very recently, this complexity bound was proved to be optimal [14,2].

While the complexity of the reachability problem is settled, its parameterized version, in fixed dimension d, is still open with a large complexity gap between the lower-bound and the upper-bound. Some recent results provided ways to decrease that gap (see for instance [1,11]) but the problem remains open. Since there exists d-dimensional VAS with finite but very large reachability sets [18], any reachability algorithm directly based on the Mayr algorithm will necessarily fail in providing a better complexity upper-bound. In fact that algorithm enumerates in some way each possible reachable configurations when the reachability set is finite.

There is another algorithm for deciding the reachability problem independent of the Mayr algorithm. In fact, in [13], we introduced a simple enumerating algorithm for deciding the reachability problem by observing that negative instances of the reachability problem can be witnessed by partitioning the set of configurations into semilinear sets called *complete separators*. Since we can decide in elementary time if a pair of semilinear sets denotes a complete separator, and the reachability problem is Ackermannian-hard, the size of such a witness is necessarily Ackermannian in the worst case.

In this paper, we take the opportunity to show how to combine papers [15] and [13] to prove that from any negative instance of the reachability problem, we can effectively compute in Ackermannian time a complete separator witnessing that property. This result prove the optimality of algorithms based on complete separators for deciding the general reachability problem. Since this paper is an invited paper at FOSSACS'24, so without any reviewing process, no new proof are given in this paper. If a proof is given, it just to be self-content. But in any case, those proofs are copy-past from [15] and [13].

Even if our result does not provide a better understanding of the complexity of the parameterized reachability problem, it shows that algorithms based on complete separators are optimal in general dimension.

2   Basic Notions

In this section, we introduce basic notions and notation.

Notation for Vectors of Integers. By $\mathbb{Z}$ we denote the set of integers, and by $\mathbb{N}$ the set $\{0, 1, 2, \dots\}$ of non-negative integers. Given $d \in \mathbb{N}$, the elements of $\mathbb{Z}^d$ are called (d-dim) *vectors*; they are denoted in bold face, and for $\mathbf{x} \in \mathbb{Z}^d$ we put $\mathbf{x} = (\mathbf{x}(1), \dots, \mathbf{x}(d))$ so that we can refer to the vector components. In this context, d is called the *dimension* of $\mathbf{x}$. We use the component-wise sum

$\mathbf{x} + \mathbf{y}$ of vectors, and their component-wise order $\mathbf{x} \leq \mathbf{y}$. For $c \in \mathbb{N}$, we put $c \cdot \mathbf{x} = (c \cdot \mathbf{x}(1), \ldots, c \cdot \mathbf{x}(d))$.

Linear and Semilinear Sets. A *set* $\mathbf{L} \subseteq \mathbb{N}^d$ is *linear* if there are d-dim vectors $\mathbf{b}$, the *basis*, and $\mathbf{p}_1, \ldots, \mathbf{p}_k$, the *periods* (for $k \in \mathbb{N}$), such that $\mathbf{L} = \{\mathbf{x} \in \mathbb{N}^d \mid \mathbf{x} = \mathbf{b} + \mathbf{u}(1) \cdot \mathbf{p}_1 + \cdots + \mathbf{u}(k) \cdot \mathbf{p}_k$ for some $\mathbf{u} \in \mathbb{N}^k\}$. In this case, by a *presentation of* $\mathbf{L}$ we mean the tuple $(\mathbf{b}, \mathbf{p}_1, \ldots, \mathbf{p}_k)$.

A *set* $\mathbf{S} \subseteq \mathbb{N}^d$ is *semilinear* if it is a finite union of linear sets, i.e. $\mathbf{S} = \mathbf{L}_1 \cup \cdots \cup \mathbf{L}_k$ where $\mathbf{L}_j$ are linear sets for all j. In this case, by a *presentation of* $\mathbf{S}$ we mean the sequence of presentations of $\mathbf{L}_1, \ldots, \mathbf{L}_k$. When we say that a *semilinear set* $\mathbf{S}$ is given, we mean that we are given a presentation of $\mathbf{S}$; when we say that $\mathbf{S}$ is *effectively constructible* in some context, we mean that there is an algorithm computing its presentation (in the respective context).

We recall that a set $\mathbf{S} \subseteq \mathbb{N}^d$ is semilinear if, and only if, it is expressible in Presburger arithmetic [4]; the respective transformations between presentations and formulas are effective and elementary. Hence if $\mathbf{S} \subseteq \mathbb{N}^d$ is semilinear, then also its complement, denoted as $\overline{\mathbf{S}}$, is semilinear, and $\overline{\mathbf{S}}$ is effectively constructible when (a presentation of) $\mathbf{S}$ is given.

Fast Growing Functions. The Grzegorczyk hierarchy [5,16] is defined thanks to a family $(F_d)_{d \in \mathbb{N}}$ of functions $F_d : \mathbb{N} \to \mathbb{N}$ such that every primitive recursive function is asymptotically bounded by some function F_d. This family is defined by $F_0(n) \stackrel{\text{def}}{=} n + 1$ and inductively by $F_{d+1}(n) \stackrel{\text{def}}{=} F_d^{n+1}(n)$ for every $n, d \in \mathbb{N}$. Observe that $F_1(n) = 2n + 1$, $F_2(n) = 2^{n+1}(n+1) - 1$, and $F_3(n)$ grows as a tower of n exponentials. It follows that F_3 is a non elementary function since it eventually exceeds any fixed iteration of the exponential function. An *Ackermannian function*, denoted as F_ω is defined thanks to the diagonal extraction $F_\omega(n) \stackrel{\text{def}}{=} F_{n+1}(n)$ for every $n \in \mathbb{N}$. This function is non primitive recursive.

Vector Addition Systems. A (d-dim) *vector addition system* (*VAS* for short) is a finite set $\mathbf{A}$ of vectors in $\mathbb{Z}^d$ called *actions*. Vectors $\mathbf{x} \in \mathbb{N}^d$ are called *configurations*, and with an action $\mathbf{a}$ we associate the binary relation $\stackrel{\mathbf{a}}{\to}$ on the configurations in $\mathbb{N}^d$ by putting $\mathbf{x} \stackrel{\mathbf{a}}{\to} \mathbf{y}$ for all $\mathbf{x}, \mathbf{y} \in \mathbb{N}^d$ such that $\mathbf{y} - \mathbf{x} = \mathbf{a}$. The relations $\stackrel{\mathbf{a}}{\to}$ are naturally extended to the relations $\stackrel{\sigma}{\to}$ for finite sequences $\sigma = \mathbf{a}_1 \ldots \mathbf{a}_k$ of actions by $\mathbf{x} \stackrel{\sigma}{\to} \mathbf{y}$ if $\mathbf{x} \stackrel{\mathbf{a}_1}{\to} \cdots \stackrel{\mathbf{a}_k}{\to} \mathbf{y}$ for all $\mathbf{x}, \mathbf{y} \in \mathbb{N}^d$.

On the set $\mathbb{N}^d$ of configurations we define the *reachability relation* $\stackrel{\mathbf{A}^*}{\longrightarrow}$: we put $\mathbf{x} \stackrel{\mathbf{A}^*}{\longrightarrow} \mathbf{y}$ if there is $\sigma \in \mathbf{A}^*$ such that $\mathbf{x} \stackrel{\sigma}{\to} \mathbf{y}$. For $\mathbf{x} \in \mathbb{N}^d$ and $\mathbf{X} \subseteq \mathbb{N}^d$ we put $\text{POST}_{\mathbf{A}}^*(\mathbf{x}) \stackrel{\text{def}}{=} \{\mathbf{y} \in \mathbb{N}^d \mid \mathbf{x} \stackrel{\mathbf{A}^*}{\longrightarrow} \mathbf{y}\}$, and $\text{POST}_{\mathbf{A}}^*(\mathbf{X}) \stackrel{\text{def}}{=} \bigcup_{\mathbf{x} \in \mathbf{X}} \text{POST}_{\mathbf{A}}^*(\mathbf{x})$. Symmetrically, for $\mathbf{y} \in \mathbb{N}^d$ and $\mathbf{Y} \subseteq \mathbb{N}^d$ we put $\text{PRE}_{\mathbf{A}}^*(\mathbf{y}) \stackrel{\text{def}}{=} \{\mathbf{x} \in \mathbb{N}^d \mid \mathbf{x} \stackrel{\mathbf{A}^*}{\longrightarrow} \mathbf{y}\}$ and $\text{PRE}_{\mathbf{A}}^*(\mathbf{Y}) \stackrel{\text{def}}{=} \bigcup_{\mathbf{y} \in \mathbf{Y}} \text{PRE}_{\mathbf{A}}^*(\mathbf{y})$. By $\mathbf{X} \stackrel{\mathbf{A}^*}{\longrightarrow} \mathbf{Y}$ we denote that $\mathbf{x} \stackrel{\mathbf{A}^*}{\longrightarrow} \mathbf{y}$ for some $\mathbf{x} \in \mathbf{X}$ and $\mathbf{y} \in \mathbf{Y}$.

The semilinear *reachability problem* takes as input a triple $(\mathbf{X}, \mathbf{A}, \mathbf{Y})$ where $\mathbf{X}, \mathbf{Y}$ are (presentations of) semi-linear sets of configurations of a VAS $\mathbf{A}$, and checks if $\mathbf{X} \stackrel{\mathbf{A}^*}{\longrightarrow} \mathbf{Y}$ hold. In the standard definition of the reachability problem the sets $\mathbf{X}, \mathbf{Y}$ are singletons; the problem is decidable [19], and it has been recently shown to be Ackermann-complete [15,14,2]. It is well-known (and easy to show) that the above more general version (the semilinear reachability problem) is tightly related to the standard version, and has thus the same complexity.

3   Separators

A *separator* is a negative instance of the semilinear reachability problem, i.e. a triple $(\mathbf{X}, \mathbf{A}, \mathbf{Y})$ where $\mathbf{X}, \mathbf{Y}$ are semilinear sets of configurations of a VAS $\mathbf{A}$ such that $\neg(\mathbf{X} \xrightarrow{\mathbf{A}^*} \mathbf{Y})$. The *domain* $\mathbf{D}$ of a separator $(\mathbf{X}, \mathbf{A}, \mathbf{Y})$ is the semilinear set $\overline{\mathbf{X} \cup \mathbf{Y}}$. Notice that $\mathbf{X}, \mathbf{D}$, and $\mathbf{Y}$ forms a partition of $\mathbb{N}^d$. When the domain is empty, the separator is said to be *complete*. Notice that a triple $(\mathbf{X}, \mathbf{A}, \mathbf{Y})$ is a *complete separator* if, and only if, $(\mathbf{X}, \mathbf{Y})$ is a partition of $\mathbb{N}^d$ into semilinear sets such that $\mathbf{y} - \mathbf{x} \neq \mathbf{a}$ for every $\mathbf{x} \in \mathbf{X}, \mathbf{y} \in \mathbf{Y}$, and $\mathbf{a} \in \mathbf{A}$. In particular this property is decidable in elementary time by encoding it as the satisfiabibility of a Presburger formula. A separator $(\mathbf{X}', \mathbf{Y}')$ is called a *completion* of a separator $(\mathbf{X}, \mathbf{Y})$ if $(\mathbf{X}', \mathbf{Y}')$ is complete, $\mathbf{X} \subseteq \mathbf{X}'$ and $\mathbf{Y} \subseteq \mathbf{Y}'$.

In [15] we proved that every separator can be effectively completed. In this paper, we show how this result can be extended with optimal complexity bounds. More formally, we prove that any separator can be completed in Ackermannian time. The Ackermannian lower-bound is immediate since the reachability problem for VAS is Ackermannian-complete and as already mentioned, we can check in elementary time if a pair of semilinear sets is a completion of a separator. The most difficult part of the result is the Ackermannian upper-bound.

4   Semi-Pseudo-Linear Sets

Given two semilinear sets $\mathbf{X}, \mathbf{Y}$ of configurations of a VAS $\mathbf{A}$, the sets $\text{POST}^*_\mathbf{A}(\mathbf{X}) \cap \mathbf{Y}$ and $\text{PRE}^*_\mathbf{A}(\mathbf{Y}) \cap \mathbf{X}$ are not semilinear in general. However, we proved in [13] that those sets are semi-pseudo-linear, a class of sets that can be tightly over-approximated by semilinear sets called *linearizations*. Linearizations are obtained by solving several instances of the semilinear reachability problem. Since in [14,2], we provided an Ackermannian upper-bounds on that decision problem, we can reasonably think that the completion of separators can be done in Ackermannian time. To prove that result, in this section we provide complexity bounds on the size of linearizations. Those linearizations will be used in the next section for completing separators in Ackermannian time.

Let us recall some definitions. A monoid $\mathbf{M}$ is a set of configurations such that $\mathbf{0} \in \mathbf{M}$, and such that $\mathbf{M} + \mathbf{M} \subseteq \mathbf{M}$. The monoid *spanned* by a set $\mathbf{P} \subseteq \mathbb{N}^d$ is the set of finite sums of vectors in $\mathbf{P}$. It is denoted as $\Sigma \mathbf{P}$. A vector $\mathbf{a} \in \mathbb{N}^d$ is called an *interior vector* of a monoid $\mathbf{M}$, if for every $\mathbf{m} \in \mathbf{M}$, there exists a natural number $n \geq 1$ such that $n\mathbf{a} \in \mathbf{m} + \mathbf{M}$.

A *pseudo-linear set* is a set $\mathbf{X} \subseteq \mathbb{N}^d$ such that there exists a linear set $\mathbf{L} = \mathbf{b} + \mathbf{M}$ where $\mathbf{M}$ is the monoid spanned by the periods of $\mathbf{L}$, such that

$\mathbf{X} \subseteq \mathbf{L}$ and such that for every finite set $\mathbf{R}$ of interior vectors of $\mathbf{M}$, there exists $\mathbf{x} \in \mathbf{X}$ such that $\mathbf{x} + \Sigma \mathbf{R} \subseteq \mathbf{X}$. In that case, the linear set $\mathbf{L}$ is called a *linearization* of $\mathbf{X}$. A semi-pseudo-linear set $\mathbf{X}$ is a finite union of pseudo-linear sets $\mathbf{X} = \mathbf{X}_1 \cup \ldots \cup \mathbf{X}_k$. In that case a semilinear set of the form $\mathbf{L}_1 \cup \ldots \cup \mathbf{L}_k$ where $\mathbf{L}_j$ is a linearization of $\mathbf{X}_j$ is a called a *linearization* of $\mathbf{X}$.

By combining the proof of [13, Theorem 6.4] with [15], we deduce the following theorem where f_d is a function of the form $F_{d+3}(Cn)$ for some constant C independent of d. In this theorem, the size in binary or in unary does not change the result and there is a lot of freedom in the definition of the size of presentations of semilinear sets and VAS.

Theorem 1. *Given two semilinear sets $\mathbf{X}$ and $\mathbf{Y}$ of configurations of a d-dim VAS $\mathbf{A}$, the sets $\mathrm{POST}_{\mathbf{A}}^*(\mathbf{X}) \cap \mathbf{Y}$ and $\mathrm{PRE}_{\mathbf{A}}^*(\mathbf{Y}) \cap \mathbf{X}$ are semi-pseudo-linear. Moreover, we can effectively compute in time $f_d(n)$ where n is the size of the input, presentations of linearizations of those sets.*

The tightness of linearization approximations can be emphasis by introducing the notion of *rank*[1] given in [13]. Formally, the *rank* of a set $\mathbf{X} \subseteq \mathbb{N}^d$, denoted as rank $\mathbf{X}$ is the minimal $r \in \{-\infty, 0, \ldots, d\}$ such that there exists a semi-linear set $\mathbf{S}$ that contains $\mathbf{X}$ of the form $\mathbf{b}_1 + \mathbf{M}_1 \cup \ldots \cup \mathbf{b}_k + \mathbf{M}_k$ where $\mathbf{M}_1, \ldots, \mathbf{M}_k$ are monoids spanned by at most r vectors. In [13], we prove that rank$(\mathbf{X}) = -\infty$ iff $\mathbf{X}$ is empty, rank$(\mathbf{X}) \le$ rank$(\mathbf{Y})$ if $\mathbf{X} \subseteq \mathbf{Y}$, and the following theorem.

Theorem 2 (Proposition 7.10 of [13]). *Let $\mathbf{S}_1, \mathbf{S}_2$ be linearizations of two non-empty semi-pseudo-linear sets $\mathbf{X}_1, \mathbf{X}_2$ with an empty intersection. We have:*

$$\text{rank}(\mathbf{S}_1 \cap \mathbf{S}_2) < \text{rank}(\mathbf{X}_1 \cup \mathbf{X}_2)$$

5 Ackermannian Completion

We show in this section who a separator $(\mathbf{X}, \mathbf{A}, \mathbf{Y})$ can be completed in Ackermannian time. We follow the algorithm introduced in [13] by first proving that if $(\mathbf{X}, \mathbf{A}, \mathbf{Y})$ is not complete, i.e. if the domain $\mathbf{D}$ is non empty, we can effectively compute a separator $(\mathbf{X}', \mathbf{A}, \mathbf{Y}')$ with a domain $\mathbf{D}'$ such that $\mathbf{X} \subseteq \mathbf{X}'$, $\mathbf{Y} \subseteq \mathbf{Y}'$, and such that rank$(\mathbf{D}') <$ rank$(\mathbf{D})$. It follows that by applying at most d times this algorithm where d is the dimension of $\mathbf{A}$, we get a complete separator.

Let n be the size of the separator $(\mathbf{X}, \mathbf{A}, \mathbf{Y})$.

The set $\mathbf{Y}'$ is obtained as follows. Since $\mathbf{D}$ is semilinear and effectively computable in elementary time, it follows from Theorem 1 that we can compute in time $f_d(E(n))$ where E is some fixed elementary function a linearization $\mathbf{U}$ of the semi-pseudo-linear set $\mathrm{POST}_{\mathbf{A}}^*(\mathbf{X}) \cap \mathbf{D}$. We introduce $\mathbf{Y}' \stackrel{\text{def}}{=} \mathbf{Y} \cup (\mathbf{D} \setminus \mathbf{U})$.

Let us prove that $(\mathbf{X}, \mathbf{A}, \mathbf{Y}')$ is a separator. By contradiction, assume that $\mathbf{X} \xrightarrow{\mathbf{A}^*} \mathbf{Y}'$. Since $\neg(\mathbf{X} \xrightarrow{\mathbf{A}^*} \mathbf{Y})$, and $\mathbf{Y}' = \mathbf{Y} \cup (\mathbf{D} \setminus \mathbf{U})$, we deduce that $\mathbf{X} \xrightarrow{\mathbf{A}^*}$

---

[1] In [13] this notion is called *dimension* but in our context, the dimension word is already used for the number of components of a vector.

$(\mathbf{D} \setminus \mathbf{U})$. However, since $\text{POST}^*_{\mathbf{A}}(\mathbf{X}) \cap \mathbf{D} \subseteq \mathbf{U}$ we get a contradiction. Hence $(\mathbf{X}, \mathbf{A}, \mathbf{Y}')$ is a separator and its domain is equal to $\mathbf{D} \cap \mathbf{U}$.

The set $\mathbf{X}'$ is obtained symmetrically. Since $\mathbf{D} \cap \mathbf{U}$ is semilinear and effectively computable in elementary time, it follows from Theorem 1 that we can compute in time $f_d(E'(f_d(E(n))))$ where E' is some fixed elementary function a linearization $\mathbf{V}$ of the semi-pseudo-linear set $\text{PRE}^*_{\mathbf{A}}(\mathbf{Y}') \cap \mathbf{D} \cap \mathbf{U}$. We introduce $\mathbf{X}' \overset{\text{def}}{=} \mathbf{X} \cup ((\mathbf{D} \cap \mathbf{U}) \setminus \mathbf{V})$.

Symmetrically, we deduce that $(\mathbf{X}', \mathbf{A}, \mathbf{Y}')$ is a separator and its domain $\mathbf{D}'$ is equal to $\mathbf{D} \cap \mathbf{U} \cap \mathbf{V}$.

Since $(\mathbf{X}, \mathbf{A}, \mathbf{Y}')$ is a separator, it follows that $\text{POST}^*_{\mathbf{A}}(\mathbf{X})$ and $\text{PRE}^*_{\mathbf{A}}(\mathbf{Y}')$ have an empty intersection. In particular the semi-pseudo-linear sets $\text{POST}^*_{\mathbf{A}}(\mathbf{X}) \cap \mathbf{D}$ and $\text{PRE}^*_{\mathbf{A}}(\mathbf{Y}') \cap \mathbf{D} \cap \mathbf{U}$ have an empty intersection. If one of those semi-pseudo-linear sets is empty then $\mathbf{D}'$ is empty and in particular $\text{rank}(\mathbf{D}') < \text{rank}(\mathbf{D})$. Otherwise, from Theorem 2 we deduce that the rank of $\mathbf{U} \cap \mathbf{V}$ is strictly bounded by the rank of the union of $\text{POST}^*_{\mathbf{A}}(\mathbf{X}) \cap \mathbf{D}$ and $\text{PRE}^*_{\mathbf{A}}(\mathbf{Y}') \cap \mathbf{D} \cap \mathbf{U}$. Since this set is included in $\mathbf{D}$, and $\mathbf{D}'$ is included in $\mathbf{U} \cap \mathbf{V}$, we deduce that $\text{rank}(\mathbf{D}') < \text{rank}(\mathbf{D})$.

By replacing E and E' by $E + E'$, we can assume without loss of generality that $E = E'$. By iterating the previous construction at most d times, we deduce that from any separator $(\mathbf{X}, \mathbf{A}, \mathbf{Y})$ of size n, we can compute in time $(f_d \circ E)^{2d}(n)$ a completion of it. We deduce the main theorem of that paper.

Theorem 3. *Separators can be completed in Ackermannian time.*

6   Conclusion

In this paper, we have shown that separators can be completed in Ackermannian time. Our computation is based on a generic algorithm given in Section 5. This algorithm can be implemented as soon as we have an oracle computing semilinear sets over-approximating the sets $\text{POST}^*_{\mathbf{A}}(\mathbf{X}) \cap \mathbf{D}$ and $\text{PRE}^*_{\mathbf{A}}(\mathbf{Y}) \cap \mathbf{D}$. If those approximations are not linearizations, the termination of the algorithm is no longer true in general. However, since its correctness is maintained, it should be interesting to benchmark such an algorithm when using heuristics for implementing oracles computing reachability set over-approximations (based on abstract interpretation, acceleration techniques, parameterized invariant, and so on).

References

1. Czerwinski, W., Jecker, I., Lasota, S., Leroux, J., Orlikowski, L.: New lower bounds for reachability in vector addition systems. In: Bouyer, P., Srinivasan, S. (eds.) 43rd IARCS Annual Conference on Foundations of Software Technology and Theoretical Computer Science, FSTTCS 2023, December 18-20, 2023, IIIT Hyderabad, Telangana, India. LIPIcs, vol. 284, pp. 35:1–35:22. Schloss Dagstuhl - Leibniz-Zentrum für Informatik (2023). https://doi.org/10.4230/LIPICS.FSTTCS.2023.35

2. Czerwinski, W., Orlikowski, L.: Reachability in vector addition systems is Ackermann-complete. In: 62nd IEEE Annual Symposium on Foundations of Computer Science, FOCS 2021, Denver, CO, USA, February 7-10, 2022. pp. 1229–1240. IEEE (2021). https://doi.org/10.1109/FOCS52979.2021.00120

3. Esparza, J., Nielsen, M.: Decidability issues for Petri nets - a survey. Bulletin of the European Association for Theoretical Computer Science **52**, 245–262 (1994)

4. Ginsburg, S., Spanier, E.H.: Semigroups, Presburger formulas and languages. Pacific Journal of Mathematics **16**(2), 285–296 (1966). https://doi.org/10.2140/pjm.1966.16.285

5. Grzegorczyk, A.: Some classes of recursive functions. Instytut Matematyczny Polskiej Akademi Nauk (1953), http://eudml.org/doc/219317

6. Hack, M.: Decidability questions for Petri nets. Ph.D. thesis, MIT (1975), http://publications.csail.mit.edu/lcs/pubs/pdf/MIT-LCS-TR-161.pdf

7. Hopcroft, J.E., Pansiot, J.J.: On the reachability problem for 5-dimensional vector addition systems. Theoritical Computer Science **8**, 135–159 (1979)

8. Karp, R.M., Miller, R.E.: Parallel program schemata. J. Comput. Syst. Sci. **3**(2), 147–195 (1969). https://doi.org/10.1016/S0022-0000(69)80011-5

9. Kosaraju, S.R.: Decidability of reachability in vector addition systems (preliminary version). In: STOC. pp. 267–281. ACM (1982). https://doi.org/10.1145/800070.802201

10. Lambert, J.: A structure to decide reachability in Petri nets. Theor. Comput. Sci. **99**(1), 79–104 (1992). https://doi.org/10.1016/0304-3975(92)90173-D

11. Lasota, S.: Improved Ackermannian lower bound for the Petri nets reachability problem. In: Berenbrink, P., Monmege, B. (eds.) 39th International Symposium on Theoretical Aspects of Computer Science, STACS 2022, March 15-18, 2022, Marseille, France (Virtual Conference). LIPIcs, vol. 219, pp. 46:1–46:15. Schloss Dagstuhl - Leibniz-Zentrum für Informatik (2022). https://doi.org/10.4230/LIPIcs.STACS.2022.46

12. Leroux, J., Schmitz, S.: Demystifying reachability in vector addition systems. In: 30th Annual ACM/IEEE Symposium on Logic in Computer Science, LICS 2015, Kyoto, Japan, July 6-10, 2015. pp. 56–67. IEEE Computer Society (2015). https://doi.org/10.1109/LICS.2015.16

13. Leroux, J.: The general vector addition system reachability problem by presburger inductive invariants. In: Proceedings of the 24th Annual IEEE Symposium on Logic in Computer Science, LICS 2009, 11-14 August 2009, Los Angeles, CA, USA. pp. 4–13. IEEE Computer Society (2009). https://doi.org/10.1109/LICS.2009.10

14. Leroux, J.: The reachability problem for Petri nets is not primitive recursive. In: 62nd IEEE Annual Symposium on Foundations of Computer Science, FOCS 2021, Denver, CO, USA, February 7-10, 2022. pp. 1241–1252. IEEE (2021). https://doi.org/10.1109/FOCS52979.2021.00121

15. Leroux, J., Schmitz, S.: Reachability in vector addition systems is primitive-recursive in fixed dimension. In: 34th Annual ACM/IEEE Symposium on Logic in Computer Science, LICS 2019, Vancouver, BC, Canada, June 24-27, 2019. pp. 1–13. IEEE (2019). https://doi.org/10.1109/LICS.2019.8785796

16. Löb, M.H., Wainer, S.S.: Hierarchies of number-theoretic functions. i. Archiv für mathematische Logik und Grundlagenforschung **13**(1), 39–51 (1970). https://doi.org/10.1007/BF01967649

17. Mayr, E.W.: An algorithm for the general petri net reachability problem. In: Proceedings of the 13th Annual ACM Symposium on Theory of Computing, May 11-13, 1981, Milwaukee, Wisconsin, USA. pp. 238–246. ACM (1981). https://doi.org/10.1145/800076.802477

18. Mayr, E.W., Meyer, A.R.: The complexity of the finite containment problem for petri nets. J. ACM **28**(3), 561–576 (1981). https://doi.org/10.1145/322261.322271
19. Mayr, E.W.: An algorithm for the general Petri net reachability problem. SIAM J. Comput. **13**(3), 441–460 (1984). https://doi.org/10.1137/0213029
20. Sacerdote, G.S., Tenney, R.L.: The decidability of the reachability problem for vector addition systems (preliminary version). In: Proceedings of the 9th Annual ACM Symposium on Theory of Computing, May 4-6, 1977, Boulder, Colorado, USA. pp. 61–76. ACM (1977). https://doi.org/10.1145/800105.803396
21. Schmitz, S.: Complexity hierarchies beyond elementary. TOCT **8**(1), 3:1–3:36 (2016). https://doi.org/10.1145/2858784
22. Schmitz, S.: The complexity of reachability in vector addition systems. ACM SIGLOG News **3**(1), 4–21 (2016). https://doi.org/10.1145/2893582.2893585

Infinite Games

Fair ω-Regular Games

Daniel Hausmann[1]*, Nir Piterman[1]*, Irmak Sağlam[2](✉)**,
and Anne-Kathrin Schmuck[2]**

[1] University of Gothenburg, Gothenburg, Sweden
{hausmann,piterman}@chalmers.se
[2] Max Planck Institute for Software Systems (MPI-SWS), Kaiserslautern, Germany
{isaglam,akschmuck}@mpi-sws.org

Abstract. We consider two-player games over finite graphs in which both players are restricted by fairness constraints on their moves. Given a two player game graph $G = (V, E)$ and a set of fair moves $E_f \subseteq E$ a player is said to play *fair* in G if they choose an edge $e \in E_f$ infinitely often whenever the source node of e is visited infinitely often. Otherwise, they play *unfair*. We equip such games with two ω-regular winning conditions α and β deciding the winner of mutually fair and mutually unfair plays, respectively. Whenever one player plays fair and the other plays unfair, the fairly playing player wins the game. The resulting games are called *fair α/β games*.

We formalize fair α/β games and show that they are determined. For fair parity/parity games, i.e., fair α/β games where α and β are given each by a parity condition over G, we provide a polynomial reduction to (normal) parity games via a gadget construction inspired by the reduction of stochastic parity games to parity games. We further give a direct *symbolic fixpoint algorithm* to solve fair parity/parity games. On a conceptual level, we illustrate the translation between the gadget-based reduction and the direct symbolic algorithm which uncovers the underlying similarities of solution algorithms for fair and stochastic parity games, as well as for the recently considered class of fair games in which only one player is restricted by fair moves.

Keywords: games on graphs, fairness, two-player games, parity games

1 Introduction

Omega-regular games are a popular abstract modelling formalism for many core computational problems in the context of correct-by-construction synthesis of reactive software or hardware. This abstract view was initiated by the seminal work of Church [8] and its independent solutions by Büchi and Landweber and Rabin [18,5]. Since then these ideas have been refined and extended for solving the *reactive synthesis problems* [17,20,14].

---

* Supported by the ERC Consolidator grant D-SynMA (No. 772459).
** Supported by the DFG project SCHM 3541/1-1.

N. Kobayashi and J. Worrell (Eds.): FoSSaCS 2024, LNCS 14574, pp. 13–33, 2024.
https://doi.org/10.1007/978-3-031-57228-9_2

However, before using any such synthesis technique, the reactive software design problem at hand needs to be abstractly modelled as a two-player game. In order for the subsequently synthesized software to be 'correct-by-construction' this game graph needs to reflect all possible interactions between involved components in an abstract manner. Building such a game graph with the 'right' level of abstraction is a known severe challenge, in particular, if the synthesized software is interacting with existing components that already possess certain behavior. Here, part of the modelling challenge amounts to finding the 'right' power of both players in the resulting abstract game to ensure that winning strategies do not fail to exist due to an unnecessarily conservative overapproximation of modeling uncertainty (or the dual problem due to underapproximation).

In this context, *fairness* has been adopted as a notion to abstractly model known characteristics of the involved components in a very concise manner. *Fairness assumptions* have been used in model checking [1] and scheduler synthesis for the classical AMBA arbiter [16] or shared resource management [6]. Notably, fairness assumptions have also gained attention in cyber-physical system design [21,15,11] and robot motion planning [9,2]. In all these applications, fairness is used as an *assumption* that the synthesized (or verified) component can rely on. In particular, if these assumptions are modelled by *transition fairness* over a two-player game arena[3] $(V_\forall, V_\exists, E)$ – i.e., by a set of fair *environment* moves $E_f \subseteq E$ (i.e., with $V_\forall$ as their domain) that need to be taken infinitely often if the source node is seen infinitely often along a play – the resulting synthesis games can be solved efficiently [4,19].

While most existing work has only looked at fairness as an *assumption* to weaken the opponent in the synthesis game, all mentioned applications also naturally allow for scenarios where multiple components with intrinsically fair behavior are interacting with each other in a non-trivial manner. For example, the ability of a concurrent process to eventually free a shared resource might depend on how fair re-allocation is implemented in other threads. On an abstract level, the formal reasoning about such scenarios requires to understand how the interactive decision making of two dependent processes is influenced by intrinsic fairness constraints imposed on their decisions. Algorithmically, these synthesis questions require fairness restrictions on both players in a game, i.e., do not restrict the domain of fair moves E_f to one player only. We refer to such games simply as *fair games*.

Motivating Example. In order to better illustrate the challenges arising from solving such fair games, consider two robots in a shared workspace with narrow passages between adjacent regions that only one robot can pass at a time. One robot (say the green one) has an ω-regular objective α that specifies desired sequences of visited regions in the workspace. The other (red) robot tries to prevent the green robot from achieving this sequence. In order to rule out trivial spoiling strategies of the red robot, both robots need to implement a tie-breaking

---

[3] Whenever we interpret players in a one-sided manner as environment and system, we choose the environment player as the $\forall$-player, as we need to take all possible environment moves into account. Similarly, the system is the $\exists$-player in this scenario.

mechanism for obstacle avoidance, i.e., they must eventually move left or right if an obstacle blocks their way.

Now consider the scenario where both robots are facing each other at a gate, as depicted in Fig. 1. While both robots block the gate from one side, neither of them can move forward, but if the green robots moves left or the red robot moves right, the other robot can take the gate to reach region A. With the mentioned requirement for tie-breaking, none of the robots is allowed to block the gate forever and both eventually have to move to the side.

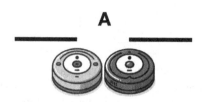

Fig. 1: Deadlock caused by fairness constraints of two robots facing a door.

Now let us assume that region A is important for both robots, hence, both robots have an incentive to enter region A first, to then move the game to an area preferable to them. However, the robot who breaks the tie first, (i.e., fulfills its fairness condition first) allows the *other* robot to enter region A first, which gives both robots the incentive to behave unfair. While it is very intuitive to make a player lose when she plays unfair and the other player plays fair, it is unclear who wins the game if both players play unfair.

To resolve this issue, we can make the objectives of the robots completely adversarial by assigning one of the players (say, green) the winner in a play where both players play unfair. In the above example, this would give the red robot the incentive to break the tie first. While this makes it harder for the red robot to spoil the objective of the green one, we might be interested in a more symmetric game which does not favor the green robot in all non-determined states of the graph. We therefore consider a second ω-regular objective β that determines the winner of (mutually) unfair plays. This results in fair games $G = (A, \alpha, \beta)$ which are determined (as shown in Sec. 3).

Contribution. Motivated by the above mentioned examples where interactive decision making of two dependent processes is influenced by intrinsic fairness constraints imposed on their decisions, this paper studies *fair games* $G = (A, \alpha, \beta)$ as their abstraction. In particular, we give solution algorithms for these games when both α and β are parity conditions induced by two different priority functions over the node set. We call such games *fair parity/parity games*.

Obviously, the previously discussed one-sided version of fair games, which we call ∀-*fair games* (as only the ∀-player (i.e., the environment) is restricted by strong transition fairness), is a special case of fair games. Both enumerative [19] and symbolic solution algorithms [4] have recently been proposed for ∀-

fair games, showing that strong transition fairness can be handled efficiently in both types of algorithms. This observation is closely related to a result for *stochastic games*, i.e., two-player games with an *additional* 'half' player that takes all its moves uniformly at random. For the purpose of qualitative analysis, such stochastic parity games have been shown in [7] to be reducible to (standard) parity games by the use of "gadgets" that turn stochastic nodes into a small sequence of ∀- and ∃-player nodes. While it is known that stochastic games can be reduced to ∀-*fair games* (and hence, fair games), it was not investigated how the different solution approaches compare. The main conceptual contribution of this paper is a unified understanding of all these solution approaches for the general class of fair games.

Concretely, our contribution is three-fold:
(1) We formalize fair games as a generalization of ∀-fair games and stochastic games such that they are determined.
(2) We show a reduction of fair parity/parity games to (standard) parity games, inspired by the gadget-based reduction of stochastic parity games to parity games in [7]. This reduction enables the use of parity game solvers over the reduced game (in particular enumerative ones such as Zielonka's algorithm [24]) and gives a gadget-based reduction of ∀-fair parity games to parity games as a corollary.
(3) We then show how our gadget construction can be used to define a *symbolic fixpoint algorithm* to solve fair parity/parity games directly (without the need for a reduction). We show the direct symbolic algorithm for ∀-fair parity games in [4] coinciding with our algorithm for this particular subclass of fair games.

With this, we believe that this paper uncovers the underlying similarities of solution algorithms for fair, ∀-fair and stochastic parity games. Further, we show how these conceptual similarities can be used to build both *enumerative* and direct *symbolic* algorithms. This is of interest as both are known to have complementary strengths, depending on how the synthesis instance is provided, and this connection was, to the best of our knowledge, not known before.

All omitted proofs are available in the extended version of this paper [10].

2   Preliminaries

We introduce infinite-duration ω-regular two-player games over finite graphs with additional *strong transition fairness conditions* on both players. For readability, we call the considered games (and their respective notions) simply *fair*.

Infinite Sequences.   We denote the set of infinite sequences over a set U by U^ω. We often view sequences $\tau = u_1 u_2 \ldots \in U^\omega$ as functions $\tau : \mathbb{N} \to U$, writing $\tau(i) = u_i$. Furthermore, we let $\mathsf{Inf}(\tau) := \{u \in U \mid \forall i.\exists j > i.\tau(j) = u\}$ denote the set of elements of U that occur infinitely often in τ. Given a function $f : U \to W$, we denote by $f(\tau) \in W^\omega$ the pointwise application of f to τ. Given a natural number n, we write $[n] := \{1, \ldots, n\}$.

Fair Game Arenas.   A *fair game arena* $A = (V_\exists, V_\forall, E, E_f)$ consists of a set of *nodes* $V = V_\exists \cup V_\forall$ that is partitioned into the sets of *existential nodes* $V_\exists$ and *universal nodes* $V_\forall$, together with a set $E \subseteq V \times V$ of *moves* that is partitioned into the set $E_f \subseteq E$ of *fair moves* and the set $E \setminus E_f$ of *normal moves*. If $E_f = \emptyset$, then we sometimes omit this component for brevity. Given a node $v \in V$ and a binary relation $R \subseteq V \times V$, we write $R(v)$ to denote the set $\{w \in V \mid (v, w) \in R\}$. We assume that E is right-total, that is, $E(v) \neq \emptyset$ for all $v \in V$. We call a node v *fair*, if it is the source node of a fair edge, i.e., $E_f(v) \neq \emptyset$ and collect all fair nodes in the set $V^{\mathsf{fair}} = \{v \in V \mid E_f(v) \neq \emptyset\}$ and define $V^{\mathsf{n}} = V \setminus V^{\mathsf{fair}}$ to be the set of nodes that are not fair ('normal nodes'). We denote $V_{\exists}^{\mathsf{fair}} = V^{\mathsf{fair}} \cap V_{\exists}$ and $V_{\forall}^{\mathsf{fair}} = V^{\mathsf{fair}} \cap V_{\forall}$.

Plays. A *play* $\tau = v_0 v_1 \ldots$ on A is an infinite sequence of nodes s.t. $v_{i+1} \in E(v_i)$ for all $i \geq 0$. Given a play $\tau = v_0 v_1 \ldots$, we define the associated sequence of moves $\tau_m = (v_0, v_1)(v_1, v_2) \ldots$. Additionally, if i is a player in $\{\exists, \forall\}$, we denote the other player by $1 - i$. We let $\mathsf{plays}(A)$ denote the set of all plays on A.

For a player $i \in \{\exists, \forall\}$, a play τ is *i-fair* if for all nodes $v \in V_i \cap \mathsf{Inf}(\tau)$ holds that $E_f|_v \subseteq \mathsf{Inf}(\tau_m)$, where $E_f|_v = \{(v, v') \in E_f \mid v' \in V\}$ denotes the set of fair edges that start at $v \in V$. Given a play τ, we write $\mathsf{fair}_i(\tau)$ to indicate that τ is *i*-fair. We call a play *mutually fair* if it is both $\exists$- and $\forall$-fair and *mutually unfair* if it is neither $\exists$- nor $\forall$-fair.

Strategies. A strategy for player $i \in \{\exists, \forall\}$ (or, an *i-strategy*) is a function $p : V^* \cdot V_i \to V$ where for each $u \cdot v \in V^* \cdot V_i$ it holds that $p(u \cdot v) \in E(v)$. A strategy p is called *positional* if $p(u \cdot v) = p(w \cdot v)$ for all $u, w \in V^*$ and $v \in V_i$.

A strategy p for player i is said to *admit* a play $\tau = v_0 v_1 \ldots$ if for all $k \in \mathbb{N}$, $v_k \in V_i$ implies $p(v_0 \ldots v_k) = v_{k+1}$. Alternatively, τ is said to be *compliant* with p. We write Σ for the set of $\exists$-strategies and Π for the set of $\forall$-strategies. Starting from a node $v \in V$, any two strategies $s \in \Sigma$ and $t \in \Pi$ induce a unique play $\mathsf{play}_v(s, t)$ in the game arena. If we do not care about the initial node of the play, we simply write $\mathsf{play}(s, t)$.

A strategy for player $i \in \{\exists, \forall\}$ is an *i-fair strategy* if every play it admits is *i*-fair. We write Σ^{fair} (resp. Π^{fair}) for the set of $\exists$-fair (resp. $\forall$-fair) strategies.

Omega-regular Winning Conditions. We consider winning conditions given by an ω-regular [22,13] language $\varphi \subseteq V^\omega$ over the node set V. In particular, we write $\varphi = \bot$ and $\varphi = \top$ to denote the trivial winning conditions $\emptyset$ and V^ω, respectively. In particular, we focus our attention to *parity* winning conditions. For a priority function $\lambda : V \to [k]$ that maps nodes of a game arena to the natural numbers bounded by k for some $k \in \mathbb{N}$, the Parity(λ) condition is given via $\varphi = \{\tau \in V^\omega \mid \max(\mathsf{Inf}(\lambda(\tau)))$ is even$\}$.

Omega-regular Games. An ω-regular game is traditionally defined via a tuple $G = (A, \alpha)$ where A is a game arena *without fair edges*, i.e. $E_f = \emptyset$ and $\alpha \subseteq V^\omega$ an ω-regular winning condition. An $\exists$-strategy $s \in \Sigma$ is said to be *winning* (for $\exists$) from a node $v \in V$, if for all $t \in \Pi$, $\mathsf{play}_v(s, t) \in \alpha$. Dually, a $\forall$-strategy $t \in \Pi$ is said to be winning (for $\forall$) from a node $v \in V$, if for all $s \in \Sigma$, $\mathsf{play}_v(s, t) \notin \alpha$. In ω-regular games, every node $v \in V$ is won by one *and only one* of the players [12,13]. This property of a game is called *determinacy*, and ω-regular games are *determined*. We denote the nodes from which $\exists$ (resp. $\forall$) has a winning strategy in G by $\mathsf{Win}_{\exists}(G)$ (resp. $\mathsf{Win}_{\forall}(G)$). When G is clear from the context, we drop the parenthesis and write $\mathsf{Win}_\exists$ and $\mathsf{Win}_\forall$ instead. Determinacy then amounts to $\mathsf{Win}_\exists \cup \mathsf{Win}_\forall = V$ and $\mathsf{Win}_\exists \cap \mathsf{Win}_\forall = \emptyset$.

Node Conventions for Figures. Throughout this paper, in all figures, the rectangular nodes represent $\forall$-player nodes and the nodes with round corners represent $\exists$-player nodes.

3   Fair Games

As already outlined in the motivating example in Sec. 1, the interpretation of winning conditions over fair games influences the characteristics of resulting winning strategies. To formalize this intuition, we will first recall a particular subclass of fair games, namely those where only one player is restricted by an additional fairness condition, in Subsec. 3.1. We will then use these games to motivate winning semantics for the general class of fair games.

3.1   Determinacy of $\forall$-Fair Games

A $\forall$-fair game is a tuple $G = (A, \alpha)$ where A is a game arena with $V^{\mathsf{fair}} \subseteq V_\forall$ (*called a $\forall$-fair game arena*), and α is an ω-regular winning condition.

In $\forall$-fair games, fairness constraints typically model known behavior of existing components that the $\exists$-player (i.e., the to be synthesized system) can rely on. This is formalized by defining that the $\exists$-player wins a $\forall$-fair game with winning condition α from node v if

$$\exists s \in \Sigma. \forall t \in \Pi^{\mathsf{fair}}. \mathsf{play}_v(s, t) \in \alpha. \tag{1a}$$

That is, $\exists$-player (or shortly, $\exists$) wins if they have a strategy that can win against all $\forall$-fair $\forall$-strategies.

Our intuition tells us that this can be converted to reasoning about general strategies for $\forall$-player (or shortly, $\forall$) by allowing $\exists$ to win whenever $\forall$ plays unfairly. In order to see this, we can look at the complement of Eq. (1a), i.e., the description of when $\forall$ wins; namely, $\forall s \in \Sigma. \exists t \in \Pi^{\mathsf{fair}}. \mathsf{play}_v(s, t) \notin \alpha$. We can replace the quantification over fair strategies with a quantification over all strategies but require that, in addition to refuting α, the resulting play be fair: $\forall s \in \Sigma. \exists t \in \Pi. \mathsf{fair}_\forall(\mathsf{play}_v(s, t)) \wedge \mathsf{play}_v(s, t) \notin \alpha$. Indeed, as we show in the extended version of this paper [10, App. A - Lem. 2], if strategy $t \in \Pi$ satisfies $\mathsf{fair}_\forall(\mathsf{play}_v(s, t))$ then we can find a fair strategy $t' \in \Pi^{\mathsf{fair}}$ with which $\mathsf{play}_v(s, t)$ is compliant. This $\forall$-fair strategy would also stop s from winning. Due to determinacy of ω-regular games, we know that the last condition is equivalent to $\exists t \in \Pi. \forall s \in \Sigma. \mathsf{fair}_\forall(\mathsf{play}_v(s, t)) \wedge \mathsf{play}_v(s, t) \notin \alpha$. In particular, this implies that t is fair. We conclude that the complement of Eq. (1) is the following equation:

$$\exists t \in \Pi^{\mathsf{fair}}. \forall s \in \Sigma. \mathsf{play}_v(s, t) \notin \alpha. \tag{1b}$$

This statement is equivalent to the determinacy of $\forall$-fair games: either $\exists$-player has a winning strategy or $\forall$-player has a winning $\forall$-fair strategy, and the two cannot be true simultaneously.

3.2   From ∀-Fair Games to Defining Determined Fair Games

Given a fair game arena A and an ω-regular objective α, a natural attempt to define winning regions in fair games would be to generalize Eq. (1) to

$$v \in \mathsf{Win}_\exists \text{ if } \exists s \in \Sigma^{\mathsf{fair}}.\forall t \in \Pi^{\mathsf{fair}}.\,\mathsf{play}_v(s,t) \in \alpha, \text{ and} \qquad (2a)$$

$$v \in \mathsf{Win}_\forall \text{ if } \exists t \in \Pi^{\mathsf{fair}}.\forall s \in \Sigma^{\mathsf{fair}}.\,\mathsf{play}_v(s,t) \notin \alpha. \qquad (2b)$$

However, in this case, $\mathsf{Win}_\exists \cup \mathsf{Win}_\forall \neq V$. Indeed, equations (2a) and (2b) are not complements of each other, that is,

$$\exists s \in \Sigma^{\mathsf{fair}}.\forall t \in \Pi^{\mathsf{fair}}.\,\mathsf{play}(s,t) \in \alpha \qquad \not\Leftrightarrow \qquad \forall t \in \Pi^{\mathsf{fair}}.\exists s \in \Sigma^{\mathsf{fair}}.\,\mathsf{play}(s,t) \in \alpha.$$

This observation makes a fair game in which winning regions are defined via Eq. (2) undetermined. The undetermined nodes $O \subseteq V$ – nodes from which none of the players has a fair winning strategy – form a separate partition of nodes, i.e., $V = \mathsf{Win}_\exists \uplus \mathsf{Win}_\forall \uplus O$. To see this, consider the following example.

Example 1. Consider the fair game arena depicted in Fig. 2 where fair edges are shown by dashed lines, $\alpha = \mathrm{Parity}(\lambda)$ and each node is labeled by its priority assigned by λ. We observe that the existential player cannot enforce reaching the even node with a ∃-fair strategy from the two middle nodes. Every ∃-fair ∃-strategy s has a counter ∀-fair ∀-strategy: choose the fair edge outgoing from the square node *after* s chooses the fair edge outgoing from the node with round corners. On the other hand, the universal player cannot prevent the play from reaching the even node with a ∀-fair strategy from these nodes for exactly the same reason. Hence, the middle two nodes are neither in $\mathsf{Win}_\exists$ nor in $\mathsf{Win}_\forall$. That is, these two nodes are undetermined; therefore they form O.

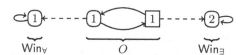

Fig. 2: A simple fair game arena discussed in Ex. 1.

In order to better understand the distinction between Equations 2a and 2b, we rely again on translation to ω-regular games. Consider the following reformulation of Eq. (2a):

$$\exists s \in \Sigma.\forall t \in \Pi.\mathsf{fair}_\exists(\mathsf{play}_v(s,t)) \wedge (\mathsf{fair}_\forall(\mathsf{play}_v(s,t)) \Rightarrow \mathsf{play}_v(s,t) \in \alpha). \qquad (3a)$$

Similarly, the following is a reformulation of Eq. (2b):

$$\exists t \in \Pi.\forall s \in \Sigma.\mathsf{fair}_\forall(\mathsf{play}_v(s,t)) \wedge (\mathsf{fair}_\exists(\mathsf{play}_v(s,t)) \Rightarrow \mathsf{play}_v(s,t) \notin \alpha). \qquad (3b^*)$$

From determinacy of ω-regular games, the negation of the latter is:

$$\exists s \in \Sigma.\forall t \in \Pi.\mathsf{fair}_\forall(\mathsf{play}_v(s,t)) \Rightarrow (\mathsf{fair}_\exists(\mathsf{play}_v(s,t)) \wedge \mathsf{play}_v(s,t) \in \alpha). \qquad (3b)$$

We formally prove the equivalences of Eqs. (2a) and (3a) and Eqs. (2b) and (3b) in [10]. It is not hard to see that the difference between Eq. (3a) and Eq. (3b) is in the way fairness is handled. Namely, in Eq. (3a) $\exists$ loses whenever she plays unfairly regardless of how $\forall$ plays. Dually, in Eq. (3b) $\exists$ wins immediately when $\forall$ plays unfairly regardless of how $\exists$ plays. It follows that determinacy can be regained by deciding the winner of the four different combinations of fairness with an ω-regular winning condition each, as summarized in the following table.

	$\mathsf{fair}_\exists(\tau)$	$\neg\mathsf{fair}_\exists(\tau)$
$\mathsf{fair}_\forall(\tau)$	$\tau \in \alpha$	$\tau \in \gamma$
$\neg\mathsf{fair}_\forall(\tau)$	$\tau \in \delta$	$\tau \in \beta$

With this generalization, we obtain (3a) if $\beta = \gamma = \bot$ and $\delta = \top$, and (3b) if $\gamma = \bot$ and $\beta = \delta = \top$.

We note that the discussion of determinacy has crucial importance to the analysis of games and the decision of how to model particular scenarios. For example, if fairness of $\forall$-player arises from physical constraints (as, e.g., in [4]) then it might make sense to consider Eq. (2b), which corresponds to $\beta = \top$. Dually, if fairness of $\exists$-player must be adhered to, then it makes sense to consider Eq. (2a), which corresponds to $\beta = \bot$. Our formulation allows to further fine tune what happens when both act unfairly by adjusting β.

Given the intuition that fairness constraints are actually additional obligations for both players, the choice of $\gamma = \bot$ and $\delta = \top$ assumed in Equations (2)-(3) is very natural. However, allowing mutually unfair plays to be decided by a different ω-regular winning condition β, allows games with more symmetric winning semantics e.g., by setting $\beta = \alpha$. We therefore restrict our attention in this paper to fair games with two winning conditions α and β while if i-player plays fairly but $(1-i)$-player plays unfairly, i-player wins, i.e., $\gamma := \bot$ and $\delta := \top$. This is formalized next.

Definition 1 (Fair Games). *A fair game $G = (A, \alpha, \beta)$ consists of a fair game arena A together with two (ω-regular) winning conditions $\alpha, \beta \subseteq \mathsf{plays}(A)$ where α and β determine the winner of mutually fair and mutually unfair plays, respectively. In fair games, a play that is i-fair and $(1-i)$-unfair is won by player i. Formally, in the fair game $G = (A, \alpha, \beta)$, $v \in \mathsf{Win}_\exists$ if and only if,*

$$\exists s \in \Sigma. \forall t \in \Pi. \mathsf{fair}_\exists(\mathsf{play}_v(s,t)) \wedge (\mathsf{fair}_\forall(\mathsf{play}_v(s,t)) \Rightarrow \mathsf{play}_v(s,t) \in \alpha)$$
$$\vee (\neg\mathsf{fair}_\exists(\mathsf{play}_v(s,t)) \wedge \neg\mathsf{fair}_\forall(\mathsf{play}_v(s,t)) \wedge \mathsf{play}_v(s,t) \in \beta) \quad (4)$$

The determinacy of fair games follows trivially from the formulation. It follows that the complement of Eq. (4) is the $\forall$ winning region, defined symmetrically by $v \in \mathsf{Win}_\forall$ if and only if

$$\exists t \in \Pi. \forall s \in \Sigma. \mathsf{fair}_\forall(\mathsf{play}_v(s,t)) \wedge (\mathsf{fair}_\exists(\mathsf{play}_v(s,t)) \Rightarrow \mathsf{play}_v(s,t) \notin \alpha)$$
$$\vee (\neg\mathsf{fair}_\forall(\mathsf{play}_v(s,t)) \wedge \neg\mathsf{fair}_\exists(\mathsf{play}_v(s,t)) \wedge \mathsf{play}_v(s,t) \notin \beta)$$

Notation. We call a fair game $G = (A, \alpha, \beta)$ a fair α/β game. Further, if α or β are given by mentioned winning conditions(e.g. $\alpha = \mathrm{Parity}(\lambda)$, $\beta = \bot$), with slight abuse of notation, we refer to the game with the name of the objectives (e.g. fair parity/$\perp$ game).

Remark 1. Stochastic games allow for an additional set V_s of stochastic game nodes that belong to neither $\exists$ nor $\forall$, and for which the stochasticity is resolved uniformly at random. It is known that for purposes of qualitative analysis (i.e., the computation of almost-sure winning strategies), stochastic games can be seen as the special case of $\forall$-fair games in which $E(v) \subseteq E_f$ holds for all stochastic nodes $v \in V_s$, and $E_f \cap E(v) = \emptyset$ for all non-stochastic nodes $v \in V_\exists \cup V_\forall$, that is, all stochastic edges are fair edges, but no non-stochastic edges are fair edges. This encoding treats stochastic branching as adversarial for the system ($\exists$-player).

3.3   Mutually Fair Strategies in Fair Parity Games

In Subsec. 3.2 and in particular in Ex. 1 we have discussed the mutually unfair plays and strategies that take such plays into account in fair α/β games. In this section, we start restricting our attention to fair parity/β games (as this will be our focus for the rest of the paper) and discuss the particularities of mutually fair strategies in such games. We will do this with the help of the games $G_1 - G_4$ depicted in Fig. 3. No mutually unfair plays exist in any of these games. This is because on all given arenas the unfair behaviour of one player makes the play trivially fair for the other. Therefore, the winning regions are independent of β.

In game G_1, both nodes are won by $\exists$. $\forall$-player loses node 3 since taking the self loop on 3 makes the play visit 3 infinitely often, however, it forces $\forall$ to play fairly, implying that they must take the edge to 4 infinitely often. Therefore, any $\forall$-fair play is won by $\exists$ since the priority 4 is seen infinitely often. Also note that if $\forall$-player decides not to play fairly, they immediately lose since all plays are trivially $\exists$-fair. The trivial winning $\exists$-strategy is depicted by red edges.

To get to game G_2, we append node 1 to the left of G_1. Here, all the nodes are won by $\forall$. This is because $\forall$-player wins node 3 by eventually taking the outgoing edge to 1 and then staying in 1 forever with the self-loop. By doing so $\forall$ evades his obligation to take the fair edges by forcing each play to see node 3 a finite number of times. One winning $\forall$-strategy is depicted by blue edges.

To get to game G_3, we append node 5 to the right of game G_1. Again, all the nodes are won by $\forall$ even though this time he cannot evade taking his fair edges. In this game $\forall$ wins due to the obligation of $\exists$ to play fairly. In a play starting from 3, $\forall$ must eventually take the outgoing edge to 4. From there on, the play will visit node 4 infinitely often, forcing $\exists$ to take his outgoing edge to 5 infinitely often. As a consequence, in every mutually fair play 5 is seen infinitely often. Therefore, the game is won by $\forall$. A winning $\forall$-strategy is depicted by blue edges on the figure, with the interpretation that blue edges from node 3 are taken alternatingly (in every sequence).

Finally, to get to game G_4, we append two nodes to game G_3. This time, all the nodes are won by $\exists$. $\exists$-player still needs to take their fair outgoing edges to 5 (and this time, also to the new node 1) infinitely often. But this time she can also take the outgoing edge to 6 infinitely often and thereby win the game. A winning

∃-strategy is depicted by red edges on the figure, again with the interpretation that red edges from node 4 are taken alternatingly (in every sequence).

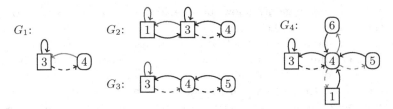

Fig. 3: Four fair parity/β games: dashed lines represent fair edges. Games G_1 and G_4 are won by ∃-player and G_2 and G_3 are won by ∀-player. In each case, a respective winning strategy is shown by colored edges. A set of colored edges represents a strategy that takes only the colored edges in the game, and whenever a source node is visited all its colored outgoing edges are taken alternatingly.

4 Reduction to Parity Games

In this section, we show how fair parity games can be reduced to parity games without fairness constraints. We show that there is a linear reduction to parity games in the case that α is a parity objective and $\beta = \top$ or $\beta = \bot$; for the case that β is a non-trivial parity objective, we show that there still is a quadratic reduction. Our reductions work by replacing each fair node in the fair game with a 3-step parity gadget. This construction is inspired by the work of Chatterjee et al. [7] where the qualitative analysis of stochastic parity games is reduced to solving parity games.

We give the formal reduction for fair parity/$\bot$ games in Subsec. 4.1 and extend it to fair parity/parity games in Subsec. 4.2. The extended version contains a discussion of the reduction for a restricted case of fair parity/$\bot$ games (fair Büchi/$\bot$ games), which can serve as a hand-holding introduction to the section.

4.1 Reduction of Fair Parity/$\bot$ Games

Let $G = (A, \mathrm{Parity}(\lambda), \bot)$ where $A = (V_\exists, V_\forall, E, E_f)$ is a fair game arena, $V = V_\exists \uplus V_\forall$ and $\lambda : V \to [2k]$ is the priority function.

The reduction to parity games replaces fair nodes $v \in V^{\mathsf{fair}}$ in G with the gadgets given in Fig. 4. Nodes $v \in V_\exists^{\mathsf{fair}}$ in G are replaced with one of the gadgets on the top (i.e. the incoming edges to v are redirected to v in the root, and the outgoing edges on the third level lead to $E(v)$ and $E_f(v)$, which are the outgoing edges and outgoing fair edges of v in G, resp.) and nodes $v \in V_\forall^{\mathsf{fair}}$ in G are replaced with one of the gadgets at the bottom. The gadgets on the left are called *existential* gadgets and the ones on the right are called *universal* gadgets, referring to the player picking the first move. Nodes in V^n are not altered.

Even though the proof works for all combinations of the gadgets (i.e. one can replace each $v \in V_\exists^{\mathsf{fair}}$ ($v \in V_\forall^{\mathsf{fair}}$) with any of the gadgets on the top (bottom)), due to space constraints we give the intuition only for the existential gadgets.

Imagine all $v \in V^{\text{fair}}$ are replaced with their existential gadgets. Within a subgame that starts at a fair node $v \in V^{\text{fair}}$, the two players intuitively interact as follows. The $\exists$-player gets to pick a number i, indicating the priorities ($2i - 1$ or $2i$) they intend to visit infinitely often in any play that visits v infinitely often. In turn, $\forall$-player gets to either pick an outgoing edge at v (for this, he pays the price of seeing the even priority $2i$), or allow $\exists$ to pick an outgoing edge (in which case he is rewarded with a visit to the odd priority $2i - 1$). Depending on the owner of v, the edge picked by $\forall$ (if $v \in V_{\exists}^{\text{fair}}$), or the edge picked by $\exists$ (if $v \in V_{\forall}^{\text{fair}}$) is required to be contained in E_f. Thus $\forall$ can insist on exploring fair edges at $V_{\exists}^{\text{fair}}$ nodes, but has to pay a price for it; dually, $\forall$ eventually has to allow $\exists$ to explore the fair edges at $V_{\forall}^{\text{fair}}$ nodes to win.

In the full reduced game defined formally in the proof of Thm. 1 below, the owner of a fair node v can *fairly* win from v by either avoiding v from some point on forever, or eventually allowing the opponent player to explore all fair edges leading out of that node. The owner wins by playing *unfairly* if and only if the opponent also plays unfairly and the owner is the $\forall$-player.

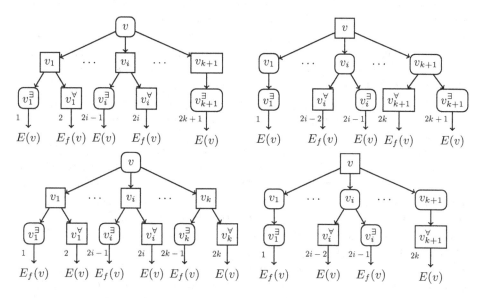

Fig. 4: Existential (left) and universal (right) gadgets for $v \in V_{\exists}^{\text{fair}}$ (top) and $v \in V_{\forall}^{\text{fair}}$ (bottom) in fair parity/$\perp$ games. For $i \in [1, k + 1]$, priorities of nodes $v_i^{\exists}$ and $v_i^{\forall}$ are given below them, priorities of nodes v_i are ignored, and the priority of v is unaltered.

Theorem 1. *Let $G = (A, Parity(\lambda), \perp)$ where $A = (V_{\exists}, V_{\forall}, E, E_f)$ is a fair game arena, $V = V_{\exists} \uplus V_{\forall}$ and $\lambda : V \to [2k]$ is the priority function. Then there exists a parity game G' on the node set V' with $V \subseteq V'$ and $|V'| \leq n(3k + 3)$ over $2k + 1$ priorities such that for $i \in \{\exists, \forall\}$, $\text{Win}_i(G) = \text{Win}_i(G') \cap V$.*

Proof (Sketch). Let $G' = (V_{\exists}', V_{\forall}', E', \Omega : V' \to [2k + 1])$ be the parity game obtained by replacing the fair nodes in G with an arbitrary combination of their corresponding existential and universal gadgets in Fig. 4. Let $V' = V_{\exists}' \cup$

$V'_\forall = V \cup V^{\text{gad}}$ where V represent the nodes coming from G and V^{gad} represent the nodes coming from the gadgets. Note that the maximum priority in G' is $\max_{\text{odd}} = 2k + 1$ which comes only from the gadget nodes V^{gad}. The maximum even priority in G' is $\max_{\text{even}} = 2k$ which can come from both V^{gad} and V. It is easy to see that $|V'| \leq n(3k + 3)$ and G' uses priorities $[2k + 1]$. To prove the correctness, we recall that the winning regions for fair parity/$\perp$ games are given via Eq. (3a), i.e. $v \in \text{Win}_\exists(G)$ if and only if

$$\exists s \in \Sigma. \forall t \in \Pi.\text{fair}_\exists(\text{play}_v(s,t)) \wedge (\text{fair}_\forall(\text{play}_v(s,t)) \Rightarrow \text{play}_v(s,t) \in \alpha). \quad (3\text{a})$$

($\Rightarrow$)   We will first show, $v \in \text{Win}_\exists(G') \cap V \Rightarrow v \in \text{Win}_\exists(G)$. To do so, we will take a (positional) winning $\exists$-strategy s' in G' and construct an $\exists$-strategy s in G such that s is $\exists$-winning in G i.e., s realizes Eq. (3a). That is, for a play ρ in G that starts from v and compliant with s Eq. (3a-s) holds.

$$\text{fair}_\exists(\rho) \wedge (\text{fair}_\forall(\rho) \Rightarrow \rho \in \alpha) \quad (3\text{a-s})$$

For this we will show the two parts of the conjunction separately. We will show (i) $\text{fair}_\exists(\rho)$, i.e. $s \in \Sigma^{\text{fair}}$, (ii) $\text{fair}_\forall(\rho) \Rightarrow \rho \in \alpha$, i.e. every $\forall$-fair play compliant with s is $\exists$-winning w.r.t. the parity condition.

Construction of the s'-subgame $G'_{s'}$: Let s' be a positional $\exists$-strategy winning every play from v in G'. We will denote the subgame of G' where $\exists$ nodes have only the outgoing edges $u \to s'(u)$ by $G'_{s'}$, and call it *the s'-subgame*. Recall that all plays that start from v in $G'_{s'}$ are $\exists$-winning.

Notation of n_u and $succ(u)$: For the existential gadgets for both $V^{\text{fair}}_\exists$ and $V^{\text{fair}}_\forall$, we call the index of the unique successor of u in $G'_{s'}$, n_u. That is, $s'(u) = u_{n_u}$. For the same gadgets, we will denote $s'(u^\exists_{n_u})$ with $succ(u)$. For the universal gadgets for both $V^{\text{fair}}_\exists$ and $V^{\text{fair}}_\forall$, we will let n_u denote the index of the rightmost child of u that is sent to its right child by s'. That is, n_u is the largest index i such that $s'(u_i) = u^\exists_i$. For the same gadgets, we will denote $s'(u^\exists_{n_u})$ with $succ(u)$.

Construction of s: We define $s : V^* \cdot V_\exists \to V$ as follows. For $u \in V^{\text{fair}}_\exists$: 1. If $n_u = k + 1$, we set $s(u) = succ(u)$. 2. Otherwise, $s(u)$ cycles through the set $\{succ(u), E_f(u)\}$ starting from $succ(u)$. For $u \in V_\exists \setminus V^{\text{fair}}_\exists$, we set $s(u) = s'(u)$.

Constraining $G'_{s'}$ with n_u: Here we will constrain $G'_{s'}$ to its subgame by limiting the choices of $\forall$-player from a u replaced by the universal gadget. For every universal gadget encountered in $G'_{s'}$, we limit the choices of $u \in V^{\text{fair}}_\forall$ to only $u \to u_{n_u}$ and $u \to u_{n_u+1}$ (if it exists). So, we remove all the other branches of u out of $G'_{s'}$. We call the remaining game $LG'_{s'}$, standing for *limited $G'_{s'}$*. Note that as $LG'_{s'}$ is a subgame of $G'_{s'}$, it is still $\exists$-winning.

$\exists$-extension: Let ρ be some play in G compliant with s. We define a play ρ' that is called *the $\exists$-extension of $\rho = u^1 u^2 \ldots$* as follows: ρ' is the play on $LG'_{s'}$ that follows ρ while 'prioritising existential nodes on $succ(u)$'. What is meant by this is, for a $u^i \in V^{\text{fair}}$, if $u^{i+1} = succ(u^i)$, then ρ' takes the unique branch in $LG'_{s'}$ that leads to u^{i+1} while passing through an existential node $(u^i)^\exists_j$. That is, regardless of which gadget u^i is replaced by, ρ' takes the branch

$$u^i \to u^i_{n_{u^i}} \to (u^i)^\exists_{n_{u^i}} \to succ(u^i) = u^{i+1} \quad (\text{branch 1})$$

On the other hand if $u^{i+1} \neq succ(u^i)$, then ρ' takes the only other branch in $LG'_{s'}$, that is (branch 2) is taken as

1. If $u^i \in V^{\text{fair}}$ is replaced by an $\exists$-gadget, then $u^i \to u^i_{n_{u^i}} \to (u^i)^{\forall}_{n_{u^i}} \to u^{i+1}$,

2. If $u^i \in V^{\text{fair}}$ is replaced by a $\forall$-gadget, then $u^i \to u^i_{n_{u^i}+1} \to (u^i)^{\forall}_{n_{u^i}+1} \to u^{i+1}$,

Note that these branches do not leave out any possible transition in ρ. That's because 1. all the successors of a $V^{\text{fair}}_{\forall}$ node are covered by one of the branches since (branch 2) leads the universal node $(u^i)^{\forall}_{n_{u^i}}$ or $(u^i)^{\forall}_{n_{u^i}+1}$, which can pick any successor of u^i. 2. all the successors of a $V^{\text{fair}}_{\exists}$ node are covered by one of the branches, since by construction of s, all the successors of u^i in ρ are in the set $\{succ(u^i)\} \cup E_f(u^i)$, where (branch 1) covers the $succ(u^i)$ successors, and (branch 2) covers the $E_f(u^i)$ successors since in this case the universal node $(u^i)^{\forall}_{n_{u^i}}$ or $(u^i)^{\forall}_{n_{u^i}+1}$ can pick any fair successor of u^i.

For $u^i \neq V^{\text{fair}}$, ρ' just takes $u^i \to u^{i+1}$.

So ρ' is well defined, and is a play in $LG'_{s'}$ that starts from v. Thus, ρ' is $\exists$-winning. Observe that if we remove the gadget nodes from ρ', we get ρ. That is, the restriction of ρ' to V, $\rho'|_V = \rho$.

(i) fair$_\exists(\rho)$: Observe that for any ρ in G compliant with s, by construction of s, the only nodes $u \in V^{\text{fair}}_{\exists}$ that ρ may not be fair on, are those for which $n_u = k+1$. So we only need to show that such nodes are seen only finitely often in ρ. Since $\rho|_V = \rho'$, that is equivalent to showing such a u cannot be seen infinitely often in its $\exists$-extension, ρ'. If it is seen infinitely often in ρ', then regardless of the gadget u is replaced with, the branch $u \to u_{k+1} \to u^{\exists}_{k+1}$ is evoked infinitely often, signalling the largest priority $2k + 1$. Therefore, ρ' is won by $\forall$-player, giving a contradiction. Therefore, we conclude ρ is $\exists$-fair.

(ii) fair$_\forall(\rho) \Rightarrow \rho \in \alpha$: Let ρ be $\forall$-fair. Look at the $\exists$-extension ρ' of ρ. Let m be the largest (even) priority in $\text{Inf}(\rho')$. Due to $\rho'|_V = \rho$, all we need to show is the existence of a $u \in \text{Inf}(\rho'|_V)$ that has priority m. Then it automatically implies that the maximum priority in $\text{Inf}(\rho)$ is m, and thus ρ is $\exists$-winning.

We will proceed with proof by contradiction and assume that the priority m appears only in $V^{\text{gad}} \cap \text{Inf}(\rho')$. Now let F be the subgame of $LG'_{s'}$ that consists of nodes and edges taken infinitely often in ρ'. Then, priority m appears in $V^{\text{gad}} \cap F$. These gadget nodes must exist in F due to nodes

- $u \in V^{\text{fair}}$ replaced by existential gadgets, and with $n_u = m \backslash 2$ (which corresponds to (branch 2)-1), or
- $u \in V^{\text{fair}}$ replaced by universal gadgets, and with $n_u = m \backslash 2 - 1$ (which corresponds to (branch 2)-2)

Note that for all such nodes u, (branch 1) of u is also in F. This is because $u \to succ(u)$ is taken infinitely often in ρ. For $u \in V^{\text{fair}}_{\exists}$, this is due to the construction of s, for $u \in V^{\text{fair}}_{\forall}$, this is due to ρ being $\forall$-fair (remember, in this case $succ(u) \in E_f(u)$).

Next, we will remove from F all priority m gadget nodes (and everything reachable only from those nodes). That is, we will prune out (branch 2) of all the nodes that bring in m priority gadget nodes to F. Due to the remaining (branch 1)s, this pruning does not cause any dead-ends. Let's call this pruned subgame of F, H. Observe once more that all plays in H are $\exists$-winning. However, the maximum priority in H is $m-1$. This is due to the remaining (branch 1)s of the pruned nodes having this priority. This implies that all infinite plays starting in H get trapped in a subgame H' of H that doesn't have nodes with priority $m-1$. Since non of the nodes in $V^{\mathsf{fair}} \cap H'$ cause a gadget node with priority m, non of their branches get pruned. That is, all nodes in H' have the same outgoing edges in H' and in F. Then any play that start in H' in F, does not leave H', making H' exactly the set of nodes seen infinitely often in ρ', i.e. $H' = F$. This contradicts our initial assumption that maximum priority seen infinitely often in ρ' being m; therefore proving ρ is $\exists$-winning.

The proof of direction ($\Leftarrow$) is similar to the proof of ($\Rightarrow$), and can be found in detail in the extended version [10].

Remark 2 (Reduction of parity/$\top$ games). In the gadgets from Fig. 4, in order to play unfairly from a $v \in V_\exists^{\mathsf{fair}}$, $\exists$-player has to take its rightmost branch and signal priority $\max_{\mathsf{odd}}$, whereas to play unfairly from $v \in V_\forall^{\mathsf{fair}}$, $\forall$-player has to take the rightmost branch and signal $\max_{\mathsf{even}}$. Since $\max_{\mathsf{odd}} > \max_{\mathsf{even}}$, this dynamic ensures mutually unfair plays are $\forall$-winning. The gadgets for a fair parity/$\top$ game with $\lambda : V \rightarrow [2k]$ can be constructed as follows with the addition of priority $2k+2$: Take the gadgets from Fig. 4. In the existential gadget for $V_\exists^{\mathsf{fair}}$ add another branch $\rightarrow v_{k+1}^\forall \rightarrow E_f(v)$ to v_{k+1} and in the universal gadget for $V_\forall^{\mathsf{fair}}$ add a rightmost branch $\rightarrow v_{k+2} \rightarrow v_{k+2}^\forall \rightarrow E_f(v)$. In the existential gadget for $V_\forall^{\mathsf{fair}}$ add a rightmost branch $\rightarrow v_{k+1} \rightarrow v_{k+1}^\exists \rightarrow E_f(v)$ and in the universal gadget for $V_\forall^{\mathsf{fair}}$ add another branch $\rightarrow v_{k+1}^\exists \rightarrow E_f(v)$ to v_{k+1}.

All the newly added gadget nodes have priority $2k+2$ and therefore $\max_{\mathsf{even}} = 2k+2 > \max_{\mathsf{odd}} = 2k+1$, which ensures that mutually unfair plays are $\exists$-winning. The correctness of the construction follows as a corollary of the reduction of fair parity/parity games given in the next section.

4.2   Reduction of Fair Parity/Parity Games

In this section, we present a quadratic reduction from fair parity/parity to parity games. So let $G = (A, \mathrm{Parity}(\lambda), \mathrm{Parity}(\Gamma))$ where $A = (V_\exists, V_\forall, E, E_f)$ is a fair game arena with $V = V_\exists \uplus V_\forall$ and priority functions $\lambda : V \rightarrow [2k]$, $\Gamma : V \rightarrow [d]$.

The reduction is based on ideas from the previous section, in particular adapting the basic structure of the introduced gadgets. However, in order to correctly treat mutually unfair plays according to the additional parity objective Γ, we annotate game nodes $v \in V$ with two memory values $p \in [d]$ and $b \in \{\exists, \forall\}$. The former is used to store the maximal priority according to Γ that the play has *recently* seen; this value is signalled (and reset after signalling) from time to time in the reduced game. The value b is used to decide (at certain nodes) whether the memory value is signalled, or not.

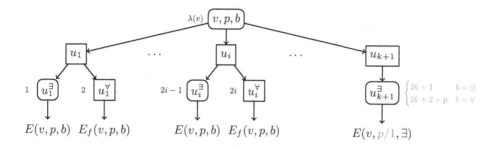

Fig. 5: Gadget for $v \in V_\exists^{\mathsf{fair}}$ in fair parity/parity games; u abbreviates (v, p, b).

It indicates the player that has last taken the rightmost branch in the gadget for one of its fair nodes. If this bit keeps flipping between $\exists$ and $\forall$ forever, then both players intuitively insist on keeping control in one of their respective fair nodes, enabling a mutually unfair play; in the reduced game, the memory content p is signalled (and then reset to 1) whenever the value flips from $\forall$ to $\exists$.

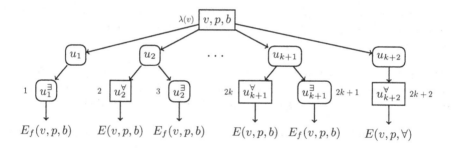

Fig. 6: Gadget for $v \in V_\forall^{\mathsf{fair}}$ in fair parity/parity games; u abbreviates (v, p, b).

Formally, the reduction is as follows. The game is based on the set $V \times [d] \times [1]$ of base nodes, where we use $[1]$ to denote $\{\exists, \forall\}$; intuitively, a node (v, p, b) from this set corresponds to $v \in V$, annotated with memory values p and b as described above. In order to succinctly refer to the combination of taking a move in G and updating the memory components, we overload notation and put

$$E(v, p, b) = \{(w, p', b) \in V \times [d] \times [1] \mid w \in E(v) \text{ and } p' = \max(p, \Gamma(v))\}$$
$$E_f(v, p, b) = \{(w, p', b) \in V \times [d] \times [1] \mid w \in E_f(v) \text{ and } p' = \max(p, \Gamma(v))\}$$

for $(v, p, b) \in V \times [d] \times [1]$. Thus a triple (w, p', b) is contained in $E(v, p, b)$ if there is a move $(v, w) \in E$ and p' is the maximum of the previous memory value p and the current priority $\Gamma(v)$ at v; in $E_f(v, p, b)$, we require $(v, w) \in E_f$ instead.

In both functions, the argument b is used to explicitly set this component of the memory to either $\exists$ or $\forall$. The reduced game consists of subgames that start at annotated nodes $u = (v, p, b) \in V \times [d] \times [1]$. In case that $v \in V^n$, the game just proceeds according to $E(v, p, b)$, with ownership of (v, p, b) determined by whether $v \in V_\exists$ or $v \in V_\forall$; this corresponds to taking a move at a normal node in G, but updating the memory component p, and keeping the component b without modifying it. For fair nodes $v \in V^{\mathsf{fair}}$, the subgame consists of three levels, and after these three steps leads back to a node from $V \times [d] \times [1]$. Fig. 5 and 6 show the subgames that start at $(v, b, p) \in V \times [d] \times [1]$ such that $v \in V^{\mathsf{fair}}_\exists$ and $v \in V^{\mathsf{fair}}_\forall$, respectively, adapting the existential gadget for $v \in V^{\mathsf{fair}}_\exists$ and the universal one for $v \in V^{\mathsf{fair}}_\forall$.

The rightmost branches in these gadgets overwrite the last component b with $\exists$ and $\forall$, respectively. The colored values in the right-most branch in the Fig. 5 gadget depend on the value of b. If $b = \forall$ (corresponding to $\forall$-player being the one that last has taken the right-most branch), then the priority $2k + 2 + p$ is signalled and the memory value p is reset to 1; if $b = \exists$ (corresponding to $\exists$-player having taken the right-most branch last), then the priority $2k + 1$ is signalled and the memory value p does not change.

Theorem 2. *Let $G = (A, Parity(\lambda), Parity(\Gamma))$ where $A = (V_\exists, V_\forall, E, E_f)$ is a fair game arena, $V = V_\exists \uplus V_\forall$ and $\lambda : V \to [2k]$ and $\Gamma : V \to [d]$ are priority functions. Then there exists a parity game G' with $6nd(k+2)$ nodes and $2k+2+d$ priorities with set $V \times [d] \times [1]$ of base nodes such that for all $v \in V$, $\exists$-player wins v in G if and only if $\exists$-player wins $(v, 1, \exists)$ in G'.*

Proof (Sketch). We construct the parity game G' following the above description, using the gadgets from Fig. 5 and 6 to treat fair nodes. The detailed construction and the correctness proof can be found in the extended version [10]. $\square$

We obtain the following bound on strategy sizes for fair parity/parity games.

Lemma 1. *Let G be a fair parity/parity game on n nodes. Then for both players the memory requirement of winning strategies in G is at most $n^2 \cdot n^n$. Furthermore, for each player a family of fair parity/$\perp$ games $(G_n)_{n \in \mathbb{N}}$ exists such that for all n, every winning strategy for the respective player requires memory at least 2^n.*

Proof (Sketch). For the upper bound, we note that in a winning i-strategy for a fair parity/parity game, as constructed in the proof of Thm. 2, the nodes in $V_i \setminus V^{\mathsf{fair}}$ have strategies with quadratic memory, but the nodes in V_i^{fair} may have to traverse all their fair successors, and possibly one more successor. In the worst case, this requires an additional local memory of $|E_f(v)| + 1 \le n$ for each $v \in V_i^{\mathsf{fair}}$, and causes an exponential blowup in the overall memory required.

For the lower bound, we consider the case for $\exists$-player; the result for $\forall$-player is obtained by switching the player's roles. Define the family $(G_n)_{n \in \mathbb{N}}$ of games by letting G_n (for $n \in \mathbb{N}$) have exactly $n+1$ nodes, one node x owned by $\forall$-player and n nodes y_i owned by $\exists$-player; let there be an edge from x to any node y_i and two *fair* edges from any node y_i back to x. Let all nodes have priority 0. Then any winning $\exists$-strategy in G_n necessarily is $\exists$-fair. There is a fair $\exists$-strategy s that uses one bit as local memory for each node $y_i \in V_\exists^{\text{fair}}$, and therefore uses memory of overall size 2^n. The claim follows since there is no $\exists$-fair strategy that uses less memory than s, which is shown by induction on n. □

5 Fixpoint Characterization of Winning Regions

In this section, we will characterize the winning regions in fair games with parity conditions by means of fixpoint expressions. Thereby we provide an alternative, symbolic route to solve such games, rather than by reducing to parity games. We start by briefly recalling details on Boolean fixpoint expressions.

Fixpoint expressions and fixpoint games. Let U be a finite set, let o be a fixed number and let $f : \mathcal{P}(U)^o \to \mathcal{P}(U)$ be a monotone function, that is, assume that whenever we have sets $X_j, Y_j \subseteq U$ such that $X_j \subseteq Y_j$ for all $1 \leq j \leq o$, then $f(X_1, \ldots, X_o) \subseteq f(Y_1, \ldots, Y_o)$. Then f and o induce the *fixpoint expression*

$$e = \eta_o X_o. \, \eta_{o-1} X_{o-1}. \, \ldots . \nu X_2. \, \mu X_1. \, f(X_1, \ldots, X_o) \tag{5}$$

where $\eta_i = \nu$ if i is even and $\eta_i = \mu$ if i is odd. We define the semantics of fixpoint expressions using parity games. Given a fixpoint expression e, the associated *fixpoint game* $G_e = (W_\exists, W_\forall, E, \text{Parity}(\kappa))$ for the priority function $\kappa : W_\exists \cup W_\forall \to [o]$ is the following parity game. We put $W_\exists = U \times \{1, \ldots, o\}$, $W_\forall = \mathcal{P}(U)^o$. Moves and priorities are defined by

$$E(v, i) = \{\overline{Z} \in W_\forall \mid v \in f(\overline{Z})\} \qquad \kappa(v, i) = i$$
$$E(\overline{Z}) = \{(v, i) \mid v \in Z_i\} \qquad \kappa(\overline{Z}) = 0$$

for $(v, i) \in W_\exists$ and $\overline{Z} = (Z_1, \ldots, Z_o) \in W_\forall$. Then we say that $v \in U$ is *contained* in e (denoted $v \in e$) if and only if $\exists$-player wins the node $(v, 1)$ in G_e.

Remark 3. The above game semantics for fixpoint expressions has been shown to be equivalent to the more traditional Knaster-Tarski semantics [3]; the cited work takes place in a more general setting and therefore uses slightly more verbose parity games.

Next we present a fixpoint characterization of the winning regions in fair games of the form $G = (A, \text{Parity}(\lambda), \perp)$ where $A = (V_\exists, V_\forall, E, E_f)$ is a fair game arena, $V = V_\exists \uplus V_\forall$ and $\lambda : V \to [2k]$ a priority function. To be able to write fixpoint expressions over such games we define monotone operators on subsets of V by putting

$$\Diamond X = \{v \in V \mid E(v) \cap X \neq \emptyset\} \qquad \Box X = \{v \in V \mid E(v) \subseteq X\}$$
$$\Diamond_f X = \{v \in V \mid E_f(v) \cap X \neq \emptyset\} \qquad \Box_f X = \{v \in V \mid E_f(v) \subseteq X\}$$

for $X \subseteq V$ and also put $\text{Cpre}(X) = (V_\exists \cap \Diamond X) \cup (V_\forall \cap \Box X)$. Then $\text{Cpre}(X)$ is the set of nodes from which $\exists$-player can force the game to reach a node from X in one step. Also, we define $C_i = \{v \in V \mid \lambda(v) = i\}$ for $1 \leq i \leq 2k$.

Using this notation, we define a function parity : $\mathcal{P}(V)^{2k} \to \mathcal{P}(V)$ by putting

$$\text{parity}(X_1, \ldots, X_{2k}) := (C_1 \cap \text{Cpre}(X_1)) \cup \ldots \cup (C_k \cap \text{Cpre}(X_{2k}))$$

for $(X_1, \ldots, X_{2k}) \subseteq \mathcal{P}(V)^{2k}$. This function is monotone and it is well-known (see e.g [23]) that the fixpoint induced by parity characterizes the winning region in parity games with priorities 1 through $2k$. This formula will still apply to 'normal' nodes V^n in the fixpoint characterization of fair parity games.

We follow the gadget constructions from Fig. 4 (using their *existential* versions) to define the following additional functions. For $1 \leq i < k$, put

$$\text{Apre}_\exists(X_i, X_{i+1}) = \Diamond X_i \cap \Box_f X_{i+1} \qquad \text{Apre}_\forall(X_i, X_{i+1}) = \Diamond_f X_i \cap \Box X_{i+1},$$

encoding nodes $(v^\forall, 2i)$ for $v \in V_\exists^{\text{fair}}$ and $v \in V_\forall^{\text{fair}}$, respectively (here, Apre stands for *alternative* predecessor function, as it encodes the additional $\forall$-choice of whether a fair edge is to be taken). Then, we let $I_p = \{i \mid i \text{ odd}, p \leq i < 2k\}$ denote the set of odd priorities that lie between p and $2k$, and put

$$\phi_{\exists,p}^{\text{fair}} = \begin{cases} \bigcup_{i \in I_p} \text{Apre}_\exists(X_i, X_{i+1}) \cup \Diamond X_{2k+1} & p \text{ is odd} \\ \bigcup_{i \in I_p} \text{Apre}_\exists(X_i, X_{i+1}) \cup \Diamond X_{2k+1} \cup \Box_f X_p & p \text{ is even,} \end{cases}$$

$$\phi_{\forall,p}^{\text{fair}} = \begin{cases} \bigcup_{i \in I_p} \text{Apre}_\forall(X_i, X_{i+1}) & p \text{ is odd} \\ \bigcup_{i \in I_p} \text{Apre}_\forall(X_i, X_{i+1}) \cup \Box X_p & p \text{ is even} \end{cases}$$

Using this notation, the winning region for the existential player in fair parity/$\bot$ games with priorities 1 through $2k$ can be characterized by the fixpoint expression induced by $2k + 1$ and the function χ that is defined to map $(X_1, \ldots, X_{2k+1}) \in \mathcal{P}(V)^{2k+1} \to \mathcal{P}(V)$ to the set

$$\chi(X_1, \ldots, X_{2k+1}) = (V^n \cap \text{parity}) \cup$$
$$(V_\exists^{\text{fair}} \cap \bigcup_{i \in [2k+1]} C_i \cap \phi_{\exists,i}^{\text{fair}}) \cup$$
$$(V_\forall^{\text{fair}} \cap \bigcup_{i \in [2k+1]} C_i \cap \phi_{\forall,i}^{\text{fair}})$$

The function χ therefore treats normal nodes from V^n in the same way as nodes in standard parity games are treated, but for fair nodes with priority i, the functions $\phi_{\exists,i}^{\text{fair}}$ and $\phi_{\forall,i}^{\text{fair}}$ are used to encode the respective gadget construction. The full fixpoint expression then is

$$e = \mu X_{2k+1}. \nu X_{2k}. \mu X_{2k-1} \ldots \nu X_2. \mu X_1. \chi(X_1, \ldots, X_{2k+1}) \qquad (6)$$

The first result of this section is that the fixpoint expression (6) characterizes the winning region of $\exists$-player in fair parity/$\bot$ games.

Theorem 3. *Let* $G = (A, Parity(\lambda), \bot)$ *where* $A = (V_\exists, V_\forall, E, E_f)$ *is a fair game arena,* $V = V_\exists \uplus V_\forall$ *and* $\lambda : V \to [2k]$ *is the priority function. Then the fixpoint expression given in* (6) *characterizes* $\text{Win}_\exists(G)$.

Proof (Sketch). The proof is by mutual transformation of winning strategies in G and in the semantic game G_e for (6). The full proof can be found in [10].

We note that for $\forall$-fair parity games ($V_\exists^{\text{fair}} = \emptyset$), Eq. (6) instantiates to the fixpoint characterization given in [4]; it follows that the parity game reductions from Sec. 4 apply to the one-sided fair parity games considered in [4] as well.

For fair parity/parity games, we obtain a similar fixpoint characterization, encoding the reduction to parity games presented in Subsec. 4.2 along the lines of Figures 5 and 6. Here, all involved functions work over (subsets of) the set $V \times [d] \times [1]$ of base nodes, consisting of game nodes that are annotated with memory values. The definition of the fixpoint expression for fair parity/parity games is straight-forward but somewhat technical since the updating and resetting mechanisms for the memory values have to be accommodated. For brevity, we refrain from elaborating the required notation and the full fixpoint expression here, and state just the main result that yields a symbolic fixpoint algorithm for fair parity/parity games; full details can be found in the extended version [10].

Theorem 4. *Let* $G = (A, Parity(\lambda), Parity(\Gamma))$ *where* $A = (V_\exists, V_\forall, E, E_f)$ *is a fair game arena,* $V = V_\exists \uplus V_\forall$ *and* $\lambda : V \to [2k]$, $\Gamma : V \to [d]$ *are priority functions. Then there is a fixpoint expression over* $V \times [d] \times [1]$ *with alternation depth* $2(k + 1) + d$ *that characterizes* $\mathsf{Win}_\exists(G)$.

Proof (Sketch). Again the proof is by mutual transformation of winning strategies in G and in the semantic game G_e for the fixpoint expression. The full proof can be found in the extended version [10].

6   Conclusion

We introduce two-player games with local transition-fairness constraints for both players, allowing two objectives α and β to decide the winner of plays in which both players play fair and both players play unfair, respectively. We show the determinacy of this class of games in the case that α and β are ω-regular objectives. In the special case that both α and β are parity conditions, there is a reduction to standard parity games with blow-up quadratic in the number of priorities used by α and β; if $\beta = \top$ or $\beta = \bot$, the reduction becomes even linear. We present both enumerative and symbolic methods to realize this reduction; in the process, we also obtain an exponential tight bound on the memory required by winning strategies in fair parity/parity games. We expect that the central idea behind the reduction generalizes from parity objectives to more general settings such as fair games in which α and β are Rabin, Streett, or even Emerson-Lei conditions, but leave this issue for future work.

Acknowledgments

We are grateful to an anonymous referee for contributing the proof of the lower bound in Lemma 1.

References

1. Aminof, B., Ball, T., Kupferman, O.: Reasoning about systems with transition fairness. In: Baader, F., Voronkov, A. (eds.) Logic for Programming, Artificial Intelligence, and Reasoning, 11th International Conference, LPAR 2004, Montevideo, Uruguay, March 14-18, 2005, Proceedings. Lecture Notes in Computer Science, vol. 3452, pp. 194–208. Springer (2004). https://doi.org/10.1007/978-3-540-32275-7_14

2. Aminof, B., Giacomo, G.D., Rubin, S.: Stochastic fairness and language-theoretic fairness in planning in nondeterministic domains. In: Beck, J.C., Buffet, O., Hoffmann, J., Karpas, E., Sohrabi, S. (eds.) Proceedings of the Thirtieth International Conference on Automated Planning and Scheduling, Nancy, France, October 26-30, 2020. pp. 20–28. AAAI Press (2020), `https://ojs.aaai.org/index.php/ICAPS/article/view/6641`

3. Baldan, P., König, B., Mika-Michalski, C., Padoan, T.: Fixpoint games on continuous lattices. Proc. ACM Program. Lang. **3**(POPL), 26:1–26:29 (2019). https://doi.org/10.1145/3290339

4. Banerjee, T., Majumdar, R., Mallik, K., Schmuck, A., Soudjani, S.: Fast symbolic algorithms for omega-regular games under strong transition fairness. TheoretiCS **2** (2023). https://doi.org/10.46298/theoretics.23.4

5. Büchi, J., Landweber, L.: Solving sequential conditions by finite-state strategies. Trans. Amer. Math. Soc. **138**, 295–311 (1969)

6. Chatterjee, K., de Alfaro, L., Faella, M., Majumdar, R., Raman, V.: Code aware resource management. Formal Methods Syst. Des. **42**(2), 146–174 (2013). https://doi.org/10.1007/s10703-012-0170-4

7. Chatterjee, K., Jurdzinski, M., Henzinger, T.: Simple stochastic parity games. In: In Proceedings of the International Conference for Computer Science Logic (CSL). pp. 100–113 (2003), `http://chess.eecs.berkeley.edu/pubs/729.html`

8. Church, A.: Application of recursive arithmetic to the problem of circuit synthesis. Journal of Symbolic Logic **28**(4) (1963)

9. D'Ippolito, N., Rodríguez, N., Sardiña, S.: Fully observable non-deterministic planning as assumption-based reactive synthesis. J. Artif. Intell. Res. **61**, 593–621 (2018). https://doi.org/10.1613/jair.5562

10. Hausmann, D., Piterman, N., Sağlam, I., Schmuck, A.K.: Fair ω-regular games (2024), extended version available at `https://arxiv.org/abs/310.13612`

11. Majumdar, R., Mallik, K., Schmuck, A.K., Soudjani, S.: Symbolic control for stochastic systems via finite parity games. Nonlinear Analysis: Hybrid Systems **51**, 101430 (2024). https://doi.org/10.1016/j.nahs.2023.101430

12. Martin, D.: Borel determinacy. Annals of Mathematics **65**, 363–371 (1975)

13. Mazala, R.: Infinite games. In: Grädel, E., Thomas, W., Wilke, T. (eds.) Automata, Logics, and Infinite Games: A Guide to Current Research. Lecture Notes in Computer Science, vol. 2500, pp. 23–42. Springer (2001)

14. Meyer, P.J., Sickert, S., Luttenberger, M.: Strix: Explicit reactive synthesis strikes back! In: 30th International Conference on Computer Aided Verification. Lecture Notes in Computer Science, vol. 10981, pp. 578–586. Springer (2018)

15. Nilsson, P., Ozay, N., Liu, J.: Augmented finite transition systems as abstractions for control synthesis. Discret. Event Dyn. Syst. **27**(2), 301–340 (2017). https://doi.org/10.1007/s10626-017-0243-z

16. Piterman, N., Pnueli, A., Sa'ar, Y.: Synthesis of reactive(1) designs. In: Proceedings of the 7th International Conference on Verification, Model Checking, and Abstract Interpretation. p. 364–380. VMCAI'06, Springer-Verlag, Berlin, Heidelberg (2006)

17. Pnueli, A., Rosner, R.: A framework for the synthesis of reactive modules. In: Proc. Intl. Conf. on Concurrency: Concurrency 88. Lecture Notes in Computer Science, vol. 335, pp. 4–17. Springer-Verlag (1988)

18. Rabin, M.: Decidability of second order theories and automata on infinite trees. Trans. Amer. Math. Soc. **141**, 1–35 (1969)

19. Sağlam, I., Schmuck, A.K.: Solving odd-fair parity games. In: 42th IARCS Annual Conference on Foundations of Software Technology and Theoretical Computer Science (2023), (to appear)

20. Schewe, S., Finkbeiner, B.: Bounded synthesis. In: 4th Int. Symp. on Automated Technology for Verification and Analysis. Lecture Notes in Computer Science, vol. 4218, pp. 245–259. Springer (2006)

21. Thistle, J.G., Malhamé, R.: Control of ω-automata under state fairness assumptions. Systems & control letters **33**(4), 265–274 (1998)

22. Thomas, W.: Languages, automata, and logic. Handbook of Formal Language Theory **III**, 389–455 (1997)

23. Walukiewicz, I.: Monadic second-order logic on tree-like structures. Theor. Comput. Sci. **275**(1-2), 311–346 (2002). https://doi.org/10.1016/S0304-3975(01)00185-2

24. Zielonka, W.: Infinite games on finitely coloured graphs with applications to automata on infinite trees. Theor. Comput. Sci. **200**(1-2), 135–183 (1998). https://doi.org/10.1016/S0304-3975(98)00009-7

Stochastic Window Mean-Payoff Games

Laurent Doyen[1][iD], Pranshu Gaba[2(✉)][iD], and Shibashis Guha[2][iD]

[1] CNRS & LMF, ENS Paris-Saclay, Gif-sur-Yvette, France
ldoyen@lmf.cnrs.fr
[2] Tata Institute of Fundamental Research, Mumbai, India
{pranshu.gaba,shibashis}@tifr.res.in

Abstract. Stochastic two-player games model systems with an environment that is both adversarial and stochastic. The environment is modeled by a player (Player 2) who tries to prevent the system (Player 1) from achieving its objective. We consider finitary versions of the traditional mean-payoff objective, replacing the long-run average of the payoffs by payoff average computed over a finite sliding window. Two variants have been considered: in one variant, the maximum window length is fixed and given, while in the other, it is not fixed but is required to be bounded. For both variants, we present complexity bounds and algorithmic solutions for computing strategies for Player 1 to ensure that the objective is satisfied with positive probability, with probability 1, or with probability at least p, regardless of the strategy of Player 2. The solution crucially relies on a reduction to the special case of non-stochastic two-player games. We give a general characterization of prefix-independent objectives for which this reduction holds. The memory requirement for both players in stochastic games is also the same as in non-stochastic games by our reduction. Moreover, for non-stochastic games, we improve upon the upper bound for the memory requirement of Player 1 and upon the lower bound for the memory requirement of Player 2.

Keywords: Stochastic games · Finitary objectives · Mean-payoff · Reactive synthesis

1 Introduction

We consider two-player turn-based stochastic games played on graphs. Games are a central model in computer science, in particular for the verification and synthesis of reactive systems [18, 11, 17]. A stochastic game is played by two players on a graph with stochastic transitions, where the players represent the system and the adversarial environment, while the objective represents the functional requirement that the synthesized system aims to satisfy with a probability p as large as possible. The vertices of the graph are partitioned into system, environment, and probabilistic vertices. A stochastic game is played in infinitely many rounds, starting by initially placing a token on some vertex. In every round, if the token is on a system or an environment vertex, then the owner of the vertex chooses a successor vertex; if the token is on a probabilistic vertex, then the

© The Author(s) 2024
N. Kobayashi and J. Worrell (Eds.): FoSSaCS 2024, LNCS 14574, pp. 34–54, 2024.
https://doi.org/10.1007/978-3-031-57228-9_3

successor vertex is chosen according to a given probability distribution. The token moves to the successor vertex, from where the next round starts. The outcome is an infinite sequence of vertices, which is winning for the system if it satisfies the given objective. The associated *quantitative satisfaction problem* is to decide, given a threshold p, whether the system can win with probability at least p. The *almost-sure problem* is the special case where $p = 1$, and the *positive problem* is to decide whether the system can win with positive probability. The almost-sure and the positive problems are referred to as the *qualitative satisfaction problems*. When the answer to these decision problems is Yes, it is useful to construct a winning strategy for the system that can be used as a model for an implementation that ensures the objective be satisfied with the given probability.

Traditional objectives in stochastic games are ω-regular such as reachability, safety, and parity objectives [11], or quantitative such as mean-payoff objectives [16, 27]. For example, a parity objective may specify that every request of the environment is eventually granted by the system, and a mean-payoff objective may specify the long-run average power consumption of the system. A well-known drawback of parity and mean-payoff objectives is that only the long-run behaviour of the system may be specified [1, 9, 21], allowing weird transient behaviour: for example, a system may grant all its requests but with an unbounded response time; or a system with long-run average power consumption below some threshold may exhibit arbitrarily long (but finite) sequences with average power consumption above the threshold. This limitation has led to considering finitary versions of those objectives [9, 23, 8], where the sequences of undesired transient behaviours must be of fixed or bounded length. Window mean-payoff objectives [8] are quantitative finitary objectives that strengthen the traditional mean-payoff objective: the satisfaction of a window mean-payoff objective implies the satisfaction of the standard mean-payoff objective. Given a length $\ell \geq 1$, the fixed window mean-payoff objective ($\mathsf{FWMP}(\ell)$) is satisfied if except for a finite prefix, from every point in the play, there exists a window of length at most ℓ starting from that point such that the mean payoff of the window is at least a given threshold. In the bounded window mean-payoff objective (BWMP), it is sufficient that there exists some length ℓ for which the $\mathsf{FWMP}(\ell)$ objective is satisfied.

Contributions. We present algorithmic solutions for stochastic games with window mean-payoff objectives, and show that the positive and almost-sure problems are solvable in polynomial time for $\mathsf{FWMP}(\ell)$ (Theorem 5), and are in NP $\cap$ coNP for BWMP (Theorem 6). The complexity result for the almost-sure problem entails that the quantitative satisfaction problem is in NP$\cap$coNP (for both the fixed and bounded version), using standard techniques for solving stochastic games with prefix-independent objectives [13]. Note that the NP $\cap$ coNP bound for the quantitative satisfaction problem matches the special case of reachability objectives in simple stochastic games [14], and thus would require a major breakthrough to be improved.

As a consequence, using the $\mathsf{FWMP}(\ell)$ objective instead of the standard mean-payoff objective provides a stronger guarantee on the system, and even with a polynomial complexity for the positive and the almost-sure problems (which is not known for mean-payoff objectives), and at no additional computational cost for the quantitative satisfaction problem. The solution relies on a reduction to non-stochastic two-player games (stochastic games without probabilistic vertices). The key result is to show that in order to win positively from some vertex of the game graph, it is necessary to win from some vertex of the non-stochastic game obtained by transforming all probabilistic vertices into adversarial vertices. While this condition, that we call the sure-almost-sure (SAS) property (Definition 1), was used to solve finitary Streett objectives [13], we follow a similar approach and generalize it to arbitrary prefix-independent objectives (Theorem 4). The bounds on the memory requirement of optimal strategies of Player 1 can also be derived from the key result, and are the same as optimal bounds for non-stochastic games. For the FWMP(ℓ) and BWMP objectives in particular, we show that the memory requirement of Player 2 is also no more than the optimal memory required by winning strategies in non-stochastic games.

As solving a stochastic game with a prefix-independent objective φ reduces to solving non-stochastic games with objective φ and showing that φ satisfies the SAS property, we show that the FWMP(ℓ) and BWMP objectives satisfy the SAS property (Lemma 4, Lemma 5) and rely on the solution of non-stochastic games with these objectives [8] to complete the reduction.

We improve the memory bounds for optimal strategies of both players in non-stochastic games. It is stated in [8] that $|V| \cdot \ell$ memory suffices for both players, where $|V|$ and ℓ are the number of vertices and the window length respectively. In [6, Theorem 2] and [19, Theorem 6.4], the bound is loosened to $\mathcal{O}(w_{\max} \cdot \ell^2)$ and $\mathcal{O}(w_{\max} \cdot \ell^2 \cdot |V|)$ for Player 1 and Player 2 respectively, where $w_{\max}$ is the maximum absolute payoff in the graph, as the original tighter bounds [8] were stated without proof. Since the payoffs are given in binary, this is exponential in the size of the input. In contrast, we tighten the bounds stated in [8]. We show that for Player 1, memory ℓ suffices (Theorem 1), and improve the bound on memory of Player 2 strategies as follows. We compute the set W of vertices from which Player 2 can ensure that the mean payoff remains negative for ℓ steps, as well as the vertices from which Player 2 can ensure that the game reaches W. These vertices are identified recursively on successive subgames of the original input game. If k is the number of recursive calls, then we show that $k \cdot \ell$ memory suffices for Player 2 to play optimally (Theorem 2). Note that $k \leq |V|$. We also provide a lower bound on the memory size for Player 2. Given $\ell \geq 2$, for every $k \geq 1$, we construct a graph with a set V of vertices such that Player 2 requires at least $k + 1 = \frac{1}{2}(|V| - \ell + 3)$ memory to play optimally (Theorem 3). This is an improvement over the result in [8] which showed that memoryless strategies do not suffice for Player 2. Our result is quite counterintuitive since given an open window (a window in which every prefix has a total payoff less than 0) that needs to be kept open for another $j \leq \ell$ steps from a vertex v, one would conjecture that it is sufficient for a Player 2 winning strategy to choose an edge from v that leads to the minimum payoff over paths of length j. Thus for every j, Player 2 should choose a fixed edge and hence memory of size ℓ should suffice. However, we show that this is not the case.

To the best of our knowledge, this work leads to the first study of stochastic games with finitary quantitative objectives.

Related work. Window mean-payoff objectives were first introduced in [8] for non-stochastic games, where solving FWMP(ℓ) was shown to be in PTIME and BWMP in NP∩coNP. These have also been studied for Markov decision processes (MDPs) in [4,3]. In [4], a threshold probability problem has been studied, while in [3], the authors studied the problem of maximising the expected value of the window mean-payoff. Positive, almost-sure, and quantitative satisfaction of BWMP in MDPs are in NP ∩ coNP [4], the same as in non-stochastic games.

Parity objectives can be viewed as a special case of mean-payoff objectives [22]. A bounded window parity objective has been studied in [9, 20, 12] where a play satisfies the objective if from some point on, there exists a bound ℓ such that from every state with an odd priority, a smaller even priority occurs within at most ℓ steps. Non-stochastic games with bounded window parity objectives can be solved in PTIME [20, 12]. Stochastic games with bounded window parity objectives have been studied in [13] where the positive and almost-sure problems are in PTIME while the quantitative satisfaction problem is in NP ∩ coNP. A fixed version of the window parity objective has been studied for two-player games and shown to be PSPACE-complete [26]. Another window parity objective has been studied in [5] for which both the fixed and the bounded variants have been shown to be in PTIME for non-stochastic games. The threshold problem for this objective has also been studied in the context of MDPs, and both fixed and bounded variants are in PTIME [4]. Finally, synthesis for *bounded* eventuality properties in LTL is 2-EXPTIME-complete [23].

Due to lack of space, some of the proofs have been omitted. A full version of the paper can be found in [15].

2   Preliminaries

Stochastic games. We consider two-player turn-based zero-sum stochastic games (or simply, stochastic games in the sequel). The two players are referred to as Player 1 and Player 2. A *stochastic game* is a weighted directed graph $\mathcal{G} = ((V, E), (V_1, V_2, V_\Diamond), \mathbb{P}, w)$, where:

- (V, E) is a directed graph with a finite set V of vertices and a set $E \subseteq V \times V$ of directed edges such that for all vertices $v \in V$, the set $E(v) = \{v' \in V \mid (v, v') \in E\}$ of out-neighbours of v is nonempty, i.e., $E(v) \neq \varnothing$ (no deadlocks);
- $(V_1, V_2, V_\Diamond)$ is a partition of V. The vertices in V_1 belong to Player 1, the vertices in V_2 belong to Player 2, and the vertices in $V_\Diamond$ are called probabilistic vertices (in figures, Player 1 vertices are shown as circles, Player 2 vertices as boxes, and probabilistic vertices as diamonds, and we use pronouns "she/her" for Player 1 and "he/him" for Player 2);

- $\mathbb{P}\colon V_\Diamond \to \mathcal{D}(V)$, where $\mathcal{D}(V)$ is the set of probability distributions over V, is a transition function that maps probabilistic vertices $v \in V_\Diamond$ to a probability distribution $\mathbb{P}(v)$ over the set $E(v)$ of out-neighbours of v such that $\mathbb{P}(v)(v') > 0$ for all $v' \in E(v)$ (i.e., all out-neighbours have nonzero probability); for the algorithmic and complexity results, we assume that probabilities are given as rational numbers.
- $w\colon E \to \mathbb{Q}$ is a *payoff function* assigning a rational payoff to every edge in the game.

Stochastic games are played in rounds. The game starts by initially placing a token on some vertex. At the beginning of a round, if the token is on a vertex v, and $v \in V_i$ for $i \in \{1,2\}$, then Player i chooses an out-neighbour $v' \in E(v)$; otherwise $v \in V_\Diamond$, and an out-neighbour $v' \in E(v)$ is chosen with probability $\mathbb{P}(v)(v')$. Player 1 receives from Player 2 the amount $w(v, v')$ given by the payoff function, and the token moves to v' for the next round. This continues ad infinitum, resulting in an infinite sequence $\pi = v_0 v_1 v_2 \cdots \in V^\omega$ such that $(v_i, v_{i+1}) \in E$ for all $i \geq 0$, called a *play*. For $i < j$, we denote by $\pi(i, j)$ the *infix* $v_i v_{i+1} \cdots v_j$ of π. Its length is $|\pi(i, j)| = j - i$, the number of edges. We denote by $\pi(0, j)$ the finite *prefix* $v_0 v_1 \cdots v_j$ of π, and by $\pi(i, \infty)$ the infinite *suffix* $v_i v_{i+1} \ldots$ of π. We denote by $\mathsf{Plays}_\mathcal{G}$ and $\mathsf{Prefs}_\mathcal{G}$ the set of all plays and the set of all prefixes in $\mathcal{G}$ respectively; the symbol $\mathcal{G}$ is omitted when it can easily be derived from the context. We denote by $\mathsf{First}(\rho)$ and $\mathsf{Last}(\rho)$ the first vertex and the last vertex of a prefix $\rho \in \mathsf{Prefs}_\mathcal{G}$ respectively. We denote by $\mathsf{Prefs}_\mathcal{G}^i$ ($i \in \{1,2\}$) the set of all prefixes ρ such that $\mathsf{Last}(\rho) \in V_i$.

Objectives. An *objective* φ is a Borel-measurable subset of $\mathsf{Plays}_\mathcal{G}$ [2]. A play $\pi \in \mathsf{Plays}_\mathcal{G}$ *satisfies* an objective φ if $\pi \in \varphi$. In a (zero-sum) stochastic game $\mathcal{G}$ with objective φ, the objective of Player 1 is φ, and the objective of Player 2 is the complement set $\overline{\varphi} = \mathsf{Plays}_\mathcal{G} \setminus \varphi$. Common examples of objectives are qualitative objectives such as reachability, safety, Büchi, and coBüchi.

An objective φ is *closed under suffixes* if for all plays π satisfying φ, all suffixes of π also satisfy φ, that is, $\pi(j, \infty) \in \varphi$ for all $j \geq 0$. An objective φ is *closed under prefixes* if for all plays π satisfying φ, for all prefixes ρ such that the concatenation $\rho \cdot \pi$ is a play in $\mathcal{G}$, i.e., $\rho \cdot \pi \in \mathsf{Plays}_\mathcal{G}$, we have that $\rho \cdot \pi \in \varphi$. An objective φ is *prefix-independent* if it is closed under both prefixes and suffixes. An objective φ is closed under suffixes if and only if the complement objective $\overline{\varphi}$ is closed under prefixes. Thus, an objective φ is prefix-independent if and only if its complement $\overline{\varphi}$ is prefix-independent.

Strategies. A (deterministic) *strategy* for Player $i \in \{1,2\}$ in a game $\mathcal{G}$ is a function $\sigma_i : \mathsf{Prefs}_\mathcal{G}^i \to V$ that maps prefixes ending in a vertex $v \in V_i$ to a successor of v. The set of all strategies of Player $i \in \{1,2\}$ in the game $\mathcal{G}$ is denoted by Λ_i. Strategies can be realised as the output of a (possibly infinite-state) Mealy machine. A *Mealy machine* is a deterministic transition system with transitions labelled by an input/output pair. Formally, a Mealy machine M is a tuple $(Q, q_0, \Sigma_i, \Sigma_o, \Delta, \delta)$ where Q is the set of states of M (the memory of the induced strategy), $q_0 \in Q$ is the initial state, Σ_i is the input alphabet, Σ_o is the output alphabet, $\Delta \colon Q \times \Sigma_i \to Q$ is a transition function that reads the current state of M and an input letter and returns the next state of M, and $\delta \colon Q \times \Sigma_i \to \Sigma_o$ is an output function that reads the current state of M and an input letter and returns an output letter. We point the reader to [15] for a description of how a strategy is defined by a Mealy machine.

The *memory size* of a strategy σ_i is the smallest number of states a Mealy machine defining σ_i can have. A strategy σ_i is *memoryless* if $\sigma_i(\rho)$ only depends on the last element of the prefix ρ, that is for all prefixes $\rho, \rho' \in \mathsf{Prefs}_{\mathcal{G}}^i$ if $\mathsf{Last}(\rho) = \mathsf{Last}(\rho')$, then $\sigma_i(\rho) = \sigma_i(\rho')$. Memoryless strategies can be defined by Mealy machines with only one state.

A play $\pi = v_0 v_1 \cdots$ is *consistent* with a strategy $\sigma_i \in \Lambda_i$ ($i \in \{1,2\}$) if $v_{j+1} = \sigma_i(\pi(0,j))$ for all $j \geq 0$ such that $v_j \in V_i$. A play π is an *outcome* of σ_1 and σ_2 if it is consistent with both σ_1 and σ_2. We denote by $\mathsf{Pr}_{\mathcal{G},v}^{\sigma_1,\sigma_2}(\varphi)$ the probability that an outcome of σ_1 and σ_2 in $\mathcal{G}$ with initial vertex v satisfies φ.

Non-stochastic two-player games. A stochastic game without probabilistic vertices (that is, with $V_\Diamond = \varnothing$) is called a *non-stochastic two-player game* (or simply, non-stochastic game in the sequel). In a non-stochastic game $\mathcal{G}$ with objective φ, a strategy σ_i is *winning* from a vertex $v \in V$ for Player i ($i \in \{1,2\}$) if every play in $\mathcal{G}$ with initial vertex v that is consistent with σ_i satisfies the objective φ. A vertex $v \in V$ is *winning* for Player i in $\mathcal{G}$ if Player i has a winning strategy in $\mathcal{G}$ from v. The set of vertices in V that are winning for Player i in $\mathcal{G}$ is the *winning region* of Player i in $\mathcal{G}$, denoted $\langle\!\langle i \rangle\!\rangle_{\mathcal{G}}(\varphi)$. If a vertex v belongs to the winning region of Player i ($i \in \{1,2\}$), then Player i is said to play *optimally* from v if she follows a winning strategy.

Subgames. Given a stochastic game $\mathcal{G} = ((V,E),(V_1,V_2,V_\Diamond),\mathbb{P},w)$, a subset $V' \subseteq V$ of vertices *induces* a subgame if (i) every vertex $v' \in V'$ has an out-neighbour in V', that is $E(v') \cap V' \neq \varnothing$, and ($ii$) every probabilistic vertex $v' \in V_\Diamond \cap V'$ has all out-neighbours in V', that is $E(v') \subseteq V'$. The induced *subgame* is $((V',E'),(V_1 \cap V', V_2 \cap V', V_\Diamond \cap V'),\mathbb{P}',w')$, where $E' = E \cap (V' \times V')$, and $\mathbb{P}'$ and w' are restrictions of $\mathbb{P}$ and w respectively to (V',E'). We denote this subgame by $\mathcal{G} \upharpoonright V'$. Let φ be an objective in the stochastic game $\mathcal{G}$. We define the restriction of φ to a subgame $\mathcal{G}'$ of $\mathcal{G}$ to be the set of all plays in $\mathcal{G}'$ satisfying φ, that is, the set $\mathsf{Plays}_{\mathcal{G}'} \cap \varphi$.

Satisfaction probability. A strategy σ_1 of Player 1 is *winning with probability p* from an initial vertex v in $\mathcal{G}$ for objective φ if $\mathsf{Pr}_{\mathcal{G},v}^{\sigma_1,\sigma_2}(\varphi) \geq p$ for all strategies σ_2 of Player 2. A strategy σ_1 of Player 1 is *positive* winning (resp., *almost-sure* winning) from v for Player 1 in $\mathcal{G}$ with objective φ if $\mathsf{Pr}_{\mathcal{G},v}^{\sigma_1,\sigma_2}(\varphi) > 0$ (resp., $\mathsf{Pr}_{\mathcal{G},v}^{\sigma_1,\sigma_2}(\varphi) = 1$) for all strategies σ_2 of Player 2. We refer to positive and almost-sure winning as *qualitative* satisfaction of φ, while for arbitrary $p \in [0,1]$, we call it *quantitative* satisfaction. We denote by $\langle\!\langle 1 \rangle\!\rangle_{\mathcal{G}}^{\mathsf{Pos}}(\varphi)$ (resp., by $\langle\!\langle 1 \rangle\!\rangle_{\mathcal{G}}^{\mathsf{AS}}(\varphi)$) the positive (resp., almost-sure) winning region of Player 1, i.e., the set of all vertices in $\mathcal{G}$ from which Player 1 has a positive (resp., almost-sure) winning strategy for $\mathcal{G}$ with objective φ. If a vertex v belongs to the positive (resp., almost-sure) winning region of Player 1, then Player 1 is said to play *optimally* from v if she follows a positive (resp., almost-sure) winning strategy from v. We omit analogous definitions for Player 2.

Positive attractors and traps. The Player i *positive attractor* $(i \in \{1, 2\})$ to $T \subseteq V$, denoted $\mathsf{PosAttr}_i(T)$, is the set of vertices in V from which Player i can ensure that the token reaches a vertex in T with positive probability. It is possible to compute the positive attractor in $\mathcal{O}(|E|)$ time [10]. In non-stochastic games, a positive attractor to a set T is the same as an attractor to the set T, which we denote by $\mathsf{Attr}_i(T)$. Computation of $\mathsf{PosAttr}_i(T)$ gives a memoryless strategy for Player i that ensures that the token reaches T with positive probability. We call such a strategy a *positive-attractor strategy* of Player i.

A *trap* for Player 1 is a set $T \subseteq V$ such that for every vertex $v \in T$, if $v \in V_1 \cup V_{\Diamond}$, then $E(v) \subseteq T$, and if $v \in V_2$, then $E(v) \cap T \neq \varnothing$. In other words, from every vertex $v \in T$, Player 2 can ensure (with probability 1) that the token never leaves T, moreover using a memoryless strategy. A trap for Player 2 can be defined analogously.

Remark 1. Let $\mathcal{G}$ be a non-stochastic game with objective φ for Player 1. If φ is closed under suffixes, then the winning region of Player 1 is a trap for Player 2. As a corollary, if φ is prefix-independent, then the winning region of Player 1 is a trap for Player 2 and the winning region of Player 2 is a trap for Player 1.

3   Window mean payoff

We consider two types of window mean-payoff objectives, introduced in [8]: (i) *fixed window mean-payoff* objective $(\mathsf{FWMP}(\ell))$ in which a window length $\ell \geq 1$ is given, and (ii) *bounded window mean-payoff* objective (BWMP) in which for every play, we need a bound on window lengths. We define these objectives below.

For a play π in a stochastic game $\mathcal{G}$, the *total payoff* of an infix $\pi(i, i+n) = v_i v_{i+1} \cdots v_{i+n}$ is defined as $\mathsf{TP}(\pi(i, i+n)) = \sum_{k=i}^{i+n-1} w(v_k, v_{k+1})$. The *mean payoff* of an infix $\pi(i, i+n)$ is defined as $\mathsf{MP}(\pi(i, i+n)) = \frac{1}{n}\mathsf{TP}(\pi(i, i+n))$. Observe that the mean payoff of an infix is nonnegative if and only if the total payoff of the infix is nonnegative. The mean payoff of a play π is defined as $\mathsf{MP}(\pi) = \liminf_{n \to \infty} \mathsf{MP}(\pi(0, n))$. Given a window length $\ell \geq 1$, a play $\pi = v_0 v_1 \cdots$ in $\mathcal{G}$ satisfies the *fixed window mean-payoff objective* $\mathsf{FWMP}_{\mathcal{G}}(\ell)$ if from every position after some point, it is possible to start an infix of length at most ℓ with a nonnegative mean payoff. Formally,

$$\mathsf{FWMP}_{\mathcal{G}}(\ell) = \{\pi \in \mathsf{Plays}_{\mathcal{G}} \mid \exists k \geq 0 \cdot \forall i \geq k \cdot \exists j \in \{1, \dots, \ell\} : \mathsf{MP}(\pi(i, i+j)) \geq 0\}.$$

We omit the subscript $\mathcal{G}$ when it is clear from the context. Note that when $\ell = 1$, the $\mathsf{FWMP}(1)$ and $\overline{\mathsf{FWMP}(1)}$ (i.e., the complement of $\mathsf{FWMP}(1)$) objectives reduce to coBüchi and Büchi objectives respectively. The following properties of $\mathsf{FWMP}(\ell)$ have been observed in [8]. For all window lengths $\ell \geq 1$, if a play

π satisfies $\mathsf{FWMP}(\ell)$, then $\mathsf{MP}(\pi) \geq 0$. In all plays satisfying $\mathsf{FWMP}(\ell)$, there exists a suffix that can be decomposed into infixes of length at most ℓ, each with a nonnegative mean payoff. Such a desirable robust property is not guaranteed by the classical mean-payoff objective, where infixes of unbounded lengths may have negative mean payoff.

As defined in [8], given a play $\pi = v_0v_1\cdots$ and $0 \leq i < j$, we say that the window $\pi(i,j)$ is *open* if the total-payoff of $\pi(i,k)$ is negative for all $i < k \leq j$. Otherwise, the window is *closed*. Given $j > 0$, we say a window is open at j if there exists an open window $\pi(i,j)$ for some $i < j$. The window starting at position i *closes* at position j if j is the first position after i such that the total-payoff of $\pi(i,j)$ is nonnegative. If the window starting at i closes at j, then for all $i \leq k < j$, the windows $\pi(k,j)$ are closed. This property is called the *inductive property of windows*.

We also have the bounded window mean-payoff objective BWMP. A play π satisfies the BWMP objective if there exists a window length $\ell \geq 1$ for which π satisfies $\mathsf{FWMP}(\ell)$, i.e.,

$$\mathsf{BWMP}_{\mathcal{G}} = \{\pi \in \mathsf{Plays}_{\mathcal{G}} \mid \exists \ell \geq 1 : \pi \in \mathsf{FWMP}(\ell)\}$$

Equivalently, a play π does not satisfy BWMP if for every suffix of π, for all $\ell \geq 1$, the suffix contains an open window of length ℓ. Note that both $\mathsf{FWMP}(\ell)$ for all $\ell \geq 1$ and BWMP are prefix-independent objectives.

Decision problems. Given a game $\mathcal{G}$, an initial vertex $v \in V$, a rational threshold $p \in [0,1]$, and an objective φ (that is either $\mathsf{FWMP}(\ell)$ for a given window length $\ell \geq 1$, or BWMP), consider the problem of deciding:

- *Positive satisfaction of φ*: whether Player 1 positively wins φ from v, i.e., whether $v \in \langle\!\langle 1 \rangle\!\rangle_{\mathcal{G}}^{\mathsf{Pos}}(\varphi)$.
- *Almost-sure satisfaction of φ*: whether Player 1 almost-surely wins φ from v, i.e., whether $v \in \langle\!\langle 1 \rangle\!\rangle_{\mathcal{G}}^{\mathsf{AS}}(\varphi)$.
- *Quantitative satisfaction of φ* (also known as *quantitative value problem* [13]): whether Player 1 wins φ from v with probability at least p, i.e., whether $\sup_{\sigma_1 \in \Lambda_1} \inf_{\sigma_2 \in \Lambda_2} \mathsf{Pr}_{\mathcal{G},v}^{\sigma_1,\sigma_2}(\varphi) \geq p$.

Note that these three problems coincide for non-stochastic games. As considered in previous works [8,3,4], the window length ℓ is usually small (typically $\ell \leq |V|$), and therefore we assume that ℓ is given in unary (while the payoff on the edges is given in binary). From determinacy of Blackwell games [24], stochastic games with window mean-payoff objectives as defined above are determined, i.e., the largest probability with which Player 1 is winning and the largest probability with which Player 2 is winning add up to 1.

Algorithms for non-stochastic window mean-payoff games. To compute the positive and almost-sure winning regions for Player 1 for $\mathsf{FWMP}(\ell)$, we recall intermediate objectives defined in [8]. The *good window* objective $\mathsf{GW}_{\mathcal{G}}(\ell)$ consists

Algorithm 1 NonStocFWMP($\mathcal{G}, \ell$)	Algorithm 2 NonStocDirFWMP($\mathcal{G}, \ell$)
In: $\mathcal{G} = ((V, E), (V_1, V_2, \varnothing), w)$ and $\ell \geq 1$	**In:** $\mathcal{G} = ((V, E), (V_1, V_2, \varnothing), w)$ and $\ell \geq 1$
Out: $\langle\!\langle 1 \rangle\!\rangle_{\mathcal{G}}(\text{FWMP}(\ell))$	**Out:** $\langle\!\langle 1 \rangle\!\rangle_{\mathcal{G}}(\text{DirFWMP}(\ell))$
1: $W_d \leftarrow$ NonStocDirFWMP($\mathcal{G}, \ell$)	1: $W_{gw} \leftarrow$ GoodWin($\mathcal{G}, \ell$)
2: **if** $W_d = \varnothing$ **then**	2: **if** $W_{gw} = V$ or $W_{gw} = \varnothing$ **then**
3: **return** $\varnothing$	3: **return** W_{gw}
4: **else**	4: **else**
5: $A \leftarrow \text{Attr}_1(W_d)$	5: $A \leftarrow \text{Attr}_2(V \setminus W_{gw})$
6: **return** $A \cup$ NonStocFWMP($\mathcal{G} \upharpoonright (V \setminus A), \ell$)	6: **return** NonStocDirFWMP($\mathcal{G} \upharpoonright (W_{gw} \setminus A), \ell$).

of all plays π in $\mathcal{G}$ such that the window opened at the first position in the play closes in at most ℓ steps:

$$\text{GW}_{\mathcal{G}}(\ell) = \{\pi \in \text{Plays}_{\mathcal{G}} \mid \exists j \in \{1, \ldots, \ell\} : \text{MP}(\pi(0, j)) \geq 0\}$$

The *direct fixed window mean-payoff* objective $\text{DirFWMP}_{\mathcal{G}}(\ell)$ consists of all plays π in $\mathcal{G}$ such that from every position in π, the window closes in at most ℓ steps:

$$\text{DirFWMP}_{\mathcal{G}}(\ell) = \{\pi \in \text{Plays}_{\mathcal{G}} \mid \forall i \geq 0 : \pi(i, \infty) \in \text{GW}_{\mathcal{G}}(\ell)\}$$

The $\text{FWMP}_{\mathcal{G}}(\ell)$ objective can be expressed in terms of $\text{DirFWMP}_{\mathcal{G}}(\ell)$:

$$\text{FWMP}_{\mathcal{G}}(\ell) = \{\pi \in \text{Plays}_{\mathcal{G}} \mid \exists k \geq 0 : \pi(k, \infty) \in \text{DirFWMP}_{\mathcal{G}}(\ell)\}$$

We refer to Algorithms 1, 2, and 3 from [8] shown here with the same numbering. They compute the winning regions for Player 1 for the $\text{FWMP}(\ell)$, $\text{DirFWMP}(\ell)$, and $\text{GW}(\ell)$ objectives in non-stochastic games respectively. The original algorithms in [8] contain subtle errors for which the fixes are known [6, 19]. For completeness, we refer the reader to [15] for counterexamples for the algorithms in [8] along with brief explanations of correctness for the modified versions.

    Algorithm 3 uses dynamic programming to compute, for all $v \in V$ and all lengths $i \in \{1, \ldots, \ell\}$, the largest payoff $C_i(v)$ that Player 1 can ensure from v within at most i steps. The winning region for $\text{GW}(\ell)$ for Player 1 consists of all vertices v such that $C_\ell(v) \geq 0$.

4   Memory requirement for non-stochastic window mean-payoff games

The memory requirement for winning strategies of Player 1 in non-stochastic games with objective $\text{FWMP}(\ell)$ is claimed to be $\mathcal{O}(|V| \cdot \ell)$ without proof [8, Lemma 7], and further "correctly stated" as $\mathcal{O}(w_{\max} \cdot \ell^2)$, where $w_{\max}$ is the maximum absolute payoff in the graph [6, Theorem 2]. We improve upon these bounds and show that memory of size ℓ suffices for a winning strategy of Player 1.

Algorithm 3 GoodWin($\mathcal{G}, \ell$)

In: $\mathcal{G} = ((V, E), (V_1, V_2, \varnothing), w)$ the non-stochastic game, and $\ell \geq 1$, the window length
Out: The set of vertices from which Player 1 wins GW(ℓ) in $\mathcal{G}$
1: **for all** $v \in V$ **do**
2:     $C_0(v) \leftarrow 0$
3:     **for all** $i \in \{1, \ldots, \ell\}$ **do**
4:         $C_i(v) \leftarrow -\infty$
5: **for all** $i \in \{1, \ldots, \ell\}$ **do**
6:     **for all** $v \in V_1$ **do**
7:         $C_i(v) \leftarrow \max_{(v,v') \in E}\{\max\{w(v, v'), w(v, v') + C_{i-1}(v')\}\}$
8:     **for all** $v \in V_2$ **do**
9:         $C_i(v) \leftarrow \min_{(v,v') \in E}\{\max\{w(v, v'), w(v, v') + C_{i-1}(v')\}\}$
10: $W_{gw} \leftarrow \{v \in V \mid C_\ell(v) \geq 0\}$
11: **return** W_{gw}

We also present a family of games with arbitrarily many vertices where Player 2 is winning and all his winning strategies require at least $\frac{1}{2}(|V| - \ell) + 3$ memory, while it was only known that memoryless strategies are not sufficient for Player 2 [8].

4.1 Memory requirement for Player 1 for FWMP objective

Upper bound on memory requirement for Player 1. We show that memory of size ℓ suffices for winning strategies of Player 1 for the DirFWMP(ℓ) objective (Lemma 1), which in turn shows that the same memory also works for the FWMP(ℓ) objective (Theorem 1).

Lemma 1. *If Player 1 wins in a non-stochastic game with objective* DirFWMP(ℓ), *then Player 1 has a winning strategy with memory of size ℓ.*

Proof (Sketch). Given a non-stochastic game $\mathcal{G}$, let W_d be the winning region of Player 1 in $\mathcal{G}$ for objective DirFWMP(ℓ). By definition, every vertex in W_d is also winning for Player 1 for the GW(ℓ) objective.

A winning strategy σ_d of Player 1 in W_d satisfies the objective GW(ℓ) by closing a window within at most ℓ steps and then restarts with the same strategy, playing for GW(ℓ) and so on. Using memory space $Q = \{1, \ldots, \ell\}$, we may store the number of steps remaining before the window must close. However, the window may close any time within ℓ steps, and the difficulty lies in detecting this independently of the history. For memory state $q = i$ and the next visited vertex being v, intuitively, the memory should be updated to $q = i - 1$ if the window did not close yet upon reaching v, and to $q = \ell$ if it did, but that depends on which path was followed to reach v (not just on v), which is not stored in the memory space.

The crux is to show that it is not always necessary for Player 1 to be able to infer when the window closes. Given the current memory state $q = i$, and the next visited vertex v, the memory update is as follows: if $C_i(v) \geq 0$ (that is, Player 1 can ensure the window from v will close within i steps), then we update to $q = i - 1$ (*decrement*) although the window may or may not have closed upon reaching v; otherwise $C_i(v) < 0$ and we update to $q = \ell - 1$ (*reset to ℓ and decrement*) and we show that in this case the window did close. Intuitively, updating to $q = i - 1$ is safe even if the window did close, because the strategy of Player 1 will anyway ensure the (upcoming) window is closed within $i - 1 < \ell$ steps. A formal description of a Mealy machine with ℓ states defining a winning strategy of Player 1 for the $\mathsf{DirFWMP}(\ell)$ objective is given in [15].   □

Theorem 1. *If Player 1 wins in a non-stochastic game $\mathcal{G}$ with objective $\mathsf{FWMP}(\ell)$, then Player 1 has a winning strategy with memory of size ℓ.*

Proof (Sketch). Since $\mathsf{FWMP}(\ell)$ is a prefix-independent objective, we have that the winning region $\langle\!\langle 1 \rangle\!\rangle_{\mathcal{G}}(\mathsf{FWMP}(\ell))$ of Player 1 is a trap for Player 2 (Remark 1), and induces a subgame, say $\mathcal{G}_0$. Let there be $k + 1$ calls to the subroutine $\mathsf{NonStocDirFWMP}$ from Algorithm 1 where $k < |V|$. We denote by $(W_i)_{i \in \{1,\dots,k\}}$ the nonempty W_d returned by the ith call to the subroutine, and let $A_i = \mathsf{Attr}_1(W_i)$. The A_i's are pairwise disjoint, and their union is $\bigcup_{i=1}^{k} A_i = \langle\!\langle 1 \rangle\!\rangle_{\mathcal{G}}(\mathsf{FWMP}(\ell))$. For $i \in \{1,\dots,k\}$, inductively define $\mathcal{G}_i$ to be the subgame induced by the complement of A_i in $\mathcal{G}_{i-1}$. Since $\mathsf{DirFWMP}(\ell)$ is closed under suffixes, for all $i \in \{1,\dots,k\}$, we have that W_i is a trap for Player 2 in $\mathcal{G}_i$ (Remark 1).

We construct a strategy σ_1^{NS} that follows the (memoryless) attractor strategy in $\bigcup_i (A_i \setminus W_i)$, and follows the winning strategy σ_d for $\mathsf{DirFWMP}(\ell)$ objective (defined in the proof of Lemma 1) in $\bigcup_i W_i$. The reader is pointed to [15] for a formal description of a Mealy machine defining the strategy σ_1^{NS}. For the correctness of the construction, the crux is to show that one of the sets W_i ($i \in \{1,\dots,k\}$) is never left from some point on. Intuitively, given the token is in A_i for some $i \in \{1,\dots,k\}$ (thus in $\mathcal{G}_i$), following σ_1^{NS}, the token will either remain in A_i, or leave the subgame $\mathcal{G}_i$ and enter A_j for a smaller index $j < i$. The result follows since this can be done at most k times.   □

Lower bound on memory requirement for Player 1. In [8], the authors show a game with $\ell = 4$ where Player 1 requires memory at least 3. This can be generalized to arbitrary ℓ to show that memory of size $\ell - 1$ may be necessary (See [15] for details).

4.2   Memory requirement for Player 2 for FWMP objective

Upper bound on memory requirement for Player 2. Now we show that for the $\overline{\mathsf{FWMP}(\ell)}$ objective, Player 2 has a winning strategy that uses memory of size at most $|V| \cdot \ell$. This has been loosely stated in [8] without a formal proof.

Theorem 2. *Let $\mathcal{G}$ be a non-stochastic game with objective $\overline{\mathsf{FWMP}(\ell)}$ for Player 2. Then, Player 2 has a winning strategy with memory size at most $|V| \cdot \ell$.*

Proof (Sketch). Since $\mathsf{FWMP}(\ell)$ is a prefix-independent objective, so is $\overline{\mathsf{FWMP}(\ell)}$. We have that $\langle\!\langle 2 \rangle\!\rangle_{\mathcal{G}}(\overline{\mathsf{FWMP}(\ell)})$ is a trap for Player 1 (Remark 1) and induces a subgame, say $\mathcal{H}_0$, of $\mathcal{G}$. Let there be $k+1$ calls to the subroutine GoodWin from Algorithm 2 (where $k < |V|$), and let $\mathcal{H}_i$ be the subgame corresponding to the ith call of the subroutine. We denote by $(W_i)_{i \in \{1,\dots,k\}}$ the complement of W_{gw} in $\mathcal{H}_i$, where W_{gw} is returned by the ith call to the subroutine, and let $A_i = \text{Attr}_2(W_i)$. The A_i's are pairwise disjoint, and their union is $\bigcup_{i=1}^{k} A_i = \langle\!\langle 2 \rangle\!\rangle_{\mathcal{G}}(\overline{\text{FWMP}(\ell)})$.

We describe a winning strategy for the $\overline{\text{FWMP}(\ell)}$ objective with memory $k \cdot \ell$, which is at most $|V| \cdot \ell$. The strategy is always in either *attractor mode* or *window-open mode*. When the game begins, it is in attractor mode. If the strategy is in attractor mode and the token is on a vertex $v \in A_i \setminus W_i$ for some $i \in \{1, \dots, k\}$, then the attractor strategy is to eventually reach W_i. If the token reaches W_i, then the strategy switches to window-open mode. Since all vertices in W_i are winning for Player 2 for the $\overline{\text{GW}(\ell)}$ objective, he can keep the window open for ℓ more steps, provided that Player 1 does not move the token out of the subgame $\mathcal{H}_i$. If, at some point, Player 1 moves the token out of the subgame $\mathcal{H}_i$ to A_j for a smaller index $j < i$, then the strategy switches back to attractor mode, this time trying to reach W_j in the bigger subgame $\mathcal{H}_j$. Otherwise, if Player 2 keeps the window open for ℓ steps, then the strategy switches back to attractor mode until the token reaches a vertex in $\bigcup_{i=1}^{k} W_i$. This strategy can be defined by a Mealy machine M_2^{NS} with states $\{1, \dots, k\} \times \{1, \dots, \ell\}$, where the first component tracks the smallest subgame $\mathcal{H}_i$ in which the window started to remain open, and the second component indicates how many more steps the window needs to be kept open for. A formal description of M_2^{NS} can be found in [15]. □

Lower bound on memory requirement for Player 2. In [8], it was shown that memoryless strategies do not suffice for Player 2. We improve upon this lower bound. Given a window length $\ell \geq 2$, for every $k \geq 1$, we construct a game $\mathcal{G}_{k,\ell}$ with $2k + \ell - 1$ vertices such that every winning strategy of Player 2 in $\mathcal{G}_{k,\ell}$ requires at least $k+1$ memory.

Theorem 3. *There exists a family of non-stochastic games $\{\mathcal{G}_{k,\ell}\}_{k \geq 1, \ell \geq 2}$ with objective $\text{FWMP}(\ell)$ for Player 1 and edge weights in $\{-1, 0, +1\}$ such that every winning strategy of Player 2 requires at least $\frac{1}{2}(|V| - \ell + 1) + 1$ memory, where $|V| = 2k + \ell - 1$.*

Proof (Sketch). Let $A = \{a_1, \dots, a_k\}$, $B = \{b_1, \dots, b_k\}$, and $C = \{c_1, \dots, c_{\ell-1}\}$ be pairwise disjoint sets. The vertices of $\mathcal{G}_{k,\ell}$ are $A \cup B \cup C$ with $V_1 = A \cup C$ and $V_2 = B$. Figure 1 shows the game $\mathcal{G}_{4,3}$. A more formal description of $\mathcal{G}_{k,\ell}$ can be found in [15].

Observe that the only open windows of length ℓ in the game $\mathcal{G}_{k,\ell}$ are sequences of the form $a_p b_r c_{\ell-1} \cdots c_1$ for all $p \leq r$. Also note that Player 2 has a winning strategy that wins starting from every vertex in the game, as Player 2 can force the token to eventually take a red edge followed by two black edges.

When the token reaches a vertex $b_r \in B$, Player 2 can either move the token to $a_r \in A$ or to $c_{\ell-1} \in C$. Depending on which vertex the token was on before reaching b_r, one of the two choices is *good* for Player 2. If the token reaches b_r

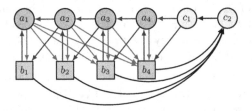

Figure 1: Game $\mathcal{G}_{4,3}$ with parameter $k = 4$ and window length $\ell = 3$. Red edges (from a_p to b_r for $p \leq r$) have payoff -1, black edges (from b_r to c_2) have payoff 0, and blue edges (the remaining edges) have payoff $+1$.

Table 1: Good choices $\chi(u, b_r)$ for all $u \in A \cup \{c_1\}$ and $b_r \in B$ in the game $\mathcal{G}_{4,3}$

$a_1b_1 \to c_2$	$a_2b_1 \to a_1$			
$a_1b_2 \to c_2$	$a_2b_2 \to c_2$	$a_3b_2 \to a_2$		
$a_1b_3 \to c_2$	$a_2b_3 \to c_2$	$a_3b_3 \to c_2$	$a_4b_3 \to a_3$	
$a_1b_4 \to c_2$	$a_2b_4 \to c_2$	$a_3b_4 \to c_2$	$a_4b_4 \to c_2$	$c_1b_4 \to a_4$

from a_p for $p \leq r$, then it is *good* for Player 2 to move the token to $c_{\ell-1} \in C$ so that the window starting at a_p remains open for ℓ steps. Otherwise, if the token reaches b_r from a_{r+1}, then it is *good* for Player 2 to move the token to a_r so that an edge with negative payoff may eventually be taken. For all $u \in A \cup \{c_1\}$, for all $b_r \in B$ such that (u, b_r) is an edge in $\mathcal{G}_{k,\ell}$, we denote by $\chi(u, b_r)$ the vertex a_r or $c_{\ell-1}$ that is good for Player 2. We list the good choices in the game $\mathcal{G}_{4,3}$ in Table 1. The columns are indexed by $u \in A \cup \{c_1\}$ and the rows are indexed by $b_r \in B$.

We show that for each column in the table, there exists a distinct memory state in every Mealy machine defining a winning strategy of Player 2. This gives a lower bound of $k + 1$ on the number of states of such a Mealy machine. Since $\mathcal{G}_{k,\ell}$ has $2k + \ell - 1$ vertices, the memory requirement of a winning strategy of Player 2 is at least $\frac{1}{2}(|V| - \ell + 1) + 1$. $\square$

Given a winning strategy σ_2^{NS} of Player 2 for the $\overline{\mathsf{FWMP}(\ell)}$ objective, the following lemma gives an upper bound on the number of steps between consecutive open windows of length ℓ in any play consistent with σ_2^{NS}. This lemma is used in Section 6, where we construct an almost-sure winning strategy of Player 2 for the $\overline{\mathsf{FWMP}(\ell)}$ objective.

Lemma 2. *Let $\mathcal{G}$ be a non-stochastic game such that $\langle\!\langle 2 \rangle\!\rangle_{\mathcal{G}}(\overline{\mathsf{FWMP}(\ell)}) = V$. Let σ_2^{NS} be a finite-memory strategy of Player 2 of memory size M that is winning for $\overline{\mathsf{FWMP}(\ell)}$ from all vertices in $\mathcal{G}$. Then, for every play π of $\mathcal{G}$ consistent with σ_2^{NS}, every infix of π of length $\mathsf{M} \cdot |V| \cdot \ell$ contains an open window of length ℓ.*

The proof is based on the pigeonhole principle and appears in [15].

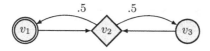

Figure 2: Büchi objective does not satisfy the SAS property in this game.

5 Reducing stochastic games to non-stochastic games

For a stochastic game $\mathcal{G}$, let $\mathcal{G}_{NS} = ((V, E), (V_1, V_2 \cup V_\Diamond, \varnothing), w)$ be the *(adversarial) non-stochastic game corresponding to* $\mathcal{G}$, obtained by changing all probabilistic vertices in $\mathcal{G}$ to Player 2 vertices. In [13], a property of finitary Streett objective was used to solve stochastic games by reducing them to non-stochastic games with the same objective. In this section, we generalize this property for arbitrary prefix-independent objectives.

Definition 1 (Sure-almost-sure (SAS) property). *A prefix-independent objective φ in a game $\mathcal{G}$ satisfies the* SAS *property if* $\langle\langle 2 \rangle\rangle_{\mathcal{G}_{NS}}(\overline{\varphi}) = V$ *implies* $\langle\langle 2 \rangle\rangle_{\mathcal{G}}^{AS}(\overline{\varphi}) = V$, *that is, if Player 2 wins the objective $\overline{\varphi}$ from every vertex in $\mathcal{G}_{NS}$, then Player 2 almost-surely wins the same objective $\overline{\varphi}$ from every vertex in $\mathcal{G}$.*

Every prefix-independent objective satisfies the converse of the SAS property since if Player 2 even wins positively from all vertices in $\mathcal{G}$, then since he controls all probabilistic vertices in $\mathcal{G}_{NS}$, he wins from all vertices in $\mathcal{G}_{NS}$ by choosing optimal successors of probabilistic vertices. We show in Section 6 that for all stochastic games $\mathcal{G}$, the objectives FWMP(ℓ) and BWMP satisfy the SAS property, while in Example 1, we show that there exists a stochastic game in which Büchi objective does not satisfy the SAS property.

Example 1. Consider the game $\mathcal{G}$ in Figure 2. The objective φ in this game is a Büchi objective: a play π satisfies the Büchi objective if π visits vertex v_1 infinitely often. Although from every vertex, with positive probability (in fact, with probability 1), a play visits v_1 infinitely often, from none of the vertices, Player 1 can ensure the Büchi objective in the non-stochastic game $\mathcal{G}_{NS}$.

Theorem 4 gives complexity bounds for solving stochastic games with objectives satisfying the SAS property in terms of the complexity of solving non-stochastic games with the same objective.

Theorem 4. *Given $\mathcal{G}$ and φ, suppose in every subgame $\mathcal{G}'$ of $\mathcal{G}$, the objective φ restricted to $\mathcal{G}'$ satisfies the* SAS *property. Let* NonStocWin$_\varphi(\mathcal{G}_{NS})$ *be an algorithm computing* $\langle\langle 1 \rangle\rangle_{\mathcal{G}_{NS}}(\varphi)$ *in $\mathcal{G}_{NS}$ in time $\mathbb{C}$. Then, the positive and almost-sure satisfaction of φ can be decided in time $\mathcal{O}(|V| \cdot (\mathbb{C} + |E|))$ and $\mathcal{O}(|V|^2 \cdot (\mathbb{C} + |E|))$ respectively.*

Moreover, for positive and almost-sure satisfaction of φ, the memory requirement for Player 1 to play optimally in stochastic games is no more than that for non-stochastic games.

Algorithm 4 PosWin$_\varphi(\mathcal{G})$	Algorithm 5 ASWin$_\varphi(\mathcal{G})$
In: $\mathcal{G} = ((V,E),(V_1,V_2,V_\Diamond),\mathbb{P},w)$ and φ	**In:** $\mathcal{G} = ((V,E),(V_1,V_2,V_\Diamond),\mathbb{P},w)$ and φ
Out: $\langle\!\langle 1 \rangle\!\rangle_{\mathcal{G}}^{\mathsf{Pos}}(\varphi)$	**Out:** $\langle\!\langle 1 \rangle\!\rangle_{\mathcal{G}}^{\mathsf{AS}}(\varphi)$
1: $W_1 \leftarrow$ NonStocWin$_\varphi(\mathcal{G}_{\mathsf{NS}})$	1: $W_2 \leftarrow V \setminus$ PosWin$_\varphi(\mathcal{G})$
2: **if** $W_1 = \varnothing$ **then**	2: **if** $W_2 = \varnothing$ **then**
3: **return** $\varnothing$	3: **return** V
4: **else**	4: **else**
5: $A_1 \leftarrow$ PosAttr$_1(W_1)$	5: $A_2 \leftarrow$ PosAttr$_2(W_2)$
6: **return** $A_1 \cup$ PosWin$_\varphi(\mathcal{G} \restriction (V \setminus A_1))$	6: **return** ASWin$_\varphi(\mathcal{G} \restriction (V \setminus A_2))$

Theorem 4 does not give bounds on the memory requirement for winning strategies of Player 2 for objective φ in the stochastic game, but we provide such bounds specifically for FWMP(ℓ) and BWMP in Section 6. We give a sketch of the proof of Theorem 4 below. The complete proof appears in [15].

The algorithms to compute the positive and almost-sure winning regions in $\mathcal{G}$, and their proofs of correctness are the same as in the case of finitary Streett objectives described in [13]. The PosWin$_\varphi$ algorithm (Algorithm 4) uses NonStocWin$_\varphi$ as a subroutine to compute $\langle\!\langle 1 \rangle\!\rangle_{\mathcal{G}}^{\mathsf{Pos}}(\varphi)$. The fact that φ satisfies the SAS property is used to show the correctness of this algorithm. The depth of recursive calls of this algorithm is bounded above by $|V|$, which gives the complexity bound. The ASWin$_\varphi$ algorithm (Algorithm 5) in turn uses PosWin$_\varphi$ as a subroutine to compute the $\langle\!\langle 1 \rangle\!\rangle_{\mathcal{G}}^{\mathsf{AS}}(\varphi)$. The depth of recursive calls of this algorithm is also bounded above by $|V|$, which gives the complexity bound. The following lemma, which is a special case of Theorem 1 in [7], is used to show the correctness of this algorithm.

Lemma 3. [7, Theorem 1] For a stochastic game $\mathcal{G}$ with prefix-independent objective φ, if $\langle\!\langle 2 \rangle\!\rangle_{\mathcal{G}}^{\mathsf{Pos}}(\varphi) = V$, then $\langle\!\langle 2 \rangle\!\rangle_{\mathcal{G}}^{\mathsf{AS}}(\varphi) = V$.

For both positive and almost-sure winning, Player 1 does not require any additional memory in the stochastic game compared to the non-stochastic game. We describe a strategy σ_1^{Pos} of Player 1 that is positive winning from all vertices in $\langle\!\langle 1 \rangle\!\rangle_{\mathcal{G}}^{\mathsf{Pos}}(\varphi)$. In each recursive call to PosWin$_\varphi$ algorithm, from every vertex in W_1, the strategy σ_1^{Pos} mimics a winning strategy of Player 1 in $\mathcal{G}_{\mathsf{NS}}$, while for vertices in $A_1 \setminus W_1$, it follows a memoryless attractor strategy to reach W_1. The same strategy is almost-sure winning for Player 1 from all vertices in $\langle\!\langle 1 \rangle\!\rangle_{\mathcal{G}}^{\mathsf{AS}}(\varphi)$.

Finally, we look at the quantitative decision problem. The quantitative satisfaction for φ can be decided in NP^B ([13, Theorem 6]), where B is an oracle deciding positive and almost-sure satisfaction problems for φ. It is not difficult to see that the quantitative satisfaction for φ can be decided in $\mathsf{NP}^B \cap \mathsf{coNP}^B$. Moreover, from the proof of [13, Theorem 6], it follows that the memory requirement of winning strategies for both players for the quantitative decision problem is no greater than that for the qualitative decision problem.

Corollary 1. *Given $\mathcal{G}$ and φ as described in Theorem 4, let B be an oracle deciding the qualitative satisfaction of φ. Then, the quantitative satisfaction of φ is in $\mathsf{NP}^B \cap \mathsf{coNP}^B$. Moreover, the memory requirement of optimal strategies for both players is no greater than that for the positive and almost-sure satisfaction of φ.*

6 Reducing stochastic window mean-payoff games: A special case

In this section, we show that for all stochastic games $\mathcal{G}$ and for all $\ell \geq 1$, the objectives $\mathsf{FWMP}_{\mathcal{G}}(\ell)$ and $\mathsf{BWMP}_{\mathcal{G}}$, which are prefix-independent, satisfy the SAS property of Definition 1. Thus, by Theorem 4, we obtain bounds on the complexity and memory requirements of Player 1 for the positive and almost-sure satisfaction of these objectives. We also show that for both these objectives, the memory requirements of Player 2 to play optimally for positive and almost-sure winning in stochastic games is no more than that of the non-stochastic games. The algorithms to compute the positive and almost-sure winning regions of Player 1 for both $\mathsf{FWMP}(\ell)$ and BWMP objectives are obtained by instantiating φ equal to $\mathsf{FWMP}(\ell)$ and BWMP in Algorithms 4 and 5. Thus, we obtain the algorithms $\mathsf{PosWin}_{\mathsf{FWMP}(\ell)}$, $\mathsf{ASWin}_{\mathsf{FWMP}(\ell)}$, $\mathsf{PosWin}_{\mathsf{BWMP}}$, and $\mathsf{ASWin}_{\mathsf{BWMP}}$.

6.1 Fixed window mean-payoff objective

We first discuss the SAS property for the $\mathsf{FWMP}(\ell)$ objective.

Lemma 4. *In stochastic games, for all $\ell \geq 1$, the $\mathsf{FWMP}(\ell)$ objective satisfies the SAS property.*

Proof (Sketch). We show that for all stochastic games $\mathcal{G}$, if $\langle\!\langle 2 \rangle\!\rangle_{\mathcal{G}_{\mathsf{NS}}}(\overline{\mathsf{FWMP}(\ell)}) = V$, then $\langle\!\langle 2 \rangle\!\rangle_{\mathcal{G}}^{\mathsf{AS}}(\overline{\mathsf{FWMP}(\ell)}) = V$. If $\langle\!\langle 2 \rangle\!\rangle_{\mathcal{G}_{\mathsf{NS}}}(\overline{\mathsf{FWMP}(\ell)}) = V$, then from Theorem 2, there exists a finite-memory strategy σ_2^{NS} (say, with memory M) of Player 2 that is winning for objective $\overline{\mathsf{FWMP}(\ell)}$ from every vertex in $\mathcal{G}_{\mathsf{NS}}$. Given such a strategy, we construct below a strategy σ_2^{AS} of Player 2 in the stochastic game $\mathcal{G}$ that is almost-sure winning for $\overline{\mathsf{FWMP}(\ell)}$ from every vertex in $\mathcal{G}$.

In $\mathcal{G}_{\mathsf{NS}}$, Player 2 controls vertices in $V_2 \cup V_\Diamond$, while in $\mathcal{G}$, Player 2 only controls vertices in V_2 and the probability function $\mathbb{P}$ determines the successors of vertices in $V_\Diamond$. While the strategy σ_2^{NS} is winning for $\overline{\mathsf{FWMP}(\ell)}$ from all vertices in $\mathcal{G}_{\mathsf{NS}}$, it may not be almost-sure winning for $\overline{\mathsf{FWMP}(\ell)}$ in $\mathcal{G}$. This is because each time the token is on a probabilistic vertex, a *deviation* occurs with positive probability, i.e., the successor chosen by the distribution is not consistent with σ_2^{NS}, resulting in a potentially worse outcome for Player 2. For example, in Figure 3, we see a stochastic game $\mathcal{G}$ and a Mealy machine M_2^{NS} defining a strategy σ_2^{NS} that is winning for Player 2 from all vertices in the non-stochastic game $\mathcal{G}_{\mathsf{NS}}$. In all outcomes in $\mathcal{G}_{\mathsf{NS}}$ that are consistent with σ_2^{NS}, the token never moves from v_6 to v_7. However, in $\mathcal{G}$, a deviation may lead the token to move along (v_6, v_7). This

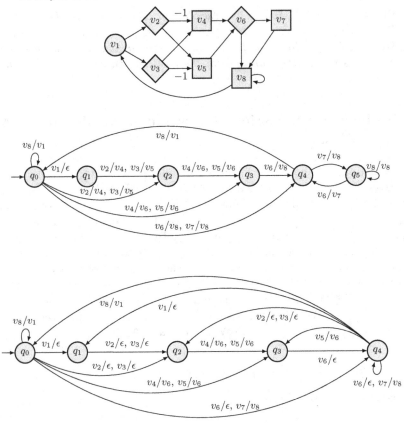

Figure 3: (top) Stochastic game $\mathcal{G}$ with objective $\overline{\mathsf{FWMP}(3)}$ for Player 2. All unlabelled edges have payoff 0. (middle) Mealy machine M_2^{NS} defining a strategy σ_2^{NS} that is winning from all vertices in $\mathcal{G}_{\mathsf{NS}}$ for $\overline{\mathsf{FWMP}(3)}$. (bottom) Part of the Mealy machine M_2^{AS} defining a reset strategy that is almost-sure winning from all vertices in $\mathcal{G}$.

results in a losing outcome for Player 2 as the token gets trapped in v_8, and subsequently no window remains open for ℓ steps. Such harmful deviations can be detected, and starting with the strategy σ_2^{NS}, we construct a strategy σ_2^{AS} that mimics σ_2^{NS} as long as harmful deviations do not occur, and *resets* otherwise, i.e., the strategy forgets the prefix of the play before the deviation. For instance, when the token moves from v_6 to v_7 in $\mathcal{G}$, the strategy resets and the play continues as if the game began from v_7. We call σ_2^{AS} a *reset strategy*. Figure 3 shows a part of a Mealy machine M_2^{AS} defining a reset strategy for the game $\mathcal{G}$. The figure contains all the reset transitions out of q_4, but the reset transitions out of q_1, q_2, and q_3 have been omitted for space. More details on how to obtain a Mealy machine that defines σ_2^{AS} from a Mealy machine that defines σ_2^{NS} without adding any new states can be found in [15].

Now, we argue that the reset strategy is almost-sure winning for Player 2 from all vertices in $\mathcal{G}$. If a play in $\mathcal{G}$ continues for $\mathsf{M} \cdot |V| \cdot \ell$ steps without deviating, then by Lemma 2, it contains an open window of length ℓ. From any point in the play, the probability that σ_2^{AS} successfully copies σ_2^{NS} for i steps (that is, no deviations occur) is at least p^i, where p is the minimum probability over all the edges in $\mathcal{G}$. It follows that from every point in the play, the probability that an open window of length ℓ occurs in the next $\mathsf{M} \cdot |V| \cdot \ell$ steps is at least $p^{\mathsf{M} \cdot |V| \cdot \ell}$. Therefore, from every position in the play, the probability that an open window of length ℓ occurs eventually is at least $\sum_{i \geq 0}(1 - p^{\mathsf{M} \cdot |V| \cdot \ell})^i \cdot p^{\mathsf{M} \cdot |V| \cdot \ell} = 1$. Thus, with probability 1, infinitely many open windows of length ℓ occur in the outcome, and the outcome satisfies $\overline{\mathsf{FWMP}(\ell)}$. Thus, all vertices in $\mathcal{G}$ are almost-sure winning for Player 2 for $\overline{\mathsf{FWMP}(\ell)}$. For all stochastic games $\mathcal{G}$, the objective $\mathsf{FWMP}(\ell)$ satisfies the SAS property. □

We now construct a strategy σ_2^{Pos} of Player 2 that is positive winning from all vertices in $\langle\langle 2 \rangle\rangle_{\mathcal{G}}^{\mathsf{Pos}}(\overline{\mathsf{FWMP}(\ell)})$. Let W_2^i and A_2^i denote the sets W_2 and A_2 computed in the ith recursive call of the $\mathsf{ASWin}_{\mathsf{FWMP}(\ell)}$ algorithm respectively. If the token is in $\bigcup_i W_2^i$, then σ_2^{Pos} mimics σ_2^{AS}; if the token is in $\bigcup_i A_2^i \setminus W_2^i$, then σ_2^{Pos} is a positive-attractor strategy to W_2^i which is memoryless. Then, σ_2^{Pos} is a positive winning strategy for Player 2 from all vertices in $\langle\langle 2 \rangle\rangle_{\mathcal{G}}^{\mathsf{Pos}}(\overline{\mathsf{FWMP}(\ell)})$. Using Theorem 4, Corollary 1, and Lemma 4, we have the following.

Theorem 5. *Given a stochastic game $\mathcal{G}$, a window length $\ell \geq 1$, and a threshold $p \in [0, 1]$, for $\mathsf{FWMP}_{\mathcal{G}}(\ell)$, the positive and almost-sure satisfaction for Player 1 are in* PTIME, *and the quantitative satisfaction is in* NP ∩ coNP. *Moreover for optimal strategies, memory of size ℓ is sufficient for Player 1 and memory of size $|V| \cdot \ell$ is sufficient for Player 2.*

6.2 Bounded window mean-payoff objective

We show that the SAS property holds for the BWMP objective for all stochastic games $\mathcal{G}$.

Lemma 5. *In stochastic games, the BWMP objective satisfies the SAS property.*

Proof (Sketch). We show that for all stochastic games $\mathcal{G}$, if $\langle\langle 2 \rangle\rangle_{\mathcal{G}_{\mathsf{NS}}}(\overline{\mathsf{BWMP}}) = V$, then $\langle\langle 2 \rangle\rangle_{\mathcal{G}}^{\mathsf{AS}}(\overline{\mathsf{BWMP}}) = V$. Since every play that satisfies $\overline{\mathsf{BWMP}}$ also satisfies $\overline{\mathsf{FWMP}(\ell)}$ for all $\ell \geq 1$, if $\langle\langle 2 \rangle\rangle_{\mathcal{G}_{\mathsf{NS}}}(\overline{\mathsf{BWMP}}) = V$, then $\langle\langle 2 \rangle\rangle_{\mathcal{G}_{\mathsf{NS}}}(\overline{\mathsf{FWMP}(\ell)}) = V$. It follows that for each $\ell \geq 1$, Player 2 has a finite-memory strategy (say, with memory M_ℓ), that is winning for the $\overline{\mathsf{FWMP}(\ell)}$ objective from all vertices in $\mathcal{G}_{\mathsf{NS}}$. For every such strategy, we construct a reset strategy σ_2^ℓ of memory size at most M_ℓ as described in the proof of Lemma 4 that is almost-sure winning for the $\overline{\mathsf{FWMP}(\ell)}$ objective from all vertices. We use these strategies to construct an infinite-memory strategy σ_2^{AS} of Player 2 that is almost-sure winning for $\overline{\mathsf{BWMP}}$ from all vertices in the stochastic game $\mathcal{G}$.

Let p be the minimum probability over all edges in the game, and for all $\ell \geq 1$, let $q(\ell)$ denote $p^{\mathsf{M}_\ell \cdot |V| \cdot \ell}$. We partition a play of the game into phases $1, 2, \ldots$ such that for all $\ell \geq 1$, the length of phase ℓ is equal to $M_\ell \cdot |V| \cdot \ell \cdot \lceil 1/q(\ell) \rceil$. We define the strategy σ_2^{AS} as follows: if the game is in phase ℓ, then σ_2^{AS} is σ_2^ℓ, the reset strategy that is almost-sure winning for $\overline{\mathsf{FWMP}(\ell)}$ in $\mathcal{G}$.

We show that σ_2^{AS} is almost-sure winning for Player 2 for $\overline{\mathsf{BWMP}}$ in $\mathcal{G}$. Let E_ℓ denote the event that phase ℓ contains an open window of length ℓ. Given a play π, if E_ℓ occurs in π for infinitely many $\ell \geq 1$, then for every suffix of π and for all $\ell \geq 1$, the suffix contains an open window of length ℓ, and π satisfies $\overline{\mathsf{BWMP}}$. For all $\ell \geq 1$, we compute the probability that E_ℓ occurs in the outcome. For all $\ell \geq 1$, we can divide phase ℓ into $\lceil 1/q(\ell) \rceil$ blocks of length $M_\ell \cdot |V| \cdot \ell$ each. If at least one of these blocks contains an open window of length ℓ, then the event E_ℓ occurs. It follows from the proof of Lemma 4 that if Player 2 follows σ_2^ℓ, then the probability that there exists an open window of length ℓ in the next $M_\ell \cdot |V| \cdot \ell$ steps is at least $q(\ell)$. Hence, the probability that none of the blocks in the phase contains an open window of length ℓ is at most $(1 - q(\ell))^{\lceil 1/q(\ell) \rceil}$. Thus, the probability that E_ℓ occurs in phase ℓ is at least $1 - (1 - q(\ell))^{\lceil 1/q(\ell) \rceil} > 1 - \frac{1}{e} \approx 0.63 > 0$. It follows that with probability 1, for infinitely many values of $\ell \geq 1$, the event E_ℓ occurs in π. □

Note that solving a non-stochastic game with the BWMP objective is in $\mathsf{NP} \cap \mathsf{coNP}$ [8]. Thus by Corollary 1, quantitative satisfaction for BWMP is in $\mathsf{NP}^{\mathsf{NP} \cap \mathsf{coNP}} \cap \mathsf{coNP}^{\mathsf{NP} \cap \mathsf{coNP}}$, which is the same as $\mathsf{NP} \cap \mathsf{coNP}$ [25].

Moreover, from [8], Player 1 has a memoryless strategy and Player 2 needs infinite memory to play optimally in non-stochastic games with the BWMP objective. From the proof of Lemma 5, by using the strategy σ_2^{AS}, Player 2 almost-surely wins $\overline{\mathsf{BWMP}}$ from all vertices in $\langle\!\langle 2 \rangle\!\rangle_{\mathcal{G}}^{\mathsf{AS}}(\overline{\mathsf{BWMP}})$. We can construct a positive winning strategy σ_2^{Pos} for Player 2 from all vertices in $\langle\!\langle 2 \rangle\!\rangle_{\mathcal{G}}^{\mathsf{Pos}}(\overline{\mathsf{BWMP}})$ in a similar manner as done for the positive winning strategy for $\overline{\mathsf{FWMP}(\ell)}$ in Section 6.1. We summarize the results in the following theorem:

Theorem 6. *Given a stochastic game $\mathcal{G}$ and a threshold $p \in [0, 1]$, for $\mathsf{BWMP}_{\mathcal{G}}$, the positive, almost-sure, and quantitative satisfaction for Player 1 are in $\mathsf{NP} \cap \mathsf{coNP}$. Moreover, a memoryless strategy suffices for Player 1, while Player 2 requires an infinite memory strategy to play optimally.*

Acknowledgement. We thank Mickael Randour for pointing out reference [6], making us aware of the bugs in the algorithms of [8] and the correct version of these algorithms as presented here. This work is partially supported by the Indian Science and Engineering Research Board (SERB) grant SRG/2021/000466 and by the Indo-French Centre for the Promotion of Advanced Research (IFCPAR).

References

1. Alur, R., Henzinger, T.A.: Finitary fairness. In: LICS. pp. 52–61. IEEE Computer Society (1994)
2. Baier, C., Katoen, J.: Principles of model checking. MIT Press (2008)

3. Bordais, B., Guha, S., Raskin, J.F.: Expected window mean-payoff. In: FSTTCS. LIPIcs, vol. 150, pp. 32:1–32:15 (2019)
4. Brihaye, T., Delgrange, F., Oualhadj, Y., Randour, M.: Life is Random, Time is Not: Markov Decision Processes with Window Objectives. Logical Methods in Computer Science **Volume 16, Issue 4** (Dec 2020)
5. Bruyère, V., Hautem, Q., Randour, M.: Window parity games: an alternative approach toward parity games with time bounds. In: GandALF. EPTCS, vol. 226, pp. 135–148 (2016)
6. Bruyère, V., Hautem, Q., Raskin, J.F.: On the complexity of heterogeneous multidimensional games. In: CONCUR. LIPIcs, vol. 59, pp. 11:1–11:15. Schloss Dagstuhl - Leibniz-Zentrum für Informatik (2016)
7. Chatterjee, K.: Concurrent games with tail objectives. Theoretical Computer Science **388**(1), 181–198 (2007)
8. Chatterjee, K., Doyen, L., Randour, M., Raskin, J.F.: Looking at mean-payoff and total-payoff through windows. Information and Computation **242**, 25–52 (2015)
9. Chatterjee, K., Henzinger, T.A.: Finitary winning in ω-regular games. In: TACAS. pp. 257–271. LNCS 3920, Springer (2006)
10. Chatterjee, K., Henzinger, T.A.: Value iteration. In: 25 Years of Model Checking - History, Achievements, Perspectives. pp. 107–138. LNCS 5000, Springer (2008)
11. Chatterjee, K., Henzinger, T.A.: A survey of stochastic ω-regular games. Journal of Computer and System Sciences **78**(2), 394–413 (2012)
12. Chatterjee, K., Henzinger, T.A., Horn, F.: Finitary winning in omega-regular games. ACM Trans. Comput. Log. **11**(1), 1:1–1:27 (2009)
13. Chatterjee, K., Henzinger, T.A., Horn, F.: Stochastic games with finitary objectives. In: MFCS. pp. 34–54. Springer Berlin Heidelberg (2009)
14. Condon, A.: The complexity of stochastic games. Information and Computation **96**(2), 203–224 (1992)
15. Doyen, L., Gaba, P., Guha, S.: Stochastic window mean-payoff games. CoRR **abs/2304.11563** (2023), https://arxiv.org/abs/2304.11563
16. Ehrenfeucht, A., Mycielski, J.: Positional strategies for mean payoff games. Int. Journal of Game Theory **8**(2), 109–113 (1979)
17. Filar, J., Vrieze, K.: Competitive Markov Decision Processes. Springer (1997)
18. Grädel, E., Thomas, W., Wilke, T. (eds.): Automata, Logics, and Infinite Games: A Guide to Current Research. LNCS 2500, Springer (2002)
19. Hautem, Q.: The Complexity of Combining Objectives in Two-Player Games. Ph.D. thesis, Université de Mons (2018)
20. Horn, F.: Faster algorithms for finitary games. In: TACAS. Lecture Notes in Computer Science, vol. 4424, pp. 472–484. Springer (2007)
21. Horn, F., Thomas, W., Wallmeier, N., Zimmermann, M.: Optimal strategy synthesis for request-response games. RAIRO Theor. Informatics Appl. **49**(3), 179–203 (2015)
22. Jurdzinski, M.: Deciding the winner in parity games is in UP ∩ co-UP. Inf. Process. Lett. **68**(3), 119–124 (1998)
23. Kupferman, O., Piterman, N., Vardi, M.Y.: From liveness to promptness. Formal Methods Syst. Des. **34**(2), 83–103 (2009)
24. Martin, D.A.: The determinacy of blackwell games. The Journal of Symbolic Logic **63**(4), 1565–1581 (1998), http://www.jstor.org/stable/2586667
25. Schöning, U.: A low and a high hierarchy within NP. Journal of Computer and System Sciences **27**(1), 14–28 (1983)
26. Weinert, A., Zimmermann, M.: Easy to win, hard to master: Optimal strategies in parity games with costs. In: CSL. LIPIcs, vol. 62, pp. 31:1–31:17 (2016)

27. Zwick, U., Paterson, M.: The complexity of mean payoff games on graphs. Theor. Comput. Sci. **158**(1&2), 343–359 (1996)

Symbolic Solution of Emerson-Lei Games for Reactive Synthesis*

Daniel Hausmann$^{(\boxtimes)}$, Mathieu Lehaut, and Nir Piterman

University of Gothenburg, Gothenburg, Sweden
hausmann@chalmers.se

Abstract. Emerson-Lei conditions have recently attracted attention due
to both their succinctness and their favorable closure properties. In the
current work, we show how infinite-duration games with Emerson-Lei
objectives can be analyzed in two different ways. First, we show that the
Zielonka tree of the Emerson-Lei condition naturally gives rise to a new
reduction to parity games. This reduction, however, does not result in
optimal analysis. Second, we show based on the first reduction (and the
Zielonka tree) how to provide a direct fixpoint-based characterization of
the winning region. The fixpoint-based characterization allows for symbolic analysis. It generalizes the solutions of games with known winning
conditions such as Büchi, GR[1], parity, Streett, Rabin and Muller objectives, and in the case of these conditions reproduces previously known
symbolic algorithms and complexity results.

We also show how the capabilities of the proposed algorithm can be
exploited in reactive synthesis, suggesting a new expressive fragment of
LTL that can be handled symbolically. Our fragment combines a safety
specification and a liveness part. The safety part is unrestricted and
the liveness part allows to define Emerson-Lei conditions on occurrences
of letters. The symbolic treatment is enabled due to the simplicity of
determinization in the case of safety languages and by using our new
algorithm for game solving. This approach maximizes the number of
steps solved symbolically in order to maximize the potential for efficient
symbolic implementations.

1   Introduction

Infinite-duration two-player games are a strong tool that has been used, notably,
for reactive synthesis from temporal specifications [38]. Many different winning
conditions are considered in the literature.

Emerson-Lei (EL) conditions [21] were initially suggested in the context of automata but are among the most general (regular) winning conditions considered
for such games. They succinctly express general liveness properties by encoding Boolean combinations of events that should occur infinitely or finitely often.
Automata and games in which acceptance or winning is defined by Emerson-Lei
conditions have garnered attention in recent years [35,40,27,25], in particular

---

* This work is supported by the ERC Consolidator grant D-SynMA (No. 772459).

N. Kobayashi and J. Worrell (Eds.): FoSSaCS 2024, LNCS 14574, pp. 55–78, 2024.
https://doi.org/10.1007/978-3-031-57228-9_4

because of their succinctness and good compositionality properties (Emerson-Lei objectives are closed under conjunction, disjunction, and negation). In this work, we show how infinite-duration two-player games with Emerson-Lei winning conditions can be solved symbolically.

It has been established that solving Emerson-Lei games is PSPACE-complete and that an exponential amount of memory may be required by winning strategies [25]. Zielonka trees are succinct tree-representations of Muller objectives [47]. They have been used to obtain tight bounds on the amount of memory needed for winning in Muller games [18], and can also be applied to analyze Emerson-Lei objectives and games. One indirect way to solve Emerson-Lei games is by transformation to equivalent parity games using later-appearance-records [25], and then solving the resulting parity games. Another, more recent, indirect approach goes through Rabin games by first extracting history-deterministic Rabin automata from Zielonka trees and then solving the resulting Rabin games [12]. Both these indirect solution methods are enumerative by nature. Here, we give a direct symbolic algorithmic solution for Emerson-Lei games. We show how the Zielonka tree allows to directly encode the game as a parity game. Furthermore, building on this reduction, we show how to construct a fixpoint equation system that captures winning in the game. As usual, fixpoint equation systems are recipes for game solving algorithms that manipulate sets of states symbolically. To the best of our knowledge, we thereby give the first description of a fully symbolic algorithm for the solution of Emerson-Lei games.

The algorithm that we obtain in this way is adaptive in the sense that the nesting structure of recursive calls is obtained directly from the Zielonka tree of the given winning objective. As the Zielonka tree is specific to the objective, this means that the algorithm performs just the fixpoint computations that are required for that specific objective. In particular, our algorithm instantiates to previously known fixpoint iteration algorithms in the case that the objective is a (generalized) Büchi, GR[1], parity, Streett, Rabin or Muller condition, reproducing previously known algorithms and complexity results. As we use fixpoint iteration, the instantiation of our algorithm to parity game solving is not directly a quasipolynomial algorithm. In the general setting, the algorithm solves unrestricted Emerson-Lei games with k colors, m edges and n nodes in time $\mathcal{O}(k! \cdot m \cdot n^k)$ and yields winning strategies with memory $\mathcal{O}(k!)$.

We apply our symbolic solution of Emerson-Lei games to the automated construction of safe systems. The ideas of synthesis of reactive systems from temporal specifications go back to the early days of computer science [14]. These concepts were modernized and connected to linear temporal logic (LTL) and finite-state automata by Pnueli and Rosner [38]. In recent years, practical applications in robotics are using this form of synthesis as part of a framework producing correct-by-design controllers [28,6,44,32,34].

A prominent way to extend the capacity of reasoning about state spaces is by reasoning *symbolically* about sets of states/paths. In order to apply this approach to reactive synthesis, different fragments of LTL that allow symbolic game analysis have been considered. Notably, the GR[1] fragment has been widely used for the applications in robotics mentioned above [37,7]. But also larger fragments are being considered and experimented with [20,19,41]. Recently, De Giacomo and Vardi suggested that similar advantages can be had by changing the usual semantics of LTL from considering infinite models to finite models (LTL_f) [22]. The complexity of the problem remains doubly-exponential, however, symbolic techniques can be applied. As models are finite, it is possible to use the classical subset construction (in contrast to Büchi determinization), which can be reasoned about symbolically. Furthermore, the resulting games have simple reachability objectives. This approach with finite models is used for applications in planning [11,10] and robotics [6].

Here, we harness our symbolic solution to Emerson-Lei games to suggest a large fragment of LTL that can be reasoned about symbolically. We introduce the *Safety and Emerson-Lei* fragment whose formulas are conjunctions $\varphi_{safety} \wedge \varphi_{EL}$ between an (unrestricted) safety condition and an (unrestricted) Emerson-Lei condition defined in terms of game states. This fragment generalizes GR[1] and the previously mentioned works in [20,19,41]. We approach safety and Emerson-Lei LTL synthesis in two steps: first, consider only the safety part and convert it to a symbolic safety automaton; second, reason symbolically on this automaton by solving Emerson-Lei games using our novel symbolic algorithm.

We show that realizability of a safety and Emerson-Lei formula $\varphi_{safety} \wedge \varphi_{EL}$ can be checked in time $2^{\mathcal{O}(m \cdot \log m \cdot 2^n)}$, where $n = |\varphi_{safety}|$ and $m = |\varphi_{EL}|$. The overall procedure therefore is doubly-exponential in the size of the safety part but only single-exponential in the size of the liveness part; notably, both the automaton determinization and game solving parts can be implemented symbolically.

We begin by recalling Emerson-Lei games and Zielonka trees in Section 2, and also prove an upper bound on the size of Zielonka trees. Next we show how to solve Emerson-Lei games by fixpoint computation in Section 3. In Section 4 we formally introduce the safety and Emerson-Lei fragment of LTL and show how to construct symbolic games with Emerson-Lei objectives that characterize realizability and that can be solved using the algorithm proposed in Section 3. Omitted proofs and further details can be found in the full version of this paper [23].

2   Emerson-Lei Games and Zielonka Trees

We recall the basics of Emerson-Lei games [25] and Zielonka trees [47], and also show an apparently novel bound on the size of Zielonka trees; previously, the main interest was on the size of winning strategies induced by Zielonka trees, which is smaller [18].

Emerson-Lei games. We consider two-player games played between the *existential player* $\exists$ and its opponent, the *universal player* $\forall$. A *game arena* $A = (V, V_\exists, V_\forall, E)$ consists of a set $V = V_\exists \uplus V_\forall$ of nodes, partitioned into sets of *existential nodes* $V_\exists$ and *universal nodes* $V_\forall$, and a set $E \subseteq V \times V$ of *moves*; we put $E(v) = \{v' \in V \mid (v, v') \in E\}$ for $v \in V$. A *play* $\pi = v_0 v_1 \ldots$ then is a sequence of nodes such that for all $i \geq 0$, $(v_i, v_{i+1}) \in E$; we denote the set of plays in A by $\mathsf{plays}(A)$. A *game* $G = (A, \alpha)$ consists of a game arena A together with an objective $\alpha \subseteq \mathsf{plays}(A)$.

A *strategy* for the existential player is a function $\sigma : V^* \cdot V_\exists \to V$ such that for all $\pi \in V^*$ and $v \in V_\exists$ we have $(v, \sigma(\pi v)) \in E$. A play $v_0 v_1 \ldots$ is said to be *compliant* with strategy f if for all $i \geq 0$ such that $v_i \in V_\exists$ we have $v_{i+1} = \sigma(v_0 \ldots v_i)$. Strategy σ is *winning* for the existential player from node $v \in V$ if all plays starting in v that are compliant with σ are contained in α; then we say that the existential player *wins* v. We denote by $W_\exists$ the *winning region* for the existential player (that is, the set of nodes that the existential player wins).

In *Emerson-Lei games*, each node is colored by a set of colors, and the objective α is induced by a formula that specifies combinations of colors that have to be visited infinitely often, or are allowed to be visited only finitely often. Formally, we fix a set C of colors and use *Emerson-Lei formulas*, that is, finite positive Boolean formulas $\varphi \in \mathbb{B}_+(\{\mathsf{Inf}\, c, \mathsf{Fin}\, c\}_{c \in C})$ over atoms of the shape $\mathsf{Inf}\, c$ or $\mathsf{Fin}\, c$, to define sets of plays. The satisfaction relation $\models$ for a set $D \subseteq C$ of colors and an Emerson-Lei formula φ (written $D \models \varphi$) is defined in the usual inductive way; D will represent the set of colors that are visited infinitely often by plays. E.g. the clauses for atoms $\mathsf{Inf}\, c$ and $\mathsf{Fin}\, c$ are

$$D \models \mathsf{Inf}\, c \Leftrightarrow c \in D \qquad\qquad D \models \mathsf{Fin}\, c \Leftrightarrow c \notin D$$

Consider a game arena $A = (V, V_\exists, V_\forall, E)$. An *Emerson-Lei condition* is given by an Emerson-Lei formula φ together with a coloring function $\gamma : V \to 2^C$ that assigns a (possibly empty) set $\gamma(v)$ of colors to each node $v \in V$. The formula φ and the coloring function γ together specify the objective

$$\alpha_{\gamma,\varphi} = \left\{ v_0 v_1 \ldots \in \mathsf{plays}(A) \,\middle|\, \{c \in C \mid \forall i.\, \exists j \geq i.\, c \in \gamma(v_j)\} \models \varphi \right\}$$

Thus a play $\pi = v_0 v_1 \ldots$ is winning for the existential player (formally: $\pi \in \alpha_{\gamma,\varphi}$) if and only if the set of colors that are visited infinitely often by π satisfies φ. Below, we will also make use of *parity games*, denoted by $(V, V_\exists, V_\forall, E, \Omega)$ where $\Omega : V \to \{1, \ldots, 2k\}$ (for $k \in \mathbb{N}$) is a priority function, assigning priorities to game nodes. The objective of the existential player then is that the maximal priority that is visited infinitely often is an even number. Parity games are an instance of Emerson-Lei games, obtained with set $C = \{p_1, \ldots, p_{2k}\}$ of colors, a coloring function that assigns exactly one color to each node and with objective

$$\mathsf{Parity}(p_1, \ldots, p_{2k}) = \bigvee\nolimits_{i \text{ even}} \left(\mathsf{Inf}\, p_i \wedge \bigwedge\nolimits_{i < j \leq 2k} \mathsf{Fin}\, p_j \right).$$

Similarly, Emerson-Lei objectives directly encode (combinations of) other standard objectives, such as Büchi, Rabin, Streett or Muller conditions:

— $\mathsf{Inf}\ f$ $\text{Büchi}(f)$
— $\bigvee_{1\le i\le k}(\mathsf{Inf}\ e_i \wedge \mathsf{Fin}\ f_i)$ $\text{Rabin}(e_1, f_1, \ldots, e_k, f_k)$
— $\bigwedge_{1\le i\le k}(\mathsf{Fin}\ r_i \vee \mathsf{Inf}\ g_i)$ $\text{Streett}(r_1, g_1, \ldots, r_k, g_k)$
— $\bigvee_{D\in\mathcal{U}}(\bigwedge_{c\in D}\mathsf{Inf}\ c \wedge \bigwedge_{d\in C\setminus D}\mathsf{Fin}\ d)$ $\text{Muller}(\mathcal{U} \subseteq 2^C)$

Zielonka Trees. We introduce a succinct encoding of the algorithmic essence of Emerson-Lei objectives in the form of so-called Zielonka trees [47,18].

Definition 1. *The* Zielonka tree *for an Emerson-Lei formula φ over set C of colors is a tuple $\mathcal{Z}_\varphi = (T, R, l)$ where $(T, R \subseteq T \times T)$ is a tree and $l : T \to 2^C$ is a labeling function that assigns sets $l(t)$ of colors to vertices $t \in T$. We denote the root of (T, R) by r. Then $\mathcal{Z}_\varphi$ is defined to be the unique tree (up to reordering of child vertices) that satisfies the following constraints.*

- *The root vertex is labeled with C, that is, $l(r) = C$.*
- *Each vertex t has exactly one child vertex t_D (labeled with $l(t_D) = D$) for each set D of colors that is maximal in $\{D' \subsetneq l(t) \mid D' \models \varphi \Leftrightarrow l(t) \not\models \varphi\}$.*

For $s, t \in T$ such that s is an ancestor of t, we write $s \le t$. Given a vertex $s \in T$, we denote its set of direct successors by $R(s) = \{t \in T \mid (s, t) \in R\}$ and the set of leafs below it by $L(s) = \{t \in T \mid s \le t$ and $R(t) = \emptyset\}$; we write L for the set of all leafs. We assume some fixed total order $\preceq$ on T that respects $\le$; this order induces a numbering of T. A vertex t in the Zielonka tree is said to be winning if $l(t) \models \varphi$, and losing otherwise. We let $T_\square$ and $T_\bigcirc$ denote the sets of winning and losing vertices in $\mathcal{Z}_\varphi$, respectively. Finally, we assign a level $\mathsf{lev}(t)$ to each vertex $t \in T$ so that $\mathsf{lev}(r) = |C|$, and $\mathsf{lev}(s') = \mathsf{lev}(s) - 1$ for all $(s, s') \in R$.

Example 2. As mentioned above, Emerson-Lei games and Zielonka trees instantiate naturally to games with, e.g., Büchi, generalized Büchi, GR[1], parity, Rabin, Streett and Muller objectives; for brevity, we illustrate this for selected examples here (more instances can be found in [23]).

1. *Generalized Büchi condition*: Given k colors $f_1, \ldots, f_k$, the winning objective $\varphi = \bigwedge_{1\le i\le k} \mathsf{Inf}\ f_i$ expresses that all colors are visited infinitely often (not necessarily simultaneously); the induced Zielonka tree is depicted below with boxes and circles representing winning and losing vertices, respectively.

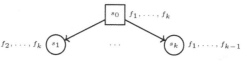

2. *Streett condition*: The vertices in the Zielonka tree for Streett condition given by $\varphi = \bigwedge_{1\le i\le k}(\mathsf{Fin}\ r_i \vee \mathsf{Inf}\ g_i)$ are identified by duplicate-free lists L of colors (each entry being r_i or g_i for some $1 \le i \le k$) that encode the vertex position in the tree. Vertex L has label $l(\mathsf{L}) = C \setminus \mathsf{L}$ and is winning if and only if $|\mathsf{L}|$ is even. Winning vertices L have one child vertex $\mathsf{L} : g_j$ for each $g_j \in C \setminus \mathsf{L}$ resulting in $|C \setminus \mathsf{L}|/2$ many child vertices. Losing vertices L have the single child vertex $\mathsf{L} : r_j$ where the last entry $\mathsf{last}(\mathsf{L})$ in L is g_j. All leafs are winning and are labeled with $\emptyset$. The tree has height $2k$ and $2(k!)$ vertices.

3. To obtain a Zielonka tree that has branching at both winning and losing vertices, we consider the objective $\varphi_{EL} = (\text{Fin } a \vee \text{Inf } b) \wedge ((\text{Fin } a \vee \text{Fin } d) \wedge \text{Inf } c)$. This property can be seen as the conjunction of a Streett pair (a, b) with two disjunctive Rabin pairs (c, a) and (c, d), altogether stating that c occurs infinitely often and a occurs finitely often or b occurs infinitely often and d occurs finitely often. Below we depict the induced Zielonka tree.

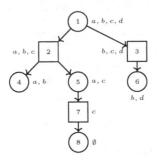

Lemma 3. *The height and the branching width of $\mathcal{Z}_\varphi$ are bounded by $|C|$ and $2^{|C|}$ respectively; the number of vertices in $\mathcal{Z}_\varphi$ is bounded by $e|C|!$ (where e is Euler's number).*

3 Solving Emerson-Lei Games

We now show how to extract from the Zielonka tree of an Emerson-Lei objective a fixpoint characterization of the winning regions of an Emerson-Lei game. Solving the game then reduces to computing the fixpoint, yielding a game solving algorithm that works by fixpoint iteration and hence is directly open to symbolic implementation. The algorithm is adaptive in the sense that the structure of its recursive calls is extracted from the Zielonka tree and hence tailored to the objective. As a stepping stone towards obtaining our fixpoint characterization, we first show how Zielonka trees can be used to reduce Emerson-Lei games to parity games that are structured into tree-like subgames.

Recall that $G = (V, V_\exists, V_\forall, E, \alpha_{\gamma,\varphi})$ is an Emerson-Lei game and that the associated Zielonka tree is $\mathcal{Z}_\varphi = (T, R, l)$ with set L of leaves, sets $T_\bigcirc$ and $T_\square$ of winning and losing vertices, respectively, and with root r. Following [18], we define the *anchor vertex* of $v \in V$ and $t \in T$ by

$$\text{anchor}(v, t) = \max_{\leq}\{s \in T \mid s \leq t \wedge \gamma(v) \subseteq l(s)\};$$

it is the lower-most ancestor of t that contains $\gamma(v)$ in its label.

A novel reduction to parity games. Intuitively, our reduction annotates nodes in G with leaves of $\mathcal{Z}_\varphi$ that act as a memory, holding information about the order in which colors have been visited. In the reduced game, the memory value $t \in L$ is updated according to a move from v to w in G by playing a subgame along the Zielonka tree. This subgame starts at the anchor vertex of v and t and the players in turn pick child vertices, with the existential player choosing the branch that is taken at vertices from $T_\bigcirc$ and the universal player choosing at vertices from $T_\square$.[1] Once this subgame reaches a leaf $t' \in L$, the memory value is updated to t' and another step of G is played. Due to the tree structure of $\mathcal{Z}_\varphi$ every play in the reduced game (walking through the Zielonka tree in the described way, repeatedly jumping from a leaf to an anchor vertex and then descending to a leaf again) has a unique topmost vertex from T that it visits infinitely often; by the definition of anchor vertices, the label of this vertex corresponds to the set of colors that is visited infinitely often by the according play of G. A parity condition can be used to decide whether this vertex is winning or losing.

Formally, we define the parity game $P_G = (V', V'_\exists, V'_\forall, E', \Omega)$, played over $V' = V \times T$, as follows. Nodes $(v, t) \in V'$ are owned by the existential player if either t is not a leaf, and it is not a winning vertex ($t \notin L$ and $t \in T_\bigcirc$), or if t is a leaf and, in G, v is owned by the existential player ($t \in L$ and $v \in V_\exists$); all other nodes are owned by the universal player. Moves and priorities are defined by

$$E'(v,t) = \begin{cases} \{v\} \times R(t) & t \notin L \\ E(v) \times \{\mathsf{anchor}(v,t)\} & t \in L \end{cases} \qquad \Omega(v,t) = \begin{cases} 2 \cdot \mathsf{lev}(t) & t \in T_\square \\ 2 \cdot \mathsf{lev}(t) + 1 & t \in T_\bigcirc \end{cases}$$

for $(v, t) \in V'$. Thus from (v, t) such that t is a leaf ($t \in L$), the owner of v picks a move $(v, w) \in E$ and the game continues with $(w, \mathsf{anchor}(v, t))$. From (v, t) such that t is not a leaf ($t \notin L$), the owner of t picks a child $t' \in R(t)$ of t in the Zielonka tree and the game continues with (v, t'), leaving the game node component v unchanged. Therefore, plays in P_G correspond to plays from G that are annotated with memory values $t \in T$ that are updated according to the colors that are visited (by moving to the anchor vertex); in addition to that, the owners of vertices in the Zielonka Tree are allowed to decide (by selecting one of the child vertices) with which colors they intend to satisfy the sub-objectives that are encoded by vertex labels. The priority function Ω then is used to identify the top-most anchor vertex s that is visited infinitely often in a play of P_G, deciding a play to be winning if and only if s is a winning vertex ($t \in T_\square$). We note that $|V'| = |V| \cdot |T| \le |V| \cdot e|C|!$ by Lemma 3.

Theorem 4. *For all $v \in V$, the existential player wins v in the Emerson-Lei game G if and only if the existential player wins (v, r) in the parity game P_G.*

This reduction yields a novel indirect method to solve Emerson-Lei games with n nodes and k colors by solving parity games with $n \cdot ek!$ nodes and $2k$ priorities; by itself, this reduction does not improve upon using later appearance records [25]. However, the game P_G consists of subgames of particular tree-like shapes. The remainder of this section is dedicated to showing how the special structure of P_G allows for direct symbolic solution by solving equivalent systems of fixpoint equations over V (rather than over the exponential-sized set V').

---

[1] Players choose from vertices where they lose, which explains the notation $T_\square$ and $T_\bigcirc$.

Fixpoint equation systems. Recall (from e.g. [4]) that a hierarchical system of fixpoint equations is given by equations

$$X_i =_{\eta_i} f_i(X_1, \ldots, X_k)$$

for $1 \leq i \leq k$, where $\eta_i \in \{\mathsf{GFP}, \mathsf{LFP}\}$ and the $f_i : \mathcal{P}(V)^k \to \mathcal{P}(V)$ are *monotone* functions, that is, $f_i(A_1, \ldots, A_k) \subseteq f_i(B_1, \ldots, B_k)$ whenever $A_j \subseteq B_j$ for all $1 \leq j \leq k$. As we aim to use fixpoint equation systems to characterize winning regions of games, it is convenient to define the semantics of equation systems also in terms of games, as proposed in [4]. For a system S of k fixpoint equations, the *fixpoint game* $G_S = (V, V_\exists, V_\forall, E, \Omega)$ is a parity game with sets of nodes $V_\exists = V \times \{1, \ldots, k\}$ and $V_\forall = \mathcal{P}(V)^k$. The set of edges E and the priority function $\Omega : V \to \{0, \ldots, 2k-1\}$ are defined, for $(v, i) \in V_\exists$ and $\bar{A} = (A_1, \ldots, A_k) \in V_\forall$, by

$$E(v, i) = \{\bar{A} \in V_\forall \mid v \in f_i(\bar{A})\} \qquad E(\bar{A}) = \{(v, i) \in V_\exists \mid v \in A_i\}$$

and by $\Omega(v, i) = 2(k - i) + \iota_i$ and $\Omega(\bar{A}) = 0$, where $\iota_i = 1$ if $\eta_i = \mathsf{LFP}$ and $\iota_i = 0$ if $\eta_i = \mathsf{GFP}$. We say that v is contained in the *solution* of variable X_i (denoted by $v \in [\![X_i]\!]$) if and only if the existential player wins the node (v, i) in G_S. In order to show containment of a node v in the solution of X_i, the existential player thus has to provide a solution $(A_1, \ldots, A_k) \in V_\forall$ for all variables such that $v \in f_i(A_1, \ldots, A_k)$; the universal player in turn can challenge a claimed solution $(A_1, \ldots, A_k)$ by picking some $1 \leq i \leq k$ and $v \in A_i$ and moving to (v, i). The game objective checks whether the dominating equation in a play (that is, the equation with minimal index among the equations that are evaluated infinitely often in the play) is a least or a greatest fixpoint equation.

Baldan et al. have shown in [4] that this game characterization is equivalent to the more traditional Knaster-Tarski-style definition of the semantics of fixpoint equation systems in terms of nested fixpoints of the involved functions f_i.

To give a flavor of the close connection between fixpoint equation systems and winning regions in games, we recall that for a given set V of nodes, the *controllable predecessor function* $\mathsf{CPre} : 2^V \to 2^V$ is defined, for $X \subseteq V$, by

$$\mathsf{CPre}(X) = \{v \in V_\exists \mid E(v) \cap X \neq \emptyset\} \cup \{v \in V_\forall \mid E(v) \subseteq X\}.$$

Example 5. Given a Büchi game $(V, V_\exists, V_\forall, E, \mathsf{Inf}\ f)$ with coloring function $\gamma : V \to 2^{\{f\}}$, the winning region of the existential player is the solution of the equation system

$$X_1 =_{\mathsf{GFP}} X_2 \qquad X_2 =_{\mathsf{LFP}} (f \cap \mathsf{CPre}(X_1)) \cup (\overline{f} \cap \mathsf{CPre}(X_2))$$

where $f = \{v \in V \mid \gamma(v) = \{f\}\}$ and $\overline{f} = V \setminus f$.

Our upcoming fixpoint characterization of winning regions in Emerson-Lei games uses the following notation that relates game nodes with anchor vertices in the Zielonka tree.

Definition 6. *For a set $D \subseteq C$ of colors, and $\bowtie \in \{\subseteq, \not\subseteq\}$ we put $\gamma_{\bowtie D}^{-1} = \{v \in V \mid \gamma(v) \bowtie D\}$. For $s, t \in T$ such that $s < t$ (that is, s is an ancestor of t in $\mathcal{Z}_\varphi$), we define*

$$\mathsf{anc}_t^s = \gamma_{\subseteq l(s)}^{-1} \cap \gamma_{\not\subseteq l(s_t)}^{-1}$$

where s_t is the child vertex of s that leads to t; we also put $\mathsf{anc}_t^t = \gamma_{\subseteq l(t)}^{-1}$.

Note that for fixed $t \in T$ and $v \in V$, there is a unique $s \in T$ such that $s \leq t$ and $v \in \mathsf{anc}_t^s$ (possibly, $s = t$); this s is the anchor vertex of t at v.

Next, we present our fixpoint characterization of winning in Emerson-Lei games, noting that it closely follows the definition of P_G.

Definition 7 (Emerson-Lei equation system). *We define the system S_φ of fixpoint equations for the objective φ by putting*

$$X_s =_{\eta_s} \begin{cases} \bigcup_{t \in R(s)} X_t & R(s) \neq \emptyset, s \in T_\bigcirc \\ \bigcap_{t \in R(s)} X_t & R(s) \neq \emptyset, s \in T_\square \\ \bigcup_{s' \leq s} \left(\mathsf{anc}_s^{s'} \cap \mathsf{CPre}(X_{s'}) \right) & R(s) = \emptyset \end{cases}$$

for $s \in T$. For every $t \in T$, we use X_t to refer to the variable X_i where i is the index of t according to $\preceq$ and similarly for η_t. Furthermore, $\eta_s = \mathsf{GFP}$ if $s \in T_\square$ and $\eta_s = \mathsf{LFP}$ if $s \in T_\bigcirc$.

Example 8. Instantiating Definition 7 to the Büchi objective $\varphi = \mathsf{Inf}\ f$ yields exactly the equation system given in Example 5. Revisiting the objectives from Example 2, we obtain the following fixpoint characterizations (further examples can be found in [23]).

1. *Generalized Büchi condition:*

$$X_{s_0} =_{\mathsf{GFP}} \bigcap_{1 \leq i \leq k} X_{s_i} \quad X_{s_i} =_{\mathsf{LFP}} (\mathsf{anc}_{s_i}^{s_0} \cap \mathsf{CPre}(X_{s_0})) \cup (\mathsf{anc}_{s_i}^{s_i} \cap \mathsf{CPre}(X_{s_i}))$$

where $\mathsf{anc}_{s_i}^{s_0} = \gamma_{\subseteq C}^{-1} \cap \gamma_{\not\subseteq C \setminus \{f_i\}}^{-1} = \{v \in V \mid f_i \in \gamma(v)\}$ and $\mathsf{anc}_{s_i}^{s_i} = \gamma_{\subseteq C \setminus \{f_i\}}^{-1}$.

2. *Streett condition:*

$$X_L =_{\eta_L} \begin{cases} \bigcap_{g_j \notin L} X_{L:g_j} & |L|\ \text{even}, |L| < 2k \\ X_{L:r_j} & |L|\ \text{odd}, \mathsf{last}(L) = g_j \\ (\mathsf{anc}_L^\emptyset \cap \mathsf{CPre}(X_\emptyset)) \cup \ldots \cup (\mathsf{anc}_L^L \cap \mathsf{CPre}(X_L)) & |L| = 2k \end{cases}$$

where $\eta_L = \mathsf{GFP}$ if $|L|$ is even and $\eta_L = \mathsf{LFP}$ if $|L|$ is odd. Here, $\mathsf{anc}_L^K = \gamma_{\subseteq C \setminus K}^{-1} \cap \gamma_{\not\subseteq C \setminus I}^{-1}$ for $K \neq L$ and $I = K_L$, and $\mathsf{anc}_L^L = \gamma_{\subseteq \emptyset}^{-1}$, both for L such that $|L| = 2k$.

3. The equation system associated to the Zielonka tree for the complex objective φ_{EL} from Example 2.3 is as follows, where we use a formula over the colors to denote the set of vertices whose label satisfies the formula. For example,

$b \wedge \neg d$ corresponds to vertices whose set of colors contains b but does not contain d.

$$X_1 =_{\mathsf{LFP}} X_2 \cup X_3 \qquad X_2 =_{\mathsf{GFP}} X_4 \cap X_5 \qquad X_3 =_{\mathsf{GFP}} X_6 \qquad X_5 =_{\mathsf{LFP}} X_7 \qquad X_7 =_{\mathsf{GFP}} X_8$$
$$X_4 =_{\mathsf{LFP}} (\neg c \wedge \neg d \cap \mathsf{Cpre}(X_4)) \cup (c \wedge \neg d \cap \mathsf{Cpre}(X_2)) \cup (d \cap \mathsf{Cpre}(X_1))$$
$$X_6 =_{\mathsf{LFP}} (\neg a \wedge \neg c \cap \mathsf{Cpre}(X_6)) \cup (\neg a \wedge c \cap \mathsf{Cpre}(X_3)) \cup (a \cap \mathsf{Cpre}(X_1))$$
$$X_8 =_{\mathsf{LFP}} (\neg a \wedge \neg b \wedge \neg c \wedge \neg d \cap \mathsf{Cpre}(X_8)) \cup (\neg a \wedge \neg b \wedge c \wedge \neg d \cap \mathsf{Cpre}(X_7)) \cup$$
$$(a \wedge \neg b \wedge \neg d \cap \mathsf{Cpre}(X_5)) \cup (b \wedge \neg d \cap \mathsf{Cpre}(X_2)) \cup (d \cap \mathsf{Cpre}(X_1)),$$

Theorem 9. *Referring to the equation system from Definition 7 and recalling that r is the root of the Zielonka tree $\mathcal{Z}_\varphi$, the solution of the variable X_r is the winning region of the existential player in the Emerson-Lei game G.*

By Theorem 4, it suffices to mutually transform winning strategies in P_G and the fixpoint game G_{S_φ} for the equation system S_φ from Definition 7.

Given the fixpoint characterization of winning regions in Emerson-Lei games with objective φ in Definition 7, we obtain a fixpoint iteration algorithm that computes the solution of Emerson-Lei games. The algorithm is by nature open to symbolic implementation. The main function is recursive, taking as input one vertex $s \in T$ of the Zielonka tree $\mathcal{Z}_\varphi$ and a list l of subsets of the set V of nodes, and returns a subset of V as result. For calls $\text{SOLVE}(s, ls)$, we require that the argument list ls contains exactly one subset $X_{s'}$ of V for each ancestor s' of s in the Zielonka tree (with $s' < s$).

Algorithm 1 $\text{SOLVE}(s, ls)$

---

if $s \in T_\bigcirc$ **then** $X_s \leftarrow \emptyset$ **else** $X_s \leftarrow V$ $\triangleright$ Initialize variable X_s for lfp/gfp
$W \leftarrow V \setminus X_s$
while $X_s \neq W$ **do** $\triangleright$ Compute fixpoint
    $W \leftarrow X_s$
    if $R(s) \neq \emptyset$ **then** $\triangleright$ Case: s is not a leaf in $\mathcal{Z}_\varphi$
        for $t \in R(s)$ **do**
           $U \leftarrow \text{SOLVE}(t, ls : W)$ $\triangleright$ Recursively solve for t
           if $s \in T_\bigcirc$ **then** $X_s \leftarrow X_s \cup U$
              else $X_s \leftarrow X_s \cap U$
        end for
    else $\triangleright$ Case: s is a leaf in $\mathcal{Z}_\varphi$
        $Y \leftarrow \emptyset$
        for $t \leq s$ **do**
           $U \leftarrow \mathsf{anc}_s^t \cap \mathsf{CPre}((ls : W)(t))$ $\triangleright$ Compute one-step attraction w.r.t. s
           $Y \leftarrow Y \cup U$
        end for
        $X_s \leftarrow Y$
    end if
end while
return X_s $\triangleright$ Return stabilized set X_s as result

---

Lemma 10. *For all $v \in V$, we have $v \in [\![X_r]\!]$ if and only if $v \in \text{SOLVE}(r, [])$.*

Proof (Sketch). The algorithm computes the solution of the equation system by standard Kleene-approximation for nested least and greatest fixpoints.

Lemma 11. *Given an Emerson-Lei game $(V, V_\exists, V_\forall, E, \alpha_{\gamma,\varphi})$ with set of colors C and induced Zielonka tree $\mathcal{Z}_\varphi$, the solution $[\![X_r]\!]$ of the equation system S_φ from Definition 7 can be computed in time $\mathcal{O}(|\mathcal{Z}_\varphi| \cdot |E| \cdot |V|^k)$, where $k \leq |C|$ denotes the height of $\mathcal{Z}_\varphi$.*

Combining Theorem 9 with Lemmas 3, 10 and 11 we obtain

Corollary 12. *Solving Emerson-Lei games with n nodes, m edges and k colors can be implemented symbolically to run in time $\mathcal{O}(k! \cdot m \cdot n^k)$; the resulting strategies require memory at most $e \cdot k!$.*

Remark 13. Strategy extraction works as follows. The algorithm computes a set $[\![X_t]\!]$ for each Zielonka tree vertex $t \in \mathcal{Z}_\varphi$. Furthermore it yields, for each non-leaf vertex $s \in T_\bigcirc$ and each $v \in [\![X_s]\!]$, a single child vertex $\text{choice}(v, s) \in R(s)$ of s such that $v \in [\![X_{\text{choice}(v,s)}]\!]$. The algorithm also yields, for each leaf vertex t and each $v \in V_\exists \cap [\![X_t]\!]$, a single game move $\text{move}(v, t)$. All these choices together constitute a winning strategy for existential player in the parity game P_G. We define a strategy for the Emerson-Lei game that uses leaves of the Zielonka tree as memory values, following the ideas used in the construction of P_G; the strategy moves, from a node $v \in V_\exists$ and having memory content m, to the node $\text{move}(v, m)$. As initial memory value we pick some leaf of $\mathcal{Z}_\varphi$ that choice associates with the initial node in G. To update memory value m according to visiting game node v, we first take the anchor vertex s of m and v. Then we pick the next memory value m to be some leaf below s that can be reached by talking the choices $\text{choice}(v, s')$ for every vertex $s' \in T_\bigcirc$ passed along the way from s to the leaf; if $s \in T_\square$, then we additionally require the following: let $q = |R(s)|$, let o be the number such that m is a leaf below the o-th child of s, and put $j = o + 1 \mod q$; then we require that m' is a leaf below the j-th child of s. By the correctness of the algorithm, the constructed strategy is a winning strategy.

Dziembowski et al. have shown that winning strategies can be extracted by using a walk through the Zielonka tree that requires memory only for the branching at winning vertices [18]. This yields, for instance, memoryless strategies for games with Rabin objectives, for which branching in the associated Zielonka trees takes place at losing vertices. Adapting the strategy extraction in our setting to this more economic method is straight-forward but notation-heavy, so we omit a more precise analysis of strategy size here.

Our algorithm hence can be implemented to run in time $2^{\mathcal{O}(k \log n)}$ for games with n nodes and $k \leq n$ colors, improving upon the bound $2^{\mathcal{O}(n^2)}$ stated in [25], where the authors only consider the case that every game node has a distinct color, implying $n = k$. We note that the later appearance record construction used in [25] is known to be hard to represent symbolically. Our fixpoint characterization generalizes previously known algorithms for e.g. parity games [8], and

Streett and Rabin games [36], recovering previously known bounds on worst-case running time of fixpoint iteration algorithms for these types of games.

While it has recently been shown that parity games can be solved in quasipoly-nomial time [9], we note that in the case of parity objectives, our algorithm is not immediately quasipolynomial. However, there are quasipolynomial methods for solving nested fixpoints [24,2] (with the latter being open to symbolic implementation); in the case of parity objectives, these more involved algorithms can be used in place of fixpoint iteration to solve our equation system and recover the quasipolynomial bound. The precise complexity of using quasipolynomial methods for solving fixpoint equation systems beyond parity conditions is subject to ongoing research.

4   Synthesis for Safety and Emerson-Lei LTL

In this section we present an application of the results from Section 3. We introduce the safety and Emerson-Lei fragment of LTL and show that synthesis for this fragment can be reasoned about symbolically. The idea for safety and Emerson-Lei LTL synthesis is twofold: first, consider only the safety part and create a symbolic arena capturing its satisfaction. Second, play a game on this arena by adding the Emerson-Lei part as a winning condition. Finally we use the results from the previous sections to solve the game symbolically.

4.1   Safety LTL and Symbolic Safety Automata

We start by defining safety LTL, symbolic safety automata, and recalling known results about those.

Definition 14 (LTL and Safety LTL [45]). *Given a non-empty set* AP *of atomic propositions, the general syntax for LTL formulas is as follows:*

$$\varphi := \top \mid \bot \mid p \mid \neg\varphi \mid \varphi_1 \wedge \varphi_2 \mid \varphi_1 \vee \varphi_2 \mid X\varphi \mid \varphi_1 U \varphi_2 \qquad p \in \mathsf{AP}.$$

Standard abbreviations are defined as follows: $\varphi_1 R \varphi_2 := \neg(\neg\varphi_1 U \neg\varphi_2)$, $F\varphi := \top U \varphi$, *and* $G\varphi := \neg F \neg \varphi$. *We define the satisfaction relation* $\models$ *for a formula* φ *and its language* $\mathcal{L}(\varphi)$ *as usual.*

An LTL formula is said to be a safety *formula if it is in negative normal form (i.e. all negations are pushed to atomic propositions) and only uses* X, R, G *as temporal operators (i.e. no* U *or* F *are allowed).*

It is a safety formula in the sense that every word that does not satisfy the formula has a finite prefix that already falsifies the formula. In other words, such a formula is satisfied as long as "bad states" are avoided forever.

Definition 15 (Symbolic Safety Automata). *A symbolic safety automaton is a tuple* $\mathcal{A} = (2^{\mathsf{AP}}, V, T, \theta_0)$ *where* V *is a set of variables,* $T(V, V', \mathsf{AP})$ *is the transition assertion, and* $\theta_0(V)$ *is the initialization assertion. A run of* $\mathcal{A}$ *on* *the word $w \in (2^{\mathsf{AP}})^\omega$ is a sequence $\rho = s_0 s_1 \ldots$ where the $s_i \in 2^V$ are variable assignments such that 1. $s_0 \models \theta_0$, and 2. for all $i \geq 0$, $(s_i, s_{i+1}, w(i)) \models T$. A word w is in $\mathcal{L}(\mathcal{A})$ if and only if there is an infinite run of $\mathcal{A}$ on w. $\mathcal{A}$ is deterministic if for all words $w \in (2^{\mathsf{AP}})^\omega$ there is at most one run of $\mathcal{A}$ on w.*

Kupferman and Vardi show how to convert a safety LTL formula into an equivalent deterministic symbolic safety automaton [30].

Lemma 16. *A safety LTL formula φ can be translated to a deterministic symbolic safety automaton $\mathcal{D}_{\mathsf{symb}}$ accepting the same language, with $|\mathcal{D}_{\mathsf{symb}}| = 2^{|\varphi|}$.*

The idea is to first convert φ to a (non-symbolic) non-deterministic safety automaton $\mathcal{N}_\varphi$, which is of size exponential of the size of the formula, and then symbolically determinize $\mathcal{N}_\varphi$ by a standard subset construction to obtain $\mathcal{D}_{\mathsf{symb}}$. Note that while the size of $\mathcal{D}_{\mathsf{symb}}$ is only exponential in the size of the formula, its state space would be double exponential when fully expanded.

Example 17. Let $\varphi = G(b \vee c) \wedge G(a \rightarrow b \vee XXb)$ be a safety LTL formula over $\mathsf{AP} = \{a, b, c\}$. An execution satisfying φ must have at least one of b or c at every step, moreover every a sees a b present at the same step or two steps afterwards.

As an intermediate step towards building the equivalent $\mathcal{D}_{\mathsf{symb}}$, we first present below a corresponding non-deterministic safety automaton $\mathcal{N}_\varphi$.

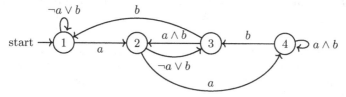

For the sake of presentation, we use Boolean combinations of AP in transitions instead of labeling them with elements of 2^{AP}, with the intended meaning that $s \xrightarrow{\psi} s' = \{s \xrightarrow{C} s' \mid C \in 2^{\mathsf{AP}}, C \models \psi\}$. We also omit the $G(b \vee c)$ part of the formula in the construction. One can simply append $\cdots \wedge (b \vee c)$ to every transition of $\mathcal{N}_\varphi$ to get back the original formula. Intuitively state 1 correspond to not seeing an a, state 2 means that a b must be seen at the next step, state 3 means that there must be a b now, and state 4 that b is needed now and next as well.

Then the symbolic safety automaton is $\mathcal{D}_{\mathsf{symb}} = (2^{\mathsf{AP}}, V, T, \theta_0)$ with:

- $V = \{v_1, v_2, v_3, v_4\}$ are the variables corresponding to the four states of $\mathcal{N}_\varphi$,
- $\theta_0 = v_1 \wedge \neg v_2 \wedge \neg v_3 \wedge \neg v_4$ asserts that only the state v_1 is initial,
- The transition assertion is $T = (v_1' \leftrightarrow (v_1 \wedge (\neg a \vee b)) \vee (v_3 \wedge b)) \wedge (v_2' \leftrightarrow (v_1 \wedge a) \vee (v_3 \wedge (a \wedge b))) \wedge (v_3' \leftrightarrow (v_2 \wedge (\neg a \vee b)) \vee (v_4 \wedge b)) \wedge (v_4' \leftrightarrow (v_2 \wedge a) \vee (v_4 \wedge (a \wedge b))) \wedge (v_1 \vee v_2 \vee v_3 \vee v_4)$.

Determinizing $\mathcal{N}_\varphi$ enumeratively would give an automaton with 9 states (see Example 23).

Remark 18. Restricting attention to safety LTL enables the two advantages mentioned above with respect to determinization. First, subset construction suffices (as observed also in [46]), avoiding the more complex Büchi determinization. Second, this construction, due to its simplicity, can be implemented symbolically. Interestingly, recent implementations of the synthesis from LTL_f [46] or from safety LTL [45] have used indirect approaches for obtaining deterministic automata. For example, by translating LTL to first order logic and applying the tool MONA to the results [45,46], or by concentrating on minimization of deterministic automata [42]. The direct construction is similar to approaches used for checking universality of nondeterministic finite automata [42] or SAT-based bounded model checking [1]. We are not aware of uses of this direct implementation of the subset construction in reactive synthesis. The worst case complexity of this part is doubly-exponential, which, just like for LTL and LTL_f, cannot be avoided [43,3].

4.2 Symbolic Games

We use *symbolic game structures* to represent a certain class of games. Formally, a *symbolic game structure* $\mathcal{G} = \langle \mathcal{V}, \mathcal{X}, \mathcal{Y}, \theta_\exists, \rho_\exists, \varphi \rangle$ consists of:

- $\mathcal{V} = \{v_1, \ldots, v_n\}$: A finite set of typed *variables* over finite domains. Without loss of generality, we assume they are all Boolean. A node s is an valuation of $\mathcal{V}$, assigning to each variable $v_i \in \mathcal{V}$ a value $s[v_i] \in \{0,1\}$. Let Σ be the set of nodes.
  We extend the evaluation function $s[\cdot]$ to Boolean expressions over $\mathcal{V}$ in the usual way. An *assertion* is a Boolean formula over $\mathcal{V}$. A node s satisfies an assertion φ denoted $s \models \varphi$, if $s[\varphi] = \textbf{true}$. We say that s is a φ-node if $s \models \varphi$.
- $\mathcal{X} \subseteq \mathcal{V}$ is a set of *input variables*. These are variables controlled by the universal player. Let $\Sigma_\mathcal{X}$ denote the possible valuations to variables in $\mathcal{X}$.
- $\mathcal{Y} = \mathcal{V} \setminus \mathcal{X}$ is a set of *output variables*. These are variables controlled by the existential player. Let $\Sigma_\mathcal{Y}$ denote the possible valuations to variables in $\mathcal{Y}$.
- $\theta_\exists(\mathcal{X}, \mathcal{Y})$ is an assertion characterizing the initial condition.
- $\rho_\exists(\mathcal{V}, \mathcal{X}', \mathcal{Y}')$ is the transition relation. This is an assertion relating a node $s \in \Sigma$ and an input value $s_\mathcal{X} \in \Sigma_\mathcal{X}$ to an output value $s_\mathcal{Y} \in \Sigma_\mathcal{Y}$ by referring to primed and unprimed copies of $\mathcal{V}$. The transition relation $\rho_\exists$ identifies a valuation $s_\mathcal{Y} \in \Sigma_\mathcal{Y}$ as a *possible output* in node s reading input $s_\mathcal{X}$ if $(s, (s_\mathcal{X}, s_\mathcal{Y})) \models \rho_\exists$, where s is the assignment to variables in $\mathcal{V}$ and $s_\mathcal{X}$ and $s_\mathcal{Y}$ are the assignment to variables in $\mathcal{V}'$ induced by $(s_\mathcal{X}, s_\mathcal{Y}) \in \Sigma$.
- φ is the winning condition, given by an LTL formula.

For two nodes s and s' of $\mathcal{G}$, s' is a *successor* of s if $(s, s') \models \rho_\exists$.
    A symbolic game structure $\mathcal{G}$ defines an arena $A_\mathcal{G}$, where $V_\forall = \Sigma$, $V_\exists = \Sigma \times \Sigma_\mathcal{X}$, and E is defined as follows:

$$E = \{(s, (s, s_\mathcal{X})) \mid s \in \Sigma \text{ and } s_\mathcal{X} \in \Sigma_\mathcal{X}\} \cup \{((s, s_\mathcal{X}), (s_\mathcal{X}, s_\mathcal{Y})) \mid (s, (s_\mathcal{X}, s_\mathcal{Y})) \models \rho_\exists\}.$$

When reasoning about symbolic game structures we ignore the intermediate visits to $V_\exists$. Indeed, they add no information as they can be deduced from the nodes in $V_\forall$ preceding and following them. Thus, a play $\pi = s_0 s_1 \ldots$ is *winning for the existential player* if σ is infinite and satisfies φ. Otherwise, σ is *winning for the universal player*.

The notion of strategy and winning region is trivially generalized from games to symbolic game structures. When needed, we treat $W_\exists$ (the set of nodes winning for the existential player) as an assertion. We define winning in the *entire* game structure by incorporating the initial assertion: a game structure $\mathcal{G}$ is said to be *won* by the existential player, if for all $s_\mathcal{X} \in \Sigma_\mathcal{X}$ there exists $s_\mathcal{Y} \in \Sigma_\mathcal{Y}$ such that $(s_\mathcal{X}, s_\mathcal{Y}) \models \theta_\exists \wedge W_\exists$.

4.3 Realizability and Synthesis

Let φ be an LTL formula over input and output variables I and O, controlled by *the environment* and *the system*, respectively (the universal and the existential player, respectively).

The reactive synthesis problem asks whether there is a strategy for the system of the form $\sigma : (2^I)^+ \to 2^O$ such that for all sequences $x_0 x_1 \cdots \in (2^I)^\omega$ we have:

$$(x_0 \cup \sigma(x_0))(x_1 \cup \sigma(x_0 x_1)) \ldots \models \varphi$$

If there is such a strategy we say that φ is *realizable* [38].

Equivalently, φ is *realizable* if the system is winning in the symbolic game $\mathcal{G}_\varphi = \langle I \cup O, I, O, \top, \top, \varphi \rangle$ with I for input variables $\mathcal{X}$ and O for output $\mathcal{Y}$.

Theorem 19. *[38] Given an LTL formula φ, the realizability of φ can be determined in doubly exponential time. The problem is 2EXPTIME-complete.*

The game $\mathcal{G}_\varphi$ above uses neither the initial condition nor the system transition. Conversely, consider a symbolic game $\mathcal{G} = \langle V, \mathcal{X}, \mathcal{Y}, \theta_\exists, \rho_\exists, \varphi \rangle$:

Theorem 20. *[7] The system wins in $\mathcal{G}$ iff $\varphi_\mathcal{G} = \theta_\exists \wedge G\rho_\exists \wedge \varphi$ is realizable.*[2][3]

4.4 Safety and Emerson-Lei Synthesis

We now define the class of LTL formulas that are supported by our technique and show how to construct appropriate games capturing their realizability problem.

For $\psi \in \mathbb{B}(\mathsf{AP})$, let $\mathsf{Inf}\,\psi := GF\psi$ and $\mathsf{Fin}\,\psi := FG\neg\psi = \neg\mathsf{Inf}\,\psi$. The *Emerson-Lei fragment* of LTL consists of all formulas that are positive Boolean combinations of $\mathsf{Inf}\,\psi$ and $\mathsf{Fin}\,\psi$ for all Boolean formulas ψ over atomic propositions. The satisfaction of such formulas depends only on the set of letters (truth assignments to propositions) appearing infinitely often in a word.

---

[2] Technically, $\rho_\exists$ contains primed variables and is not an LTL formula. This can be easily handled by using the next operator X. We thus ignore this issue.

[3] We note that Bloem et al. consider more general games, where the environment also has an initial assertion and a transition relation. Our games are obtained from theirs by setting the initial assertion and the transition relation of the environment to true.

Remark 21. The Emerson-Lei fragment easily accommodates various liveness properties that cannot be encoded in smaller fragments such as GR[1]. One prominent example for this is the property of *stability* (as encoded by LTL formulas of the shape *FG p*), which appears frequently as a guarantee in usage of synthesis for robotics and control (see, e.g., the work of Ehlers [19] and Ozay [32]), and commonly is approximated in GR[1] but, as a guarantee or as part of a specification, cannot be captured exactly in the game context. Another important example is *strong fairness* (as encoded by LTL formulas of the shape $\bigwedge_i(GF\ r_i \to GF\ g_i)$) which allows to capture the exact relation between cause and effect. Particularly, in GR[1] only if *all* "resources" are available infinitely often there is an obligation on the system to supply *all* its "guarantees". In contrast, strong fairness allows to connect particular resources to particular supplied guarantees. Ongoing studies on fairness assumptions that arise from the abstraction of continuous state spaces to discrete state spaces [32,33] provide further examples of fairness assumptions that can be expressed in EL but not in GR[1]. Emerson-Lei liveness allows free combination of all properties mentioned above and more.

Definition 22. *The* Safety and Emerson-Lei fragment *is the set of formulas of the form* $\varphi = \varphi_{\text{safety}} \wedge \varphi_{\text{EL}}$, *where* φ_{safety} *is a safety formula and* φ_{EL} *is in the Emerson-Lei fragment.*

We assume a partition $\mathsf{AP} = I \uplus O$ where I is a set of *input propositions* and O a set of *output propositions*, both non-empty. Let $\varphi = \varphi_{\text{safety}} \wedge \varphi_{\text{EL}}$ be a safety and Emerson-Lei formula over AP, and let $\mathcal{D}_{\text{symb}} = (2^{\mathsf{AP}}, V, T, \theta_0)$ be the symbolic deterministic safety automaton associated to φ_{safety}. We construct $G_\varphi = \langle V \uplus \mathsf{AP}, I, O \uplus V, \theta_0, T, \varphi_{\text{EL}} \rangle$, thus $\mathcal{X} = I$ and $\mathcal{Y} = O \uplus V$.

Example 23. Let $\varphi_{\text{safety}} = G(b \vee c) \wedge G(a \to b \vee XXb)$, our running safety example from Example 17 with its associated symbolic deterministic automaton. Partition AP into $I = \{a\}$ and $O = \{b, c\}$. We depict the arena of the game G_φ (independent of the formula φ_{EL} that is yet to be defined) in Figure 1.

To keep the illustration readable and keep it from getting too large, a few modifications to the formal arena definition have been made. First, c labels on edges have been omitted: every transition labeled with b represent two transitions with sets $\{b\}$ and $\{b, c\}$, while transitions labeled with $\neg b$ stand for a single transition with set $\{c\}$ (due to the $G(b \vee c)$ requirement forbidding $\emptyset$). Similarly, existential nodes have been omitted when all choices for the existential player lead to the same destination. Instead, the universal and existential moves have been combined in one transition: $a; *$ for an a followed by some existential move, and $a; b$ for when an a requires the existential player to play b (with or without c, as above). Finally, states are only labeled with variables from V and not AP, the latter is used to label edges instead. For a fully state-based labeling arena, states would have to store the last move, leading to various duplicate states.

Note that this game arena is given only for illustration purposes, as we want to solve the symbolic game without explicitly enumerating all its states and transitions like here.

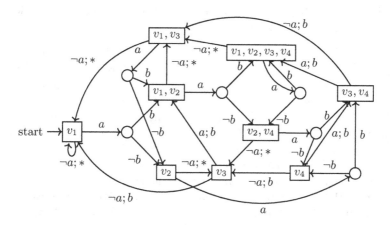

Fig. 1. Game arena for G_φ

Lemma 24. *The system wins G_φ if and only if φ is realizable.*

Next we detail how to solve the symbolic game G_φ by using the results from Section 3.

Lemma 25. *Given a symbolic game $G = \langle \mathcal{V}, \mathcal{X}, \mathcal{Y}, \theta_\exists, \rho_\exists, \varphi \rangle$ such that φ is an Emerson-Lei formula with set of colors*

$$C = \{\psi \in \mathbb{B}(\mathsf{AP}) \mid \psi \text{ is a subformula of } \varphi\},$$

the winning region $W_\exists$ of G is characterized by the equation system from Definition 7, using the assertion

$$\mathsf{CPre}(S) = \forall s_\mathcal{X} \in \Sigma_\mathcal{X}. \exists s_\mathcal{Y} \in \Sigma_\mathcal{Y}. S' \wedge (v, s_\mathcal{X}, s_\mathcal{Y}) \models \rho_\exists.$$

The proof of this lemma is by straightforward adaptation of the proof of Theorem 9 to the symbolic setting, following the relation between symbolic game structures and game arenas described above.

Finally, this gives us a procedure to solve the synthesis problem for safety and Emerson-Lei LTL.

Theorem 26. *The realizability of a formula $\varphi = \varphi_{\mathsf{safety}} \wedge \varphi_{EL}$ of the Safety and Emerson-Lei fragment of LTL can be checked in time $2^{\mathcal{O}(m \cdot \log m \cdot 2^n)}$, where $n = |\varphi_{\mathsf{safety}}|$ and $m = |\varphi_{EL}|$. Realizable formulas can be realized by systems of size at most $2^{2^n} \cdot e \cdot m!$.*

Proof. Using the construction described in this section, we obtain the symbolic game G_φ of size $q = 2^{2^n}$ with winning condition φ_{EL}, using at most m colors; by Theorem 24, this game characterizes realizibility of the formula. Using the results from the previous section, G_φ can be solved in time $\mathcal{O}(m! \cdot q^2 \cdot q^m) \in \mathcal{O}(2^{m \log m} \cdot 2^{(m+2)2^n}) \in 2^{\mathcal{O}(m \cdot \log m \cdot 2^n)}$, resulting in winning strategies with memory at most $e \cdot m!$.

Both the automata determinization and the game solving can be implemented symbolically.

Example 27. To illustrate the overall synthesis method, we consider the game that is obtained by combining the game arena $G_{\varphi_{\text{safety}}}$ from Example 23 with the winning objective $\varphi_{EL} = (\text{Fin } a \vee \text{Inf } b) \wedge (\text{Fin } a \vee \text{Fin} d) \wedge \text{Inf } c$ from Example 2.3, where we instantiate the label d to nodes satisfying $b \wedge c$ thus creating a game-specific dependency between the colors. Solving this game amounts to solving the equation system shown in Example 8.3. However, with the interpretation of $d = b \wedge c$, some of the conditions become simpler. For example, $\neg a \wedge \neg b \wedge \neg c \wedge \neg d$ becomes $\neg a \wedge \neg b \wedge \neg c$ and $b \wedge \neg d$ becomes $b \wedge \neg c$. It turns out that the system player wins the node v_1. Intuitively, the system can play $\{c\}$ whenever possible and thereby guarantee satisfaction of φ_{EL}. We extract this strategy from the computed solution of the equation system in Example 2.3 as described in Remark 13. E.g. for partial runs π that end in v_1 and for which the last leaf vertex in the induced walk ρ_π through $\mathcal{Z}_\varphi$ is the vertex 8, the system can react by playing $\{b\}$, $\{c\}$, or even $\{b, c\}$ whenever the environment plays $\emptyset$. The move $\{b\}$ continues the induced walk ρ_π through vertex 2 to the leaf vertex 5; similarly, the move $\{b, c\}$ continues ρ_π through the vertex 1 to the leaf vertex 6. The strategy construction gives precedence to the choice that leads through the lowest vertex in the Zielonka tree, which in this case means picking the move $\{c\}$ that continues ρ_π through the vertex 7 to the leaf 8. Proceeding similarly for all other combinations of game nodes and vertices in the Zielonka tree, one obtains a strategy σ for the system that always outputs singleton letters, giving precedence to $\{c\}$ whenever possible. To see that σ is a winning strategy, let π be a play that is compatible with σ. If π eventually loops at v_1 forever, then s_π is the existential vertex 7 and the existential player wins the play since it satisfies both Fin a and Inf c. Any other play π satisfies Inf a, Inf b and Inf c since all cycles that are compatible with σ (excluding the loop at v_1) contain at least one a-edge, at least one b-edge and also at least one c-edge that is prescribed by the strategy σ. For these plays, ρ_π eventually reaches the vertex 2. Since the system always plays singleton letters (so that π in particular satisfies $\text{Fin}(b \wedge c)$), the vertex 1 is not visited again by ρ_π, once vertex 2 has been reached. Hence the dominating vertex for such plays is $s_\pi = 2$, an existential vertex.

4.5   Synthesis Extensions and Optimizations

We have chosen to use safety-LTL as the safety part of the Safety-EL fragment to showcase the options opened by having symbolic algorithms for the analysis of very expressive liveness conditions. The crucial feature of the safety fragment is the ability to convert that part of the specification to a symbolic deterministic automaton. It is important to note that *every* fragment of LTL (or ω-regular in general) that can be easily converted to a symbolic deterministic automaton can be incorporated and handled with the same machinery. For example, it was suggested to extend the expressiveness of GR[1] by including deterministic automata in the safety part of the game and referring to their states in the liveness part [7]. Past LTL [31] can be handled in the same way in that it is incorporated for GR[1] [7]. An extreme example is GR-EBR, where safety parts are allowed to use bounded future and pure past, which still allows the symbolic treatment [15]. All of these alternatives can be incorporated in the safety part with no changes to our overall methodology. Unlike previous cases, if there is an easy translation to deterministic symbolic automata *with a non-trivial winning condition*, these can be incorporated as well with the EL part extended to handle their winning condition as well. We could consider also extensions to the liveness parts. For example, by using past LTL or reference to states of additional symbolic deterministic automata. The Boolean state formulas appearing as part of the EL condition can be replaced by formulas allowing one usage of the next operator, as in [39,19]. The generalization to handle transition-based EL games, which would be required in that case, rather than state-based EL games is straight-forward.

As the formulas we consider are conjunctions, optimizations can be applied to both conjuncts independently. This subsumes, for example, analyzing the winning region in a safety game prior to the full analysis [29,7,5], reductions in the size of nondeterministic automata [17], or symbolic minimization of deterministic automata [16].[4]

5 Conclusions and Future Work

We provide a symbolic algorithm to solve games with Emerson-Lei winning conditions. Our solution is based on an encoding of the Zielonka tree of the winning condition in a system of fixpoint equations. In case of known winning conditions, our algorithm recovers known algorithms and complexity results. As an application of this algorithm, we suggest an expressive fragment of LTL for which realizability can be reasoned about symbolically. Formulas in our fragment are conjunctions between an LTL safety formula and an Emerson-Lei liveness condition. This fragment is more general than, e.g., GR[1].

In the future, we believe that analysis of the Emerson-Lei part can reduce the size of Zielonka trees (and thus the symbolic algorithm). This can be done either through analysis and simplification of the LTL formula, e.g., [26], by means of alternating-cycle decomposition [12,13], or by analyzing the semantic meaning of colors. We would also like to implement the proposed overall synthesis method.

References

1. Armoni, R., Egorov, S., Fraer, R., Korchemny, D., Vardi, M.Y.: Efficient LTL compilation for sat-based model checking. In: International Conference on Computer-Aided Design. pp. 877–884. IEEE Computer Society (2005). https://doi.org/10.1109/ICCAD.2005.1560185

---

[4] Notice that explicit minimization as done, e.g., in [30] would require to explicitly construct the potentially doubly exponential deterministic automaton, nullifying the entire effort to keep all analysis symbolic.

2. Arnold, A., Niwinski, D., Parys, P.: A quasi-polynomial black-box algorithm for fixed point evaluation. In: Baier, C., Goubault-Larrecq, J. (eds.) 29th EACSL Annual Conference on Computer Science Logic, CSL 2021, January 25-28, 2021, Ljubljana, Slovenia (Virtual Conference). LIPIcs, vol. 183, pp. 9:1–9:23. Schloss Dagstuhl - Leibniz-Zentrum für Informatik (2021). https://doi.org/10.4230/LIPIcs.CSL.2021.9

3. Artale, A., Geatti, L., Gigante, N., Mazzullo, A., Montanari, A.: Complexity of safety and cosafety fragments of linear temporal logic. In: Proceedings of the AAAI Conference on Artificial Intelligence. vol. 37, pp. 6236–6244 (2023)

4. Baldan, P., König, B., Mika-Michalski, C., Padoan, T.: Fixpoint games on continuous lattices. Proc. ACM Program. Lang. 3(POPL), 26:1–26:29 (2019). https://doi.org/10.1145/3290339

5. Bansal, S., Giacomo, G.D., Stasio, A.D., Li, Y., Vardi, M.Y., Zhu, S.: Compositional safety LTL synthesis. In: 14th International Conference on Verified Software, Theories, Tools and Experiments. Lecture Notes in Computer Science, vol. 13800, pp. 1–19. Springer (2022). https://doi.org/10.1007/978-3-031-25803-9_1

6. Bhatia, A., Maly, M.R., Kavraki, L.E., Vardi, M.Y.: Motion planning with complex goals. IEEE Robotics Autom. Mag. 18(3), 55–64 (2011). https://doi.org/10.1109/MRA.2011.942115

7. Bloem, R., Jobstmann, B., Piterman, N., Pnueli, A., Sa'ar, Y.: Synthesis of reactive(1) designs. J. Comput. Syst. Sci. 78(3), 911–938 (2012). https://doi.org/10.1016/j.jcss.2011.08.007

8. Bruse, F., Falk, M., Lange, M.: The fixpoint-iteration algorithm for parity games. In: International Symposium on Games, Automata, Logics and Formal Verification, GandALF 2014. EPTCS, vol. 161, pp. 116–130 (2014). https://doi.org/10.4204/EPTCS.161.12

9. Calude, C., Jain, S., Khoussainov, B., Li, W., Stephan, F.: Deciding parity games in quasipolynomial time. In: Theory of Computing, STOC 2017. pp. 252–263. ACM (2017)

10. Camacho, A., McIlraith, S.A.: Learning interpretable models expressed in linear temporal logic. In: Twenty-Ninth International Conference on Automated Planning and Scheduling. pp. 621–630. AAAI Press (2019). https://doi.org/10.1609/icaps.v29i1.3529

11. Camacho, A., Triantafillou, E., Muise, C.J., Baier, J.A., McIlraith, S.A.: Nondeterministic planning with temporally extended goals: LTL over finite and infinite traces. In: Thirty-First AAAI Conference on Artificial Intelligence. pp. 3716–3724. AAAI Press (2017). https://doi.org/10.1609/aaai.v31i1.11058

12. Casares, A., Colcombet, T., Lehtinen, K.: On the size of good-for-games rabin automata and its link with the memory in muller games. In: Bojanczyk, M., Merelli, E., Woodruff, D.P. (eds.) International Colloquium on Automata, Languages, and Programming, ICALP 2022. LIPIcs, vol. 229, pp. 117:1–117:20. Schloss Dagstuhl - Leibniz-Zentrum für Informatik (2022). https://doi.org/10.4230/LIPIcs.ICALP.2022.117

13. Casares, A., Duret-Lutz, A., Meyer, K.J., Renkin, F., Sickert, S.: Practical applications of the alternating cycle decomposition. In: Fisman, D., Rosu, G. (eds.) Tools and Algorithms for the Construction and Analysis of Systems - 28th International Conference, TACAS 2022, Held as Part of the European Joint Conferences on Theory and Practice of Software, ETAPS 2022, Munich, Germany,

April 2-7, 2022, Proceedings, Part II. Lecture Notes in Computer Science, vol. 13244, pp. 99–117. Springer (2022). https://doi.org/10.1007/978-3-030-99527-0_6

14. Church, A.: Logic, arithmetic, and automata. In: International Congress of Mathematicians. Institut Mittag-Leffler, Sweden (1963)

15. Cimatti, A., Geatti, L., Gigante, N., Montanari, A., Tonetta, S.: Fairness, assumptions, and guarantees for extended bounded response ltl+p synthesis. Software and System Modeling (2023). https://doi.org/10.1007/s10270-023-01122-4

16. D'Antoni, L., Veanes, M.: Minimization of symbolic automata. In: Symposium on Principles of Programming Languages (POPL). pp. 541–554. ACM (2014). https://doi.org/10.1145/2535838.2535849

17. Duret-Lutz, A., Renault, E., Colange, M., Renkin, F., Aisse, A.G., Schlehuber-Caissier, P., Medioni, T., Martin, A., Dubois, J., Gillard, C., Lauko, H.: From spot 2.0 to spot 2.10: What's new? In: 34th International Conference on Computer Aided Verification. Lecture Notes in Computer Science, vol. 13372, pp. 174–187. Springer (2022). https://doi.org/10.1007/978-3-031-13188-2_9

18. Dziembowski, S., Jurdzinski, M., Walukiewicz, I.: How much memory is needed to win infinite games? In: 12th Annual IEEE Symposium on Logic in Computer Science. pp. 99–110. IEEE Computer Society (1997). https://doi.org/10.1109/LICS.1997.614939

19. Ehlers, R.: Generalized rabin(1) synthesis with applications to robust system synthesis. In: Third International Symposium on NASA Formal Methods. Lecture Notes in Computer Science, vol. 6617, pp. 101–115. Springer (2011). https://doi.org/10.1007/978-3-642-20398-5_9

20. Ehlers, R.: Unbeast: Symbolic bounded synthesis. In: 17th International Conference on Tools and Algorithms for the Construction and Analysis of Systems. Lecture Notes in Computer Science, vol. 6605, pp. 272–275. Springer (2011). https://doi.org/10.1007/978-3-642-19835-9_25

21. Emerson, E.A., Lei, C.: Modalities for model checking: Branching time logic strikes back. Sci. Comput. Program. 8(3), 275–306 (1987). https://doi.org/10.1016/0167-6423(87)90036-0

22. Giacomo, G.D., Vardi, M.Y.: Synthesis for LTL and LDL on finite traces. In: Yang, Q., Wooldridge, M.J. (eds.) Twenty-Fourth International Joint Conference on Artificial Intelligence. pp. 1558–1564. AAAI Press (2015)

23. Hausmann, D., Lehaut, M., Piterman, N.: Symbolic solution of Emerson-Lei games for reactive synthesis. CoRR abs/2305.02793 (2023), https://arxiv.org/abs/2305.02793

24. Hausmann, D., Schröder, L.: Quasipolynomial computation of nested fixpoints. In: Groote, J.F., Larsen, K.G. (eds.) Tools and Algorithms for the Construction and Analysis of Systems - 27th International Conference, TACAS 2021, Held as Part of the European Joint Conferences on Theory and Practice of Software, ETAPS 2021, Luxembourg City, Luxembourg, March 27 - April 1, 2021, Proceedings, Part I. Lecture Notes in Computer Science, vol. 12651, pp. 38–56. Springer (2021). https://doi.org/10.1007/978-3-030-72016-2_3

25. Hunter, P., Dawar, A.: Complexity bounds for regular games. In: 30th International Symposium on Mathematical Foundations of Computer Science. Lecture Notes in Computer Science, vol. 3618, pp. 495–506. Springer (2005). https://doi.org/10.1007/11549345_43

26. John, T., Jantsch, S., Baier, C., Klüppelholz, S.: Determinization and limit-determinization of Emerson-Lei automata. In: 19th International Symposium on Automated Technology for Verification and Analysis. Lecture Notes in Computer Science, vol. 12971, pp. 15–31. Springer (2021). https://doi.org/10.1007/978-3-030-88885-5_2

27. John, T., Jantsch, S., Baier, C., Klüppelholz, S.: From emerson-lei automata to deterministic, limit-deterministic or good-for-mdp automata. Innov. Syst. Softw. Eng. 18(3), 385–403 (2022). https://doi.org/10.1007/s11334-022-00445-7

28. Kress-Gazit, H., Fainekos, G.E., Pappas, G.J.: Temporal-logic-based reactive mission and motion planning. IEEE Trans. Robotics 25(6), 1370–1381 (2009). https://doi.org/10.1109/TRO.2009.2030225

29. Kugler, H., Segall, I.: Compositional synthesis of reactive systems from live sequence chart specifications. In: 15th International Conference on Tools and Algorithms for the Construction and Analysis of Systems. Lecture Notes in Computer Science, vol. 5505, pp. 77–91. Springer (2009). https://doi.org/10.1007/978-3-642-00768-2_9

30. Kupferman, O., Vardi, M.Y.: Model checking of safety properties. Formal methods in system design 19(3), 291–314 (2001). https://doi.org/10.1023/A:1011254632723

31. Lichtenstein, O., Pnueli, A., Zuck, L.D.: The glory of the past. In: Conference on Logics of Programs. Lecture Notes in Computer Science, vol. 193, pp. 196–218. Springer (1985). https://doi.org/10.1007/3-540-15648-8_16

32. Liu, J., Ozay, N., Topcu, U., Murray, R.M.: Synthesis of reactive switching protocols from temporal logic specifications. IEEE Trans. Autom. Control. 58(7), 1771–1785 (2013). https://doi.org/10.1109/TAC.2013.2246095

33. Majumdar, R., Schmuck, A.: Supervisory controller synthesis for nonterminating processes is an obliging game. IEEE Trans. Autom. Control. 68(1), 385–392 (2023). https://doi.org/10.1109/TAC.2022.3143108

34. Moarref, S., Kress-Gazit, H.: Automated synthesis of decentralized controllers for robot swarms from high-level temporal logic specifications. Auton. Robots 44(3-4), 585–600 (2020). https://doi.org/10.1007/s10514-019-09861-4

35. Müller, D., Sickert, S.: LTL to deterministic emerson-lei automata. In: Bouyer, P., Orlandini, A., Pietro, P.S. (eds.) Proceedings Eighth International Symposium on Games, Automata, Logics and Formal Verification, GandALF 2017, Roma, Italy, 20-22 September 2017. EPTCS, vol. 256, pp. 180–194 (2017). https://doi.org/10.4204/EPTCS.256.13

36. Piterman, N., Pnueli, A.: Faster solutions of rabin and streett games. In: 21th IEEE Symposium on Logic in Computer Science (LICS 2006), 12-15 August 2006, Seattle, WA, USA, Proceedings. pp. 275–284. IEEE Computer Society (2006). https://doi.org/10.1109/LICS.2006.23

37. Piterman, N., Pnueli, A., Sa'ar, Y.: Synthesis of reactive(1) designs. In: 7th International Conference on Verification, Model Checking, and Abstract Interpretation. Lecture Notes in Computer Science, vol. 3855, pp. 364–380. Springer (2006). https://doi.org/10.1007/11609773_24

38. Pnueli, A., Rosner, R.: On the synthesis of a reactive module. In: Sixteenth ACM Symposium on Principles of Programming Languages. pp. 179–190. ACM Press (1989). https://doi.org/10.1145/75277.75293

39. Raman, V., Piterman, N., Finucane, C., Kress-Gazit, H.: Timing semantics for abstraction and execution of synthesized high-level robot control. IEEE Trans. Robotics **31**(3), 591–604 (2015). https://doi.org/10.1109/TRO.2015.2414134

40. Renkin, F., Duret-Lutz, A., Pommellet, A.: Practical "paritizing" of emerson-lei automata. In: Hung, D.V., Sokolsky, O. (eds.) Automated Technology for Verification and Analysis - 18th International Symposium, ATVA 2020, Hanoi, Vietnam, October 19-23, 2020, Proceedings. Lecture Notes in Computer Science, vol. 12302, pp. 127–143. Springer (2020). https://doi.org/10.1007/978-3-030-59152-6_7

41. Sohail, S., Somenzi, F.: Safety first: a two-stage algorithm for the synthesis of reactive systems. Int. J. Softw. Tools Technol. Transf. **15**(5-6), 433–454 (2013). https://doi.org/10.1007/s10009-012-0224-3

42. Tabakov, D., Vardi, M.Y.: Experimental evaluation of classical automata constructions. In: 12th International Conference on Logic for Programming, Artificial Intelligence, and Reasoning. Lecture Notes in Computer Science, vol. 3835, pp. 396–411. Springer (2005). https://doi.org/10.1007/11591191_28

43. Vardi, M.Y., Stockmeyer, L.J.: Improved upper and lower bounds for modal logics of programs: Preliminary report. In: Proceedings of the 17th Annual ACM Symposium on Theory of Computing. pp. 240–251. ACM (1985)

44. Wongpiromsarn, T., Topcu, U., Murray, R.M.: Receding horizon temporal logic planning. IEEE Trans. Autom. Control. **57**(11), 2817–2830 (2012). https://doi.org/10.1109/TAC.2012.2195811

45. Zhu, S., Tabajara, L.M., Li, J., Pu, G., Vardi, M.Y.: A symbolic approach to safety LTL synthesis. In: 13th International Haifa Verification Conference: Hardware and Software - Verification and Testing. Lecture Notes in Computer Science, vol. 10629, pp. 147–162. Springer (2017). https://doi.org/10.1007/978-3-319-70389-3_10

46. Zhu, S., Tabajara, L.M., Pu, G., Vardi, M.Y.: On the power of automata minimization in temporal synthesis. In: Proceedings 12th International Symposium on Games, Automata, Logics, and Formal Verification. EPTCS, vol. 346, pp. 117–134 (2021). https://doi.org/10.4204/EPTCS.346.8

47. Zielonka, W.: Infinite games on finitely coloured graphs with applications to automata on infinite trees. Theor. Comput. Sci. **200**(1-2), 135–183 (1998). https://doi.org/10.1016/S0304-3975(98)00009-7

78   D. Hausmann et al.

Parity Games on Temporal Graphs

Pete Austin$^{(\boxtimes)}$ [ID], Sougata Bose [ID], and Patrick Totzke [ID]

University of Liverpool, Liverpool, UK
sgpausti@liverpool.ac.uk

Abstract. Temporal graphs are a popular modelling mechanism for dynamic complex systems that extend ordinary graphs with discrete time. Simply put, time progresses one unit per step and the availability of edges can change with time.

We consider the complexity of solving ω-regular games played on temporal graphs where the edge availability is ultimately periodic and fixed a priori.

We show that solving parity games on temporal graphs is decidable in PSPACE, only assuming the edge predicate itself is in PSPACE. A matching lower bound already holds for what we call *punctual* reachability games on static graphs, where one player wants to reach the target at a given, binary encoded, point in time. We further study syntactic restrictions that imply more efficient procedures. In particular, if the edge predicate is in P and is monotonically increasing for one player and decreasing for the other, then the complexity of solving games is only polynomially increased compared to static graphs.

Keywords: Temporal graphs · Reachability Games · Complexity · Timed automata

1 Introduction

Temporal graphs are graphs where the edge relation changes over time. They are often presented as a sequence $G_0, G_1, \ldots$ of graphs over the same set of vertices. We find it convenient to define them as pairs $G = (V, E)$ consisting of a set V of vertices and associated edge availability predicate $E : V^2 \to 2^{\mathbb{N}}$ that determines at which integral times a directed edge can be traversed. This model has been used to analyse dynamic networks and distributed systems in dynamic topologies, such as gossiping and information dissemination [36,24]. There is also a large body of work that considers temporal generalisations of various graph-theoretic notions and properties [32,14,10]. Related algorithmic questions include graph colouring [30], exploration [12], travelling salesman [33], maximum matching [29], and vertex-cover [2]. The edge relation is often deliberately left unspecified and sometimes only assumed to satisfy some weak assumptions about connectedness, frequency, or fairness to study the worst or average cases in uncontrollable environments. Depending on the application, one distinguishes between "online" questions, where the edge availability is revealed stepwise, as

N. Kobayashi and J. Worrell (Eds.): FoSSaCS 2024, LNCS 14574, pp. 79–98, 2024.
https://doi.org/10.1007/978-3-031-57228-9_5

opposed to the "offline" variant where all is given in advance. We refer to [17,31] for overviews of temporal graph theory and its applications.

Two player zero-sum verification games on directed graphs play a central role in formal verification, specifically the reactive synthesis approach [34]. Here, a controllable system and an antagonistic environment are modeled as a game in which two opposing players jointly move a token through a graph. States are either owned by Player 1 (the system) or Player 2 (the environment), and the owner of the current state picks a valid successor. Such a play is won by Player 1 if, and only if, the constructed path satisfies a predetermined *winning condition* that models the desired correctness specification. The winning condition is often given either in a temporal logic such as Linear Temporal Logic (LTL) [35], or directly as ω-automaton whose language is the set of infinite paths considered winning for Player 1. The core algorithmic problem is solving games: to determine which player has a strategy to force a win, and if so, how.

Determining the complexity of solving games on static graphs has a long history and continues to be an active area of research. We refer to [1,13] for introductions on the topic and recall here only that solving reachability games, where Player 1 aims to eventually reach a designated target state, is complete for polynomial time. The precise complexity of solving parity games is a longstanding open question. It is known to be in UP∩coUP [22], and so in particular in NP and coNP, and recent advances have led to quasi-polynomial time algorithms [6,23,26,9,25].

Related Work. Periodic temporal graphs were first studied by Floccchini, Mans, and Santoro in [14], where they show polynomial bounds on the length of explorations (paths covering all vertices). Recently, De Carufel, Flocchini, Santoro, and Simard [10] study Cops & Robber games on periodic temporal graphs. They provide an algorithm for solving one-cop games that is only quadratic in the number of vertices and linear in the period.

Games on temporal graphs with maximal age, or period of some absolute value K given in binary are games on exponentially succinctly presented arenas. Unfolding them up to time K yields an ordinary game on the exponential sized graph which allows to transfer upper bounds, that are not necessarily optimal. In a similar vein, Avni, Ghorpade, and Guha [4] have recently introduced types of games on exponentially succinct arenas called pawn games. Similar to our results, their findings provide improved PSPACE upper bounds for reachability games.

Parity games on temporal graphs are closely related to timed-parity games, which are played on the configuration graphs of timed automata [3]. However, the time in temporal graphs is discrete as opposed to the continuous time semantics in timed automata. Solving timed parity games is complete for EXP[28,8] and the lower bound already holds for reachability games on timed automata with only two clocks [21]. Unfortunately, a direct translation of (games on) temporal graphs to equivalent timed automata games requires at least two clocks: one to hold the global time used to check the edge predicate and one to ensure that time progresses one unit per step.

Contributions. We study the complexity of solving parity games on temporal graphs. As a central variant of independent interest are what we call *punctual reachability games*, that are played on a static graph and player wants to reach a target vertex at a given binary encoded time. We show that solving such games is already hard for PSPACE, which provides a lower bound for all temporal graph games we consider.

As our second, and main result, we show how to solve parity games on (ultimately) periodic temporal graphs. The difficulty to overcome here is that the period may be exponential in the number of vertices and thus a naïvely solving the game on the unfolding only yields algorithms in exponential space. Our approach relies on the existence of polynomially sized summaries that can be verified in PSPACE using punctual reachability games.

We then provide a sufficient syntactic restriction that avoids an increased complexity for game solving. In particular, if the edge predicate is in polynomial time and is monotonically increasing for one player and decreasing for the other, then the cost of solving reachability or parity games on temporal graphs increases only polynomially in the number of vertices compared to the cost of solving these games on static graphs.

None of our upper bounds rely on any particular representation of the edge predicate. Instead, we only require that the representation ensures that checking membership (if an edge is traversable at a given time) has suitably low complexity. That is, our approach to solve parity games only requires that the edge predicate is in PSPACE, and polynomial-time verifiable edge predicates suffice to derive P-time upper bounds for monotone reachability games. These conditions are met for example if the edge predicate is defined as semilinear set given as an explicit union of linear sets (NP in general and in P for singleton sets of periods), or by restricted Presburger formulae: the quantifier-free fragment is in P, the existential fragment is in NP but remains in P if the number of variables is bounded [37]. See for instance [15] and contained references.

The rest of the paper is structured as follows. We recall the necessary notations in Section 2 and then discuss reachability games in Section 3. Section 4 presents the main construction for solving parity games and finally, in Section 5, we discuss improved upper bounds for monotone temporal graphs.

2   Preliminaries

Definition 1 (Temporal Graphs). *A temporal graph $G = (V, E)$ is a directed graph where V are vertices and $E : V^2 \to 2^{\mathbb{N}}$ is the edge availability relation that maps each pair of vertices to the set of times at which the respective directed edge can be traversed. If $i \in E(s, t)$ we call t an i-successor of s and write $s \xrightarrow{i} t$.*

The horizon *of a temporal graph is $h(G) = \sup_{s,t \in V}(E(s,t))$, the largest finite time at which any edge is available, or ∞ if no such finite time exists. A temporal graph is* finite *if $h(G) \in \mathbb{N}$ i.e., every edge eventually disappears forever. A temporal graph is* periodic *with period $K \in \mathbb{N}$ if for all nodes $s, t \in V$ it holds that $E(s, t) = E(s, t) + K \cdot \mathbb{N}$. We call G static if it has period 1.*

Naturally, one can unfold a temporal graph into its *expansion* up to some time $T \in \mathbb{N} \cup \{\infty\}$, which is the graph with nodes $V \times \{0, 1, \ldots, T\}$ and directed edges $(s, i) \rightarrow (t, i + 1)$ iff $i \in E(s, t)$.

In order for algorithmic questions to be interesting, we assume that temporal graphs are given in a format that is more succinct than the expansion up to their horizon or period. We only require that the representation ensures that checking if an edge is traversable at a given time can be done reasonably efficiently.

We will henceforth use formulae in the existential fragment of Presburger arithmetic, the first-order theory over natural numbers with equality and addition. That is, the $\exists$PA formula $\Phi_{s,t}(x)$ with one free variable x represents the set of times at which an edge from s to t is available as $E(s, t) = \{n \mid \Phi_{s,t}(n) \equiv true\}$. We use common syntactic sugar including inequality and multiplication with (binary encoded) constants. For instance, $\Phi_{s,t}(x) \stackrel{\text{def}}{=} 5 \leq x \wedge x \leq 10$ means the edge is available at times $\{5, 6, 7, 8, 9, 10\}$; and $\Phi_{s,t}(x) \stackrel{\text{def}}{=} \exists y.(x = y \cdot 7) \wedge \neg(x \leq 100)$ means multiples of 7 greater than 100.

Definition 2 (Parity Games). *A parity game is a zero-sum game played by two opposing players on a directed graph. Formally, the game is given by a game graph $G = (V, E)$, a partitioning $V = V_1 \uplus V_2$ of vertices into those owned by Player 1 and Player 2 respectively, and a colouring $col : V \rightarrow C$ of vertices into a finite set $C \subsetneq \mathbb{N}$ of colours.*

The game starts with a token on an initial vertex $s_0 \in V$ and proceeds in turns where in round i, the owner of the vertex occupied by the token moves it to some successor. This way both players jointly agree on an infinite path $\rho = s_0 s_1 \ldots$ called a play. A play is winning for Player 1 if $\max\{c \mid \forall i \exists j. col(s_j) = c\}$, the maximum colour seen infinitely often, is even.

A strategy for Player i is a recipe for how to move. Formally, it is a function $\sigma_i : V^* V_i \rightarrow V$ from finite paths ending in a vertex s in V_i to some successor. We call σ positional if $\sigma(\pi s) = \sigma(\pi' s)$ for any two prefixes $\pi, \pi' \in V^*$. A strategy is winning from vertex s if Player i wins every play that starts in vertex s and during which all decisions are made according to σ.

We call a vertex s winning for Player i if there exists a winning strategy from s, and call the subset of all such vertices the *winning region* for that player. Parity games enjoy the following property (See [13, Theorem 15] for details).

Proposition 1. *Parity games are uniformly positionally determined: For every game $(V = V_1 \uplus V_2, E, col)$ there is a pair σ_1, σ_2 of positional strategies so that σ_i is winning for Player i from every vertex in the winning region of Player i.*

A *temporal parity game* is a parity game played on the infinite expansion of a temporal graph $G = (V, E)$, where the ownership and colouring of vertices are given with respect to the underlying directed graph $V = V_1 \uplus V_2$ and $col : V \rightarrow C$. The ownership and colouring are lifted to the expansion so that vertices in $V_i \times \mathbb{N}$ are owned by Player i and vertex (s, n) has colour $col(s)$.

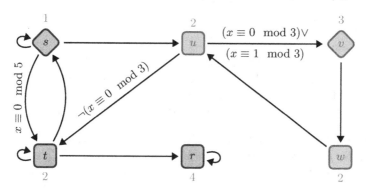

Fig. 1: An example of a temporal parity game. Player 1 controls the diamond vertices $V_1 = \{s, v\}$ and Player 2 controls square vertices $V_2 = \{r, t, u, w\}$. Edge labels are Presburger formulae constraints denoting when an edge is available; edges without constraints are always available. The grey label next to each node denotes its colour. E.g., $col(s) = 1 \in C = \{1, 2, 3, 4\}$.

Example 1. Consider the temporal parity game shown in Fig. 1. We will draw Player 1 states as diamond and those controlled by Player 2 as squares and sometimes write modulo expressions to define the edge availability. For example, the constraint on the edge from u to v can be written as the ∃PA-formula as $\exists y.(x = 3y) \vee (x = 3y + 1)$ and so this edge is available at times $0, 1, 3, 4, 6, \ldots$. The temporal graph underlying this game has period 15.

Player 1 has a winning strategy starting from (s, i) in the expansion by staying in state s until time $i' \geq i$ with $i' \equiv 0 \mod 5$ and then following the edge to $(t, i' + 1)$. If Player 2 ever chooses to move to r, he is trapped in an even-coloured cycle; if he stays in t forever, then the resulting game sees only colour 2 and is losing for him. Otherwise, if the game continues at $(s, i' + 2)$, Player 1 repeats as above (and wins plays that see both states s and t. The example shows that Player 1 s strategies depend on the time and are not positional in the vertices alone, even if the winning set has period 1. Indeed, the only possible vertex-positional strategy (cycle in s) is losing.

The vertices $\{s, t\}$ shaded in blue represent the vertex from which Player 1 can win starting at any time, following the strategy described above. From the vertices shaded in red, Player 2 can win starting at certain times. For example, Player 2 has a winning strategy from (u, i) if, and only if, $i \equiv 0 \mod 3$ or $i \equiv 1 \mod 3$ by moving to $(v, i + 1)$. Notice that this edge is not available, and thus Player 2 is forced to move to t at times $x \equiv 2 \mod 3$. In particular therefore, Player 1 wins from $(v, 0)$. The winning region for Player 1 is $\{(s, k), (t, k), (r, k), (u, 3k + 2), (v, 3k), (w, 3k + 1) \mid k \in \mathbb{N}\}$.

The algorithmic question we consider is determining the set of vertices from which Player 1 wins starting at time 0.

3   Reachability Games

We discuss a variant of temporal games that turns out to be central both for upper and lower bounds for solving games on temporal graphs.

We call these *punctual reachability games*, which are played on a static graph and Player 1 aims to reach the target precisely at a target time.

Definition 3. *A punctual reachability game $G = (V, E, s_0, F)$ is a game played on a static graph with vertices $V = V_1 \uplus V_2$, edges $E \subseteq V^2$, an initial state s_0 and set of target vertices $F \subseteq V$. An additional parameter is a target time $T \in \mathbb{N}$ given in binary. Player 1 wins a play if and only if a vertex in F is reached at time T.*

Punctual reachability games are really just a reformulation of the membership problem for alternating finite automata (AFA) [7] over a unary input alphabet. Player 1 wins the punctual reachability game with target T if, and only if, the word a^T is accepted by the AFA described by the game graph. Checking if a given unary word a^T is accepted by an AFA is complete for polynomial time if T is given in unary [20]. We first observe that it is PSPACE-hard if T is given in binary. We write in the terminology of punctual reachability games but the main argument is by reduction from the emptiness problem for unary AFA, which is PSPACE-compete [18,19]. We rely on the fact that the shortest word accepted by an AFA is at most exponential in the number of states.

Lemma 1. *Let $G = (V, E, s_0, F)$ be a reachability game on a static graph. If there exist $T \in \mathbb{N}$ so that Player 1 wins the punctual reachability game at target time T, then there exists some such $T \le 2^{|V|}$.*

Proof. Assume towards contradiction that $T \ge 2^{|V|}$ is the smallest number such that Player 1 wins the punctual reachability game and consider some winning strategy σ. For any time $k \le T$ we can consider the set $S_k \subseteq V$ of vertices occupied on any branch of length k on σ. By the pigeonhole principle, we observe $k < k' \le T$ with $S_k = S_{k'}$, which allows to create a strategy σ' that follows σ until time k, then continues (and wins) according to σ as if it had just seen a length k' history leading to the same vertex. This shows that there exists a winning strategy for target time $T - (k - k')$, which contradicts the assumption. □

A lower bound for solving punctual reachability games is now immediate.

Lemma 2. *Solving punctual reachability games with target time T encoded in binary is PSPACE-hard.*

Proof. We reduce the non-emptiness problem of AFA over unary alphabets. In our terminology this is the decision problem if, for a given a reachability game $G = (V, E, s_0, F)$ there exists some $T \in \mathbb{N}$ so that Player 1 wins the punctual reachability game at target time T. This problem is PSPACE-complete [18].

By Lemma 1, positive instances can be witnessed by a small target $T \leq 2^{|V|}$ and so we know that it is PSPACE-hard to determine the existence of such a small target time that allows Player 1 to win.

Consider now the punctual reachability game G' that extends G by a new initial vertex s_0' that is owned by Player 1 and which has a self-loop as well as an edge to the original initial vertex s_0 with target time $T' \stackrel{\text{def}}{=} 2^{|V|}$. In G', Player 1 selects some number $T \leq T'$ by waiting in the initial vertex for $T' - T$ steps and then starts the game G with the target time T. Therefore, Player 1 wins in G' for target T' if, and only if, she wins for some $T \leq 2^{|V|}$ in G.   □

Corollary 1. *Solving reachability games on finite temporal graphs is* PSPACE-*hard.*

Proof. We reduce the punctual reachability game with target T to an ordinary reachability game on a finite temporal graph. This can be done by introducing a new vertex u as the only target vertex, so that it is only reachable via edges from vertices in F at time exactly T. That is $E(s, u) \stackrel{\text{def}}{=} \{T\}$ and $E(s, t) = [0, T]$ for all $s, t \in V \setminus \{u\}$. Now Player 1 wins the reachability game for target u if, and only if, she wins the punctual reachability game with target F at time T.   □

A matching PSPACE upper bound for solving punctual reachability games, as well as reachability games on finite temporal graphs can be achieved by computing the winning region backwards as follows.[1] For any game graph with vertices $V = V_1 \uplus V_2$, set $S \subseteq V$ and $i \in \{1, 2\}$, let $Pre_i(S) \subseteq V$ denote the set of vertices from which Player i can force to reach S in one step.

$$Pre_i(S) \stackrel{\text{def}}{=} \{v \in V_i \mid \exists (v, v') \in E.v' \in S\} \cup \{v \in V_{1-i} \mid \forall (v, v') \in E.v' \in S\}$$

A straightforward induction on the duration T shows that Player i wins the punctual reachability game with target time T from vertex s if, and only if $s \in Pre_i^T(F)$, the T-fold iteration of Pre_i applied to the target set F.

Notice that knowledge of $Pre_i(S)$ is sufficient to compute $Pre_i^{k+1}(S)$. By iteratively unfolding the definition of Pre_i^k, we can compute $Pre_i^T(F)$ from $Pre_i^0(F) = F$ in polynomial space.[2] Together with Lemma 2 we conclude the following.

Theorem 1. *Solving punctual reachability games with target time T encoded in binary is* PSPACE-*complete.*

---

[1] For readers familiar with reachability games, the notion $Pre_1(S)$ above is very similar to, but not the same as the k-step attractor of S: The former contains states from which Player 1 can force to see the target in *exactly* k steps, whereas the latter contains those where the target is reachable in k *or fewer* steps.

[2] To be precise, naïvely unfolding the definition requires $\mathcal{O}(T + |V|^2)$ time, exponential in (the binary encoded input) T, and $\mathcal{O}(|V| + \log(T))$ space to memorise the current set $Pre_k \subseteq V$ as well as the time $k \leq T$ in binary.

The same approach works for reachability games on finite *temporal* graphs if applied to the expansion up to horizon $h(G)$, leading to the same time and space complexity upper bounds. The only difference is that computing $Pre_1^k(F \times \{T\})$ requires to check edge availability at time $T - k$.

Theorem 2. *Solving reachability games on finite temporal graphs is* PSPACE-*complete.*

Proof. Consider a temporal game with vertices $V = V_1 \uplus V_2$, edges $E : V^2 \to 2^{\mathbb{N}}$ target vertices $F \subseteq V$ and where $T = h(G)$ is the latest time an edge is available. We want to check if starting in an initial state s_0 at time 0, Player 1 can force to reach F at time T. In other words, for the game played on the expansion up to time T we want to decide if $(s_0, 0)$ is contained in $Pre_1^T(F \times \{T\})$.

By definition of the expansion, we have $Pre_1(S \times \{n\}) \subseteq V \times \{n-1\}$ for all $S \subseteq V$ and $n \leq T$. Since we can check the availability of an edge at time n in polynomial space, we can iteratively compute $Pre_1^n(F \times \{T\})$ backwards, starting with $Pre_1^0(F \times \{T\}) = F \times \{T\}$, and only memorising the current iteration $n \leq T$ and a set $W_n \subseteq V$ representing $Pre_1^n(F \times \{T\}) = W_n \times \{T - n\}$. $\square$

4   Parity Games

We consider Parity games played on periodic temporal graphs. As input we take a temporal graph $G = (V, E)$ with period K, a partitioning $V = V_1 \uplus V_2$ of the vertices, as well as a colouring $col : V \to C$ that associates a colour out of a finite set $C \subset \mathbb{N}$ of colours to every state.

It will be convenient to write $col(\pi) \overset{\text{def}}{=} \max\{col(s_i) \mid 0 \leq i \leq k\}$ for the maximal colour of any vertex visited along a finite path $\pi = (s_0, 0)(s_1, 1) \ldots (s_k, k)$. The following relations R_s^σ capture the guarantees provided by a strategy σ if followed for one full period from vertex s.

Definition 4. *For a strategy σ and vertex $s \in V$ define $R_s^\sigma \subseteq V \times C$ be the relation containing $(t, c) \in R_s^\sigma$ if, and only if, there exists a finite play $\pi = (s, 0) \ldots (t, K)$ consistent with σ, that starts in s at time 0, ends in t at time K, and the maximum colour seen on the way is $col(\pi) = c$. We call R_s^σ the summary of s with respect to strategy σ.*
A relation $B \subseteq V \times C$ is s-realisable if there is a strategy σ with $B = R_s^\sigma$.

Example 2. Consider the game in Fig. 2 where vertex $u \in V_2$ has colour 2 and all other vertices have colour 1. The graph has period $K = 2$. The relations $\{(t, 1)\}$ and $\{(t, 2), (t', 2)\}$ are s-realisable, as witnessed by the strategies $\sigma(s) = t$ and $\sigma(s) = u)$, respectively. However, $\{(t, 2)\}$ is not s-realisable as no Player 1 strategy guarantees to visit s then u then t.

Lemma 3. *Checking s-realisability is in* PSPACE. *That is, one can verify in polynomial space for a given temporal Parity game, state $s \in V$ and relation $B \subseteq V \times C$ whether B is s-realisable.*

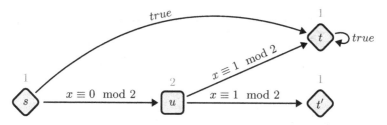

Fig. 2: The game from Example 2. Labels on vertices and edges denote colours and available times, respectively. The graph has period 2. In two rounds, Player 1 can force to end in t having seen colour 1, or in either t or t' but having seen a better colour 2.

Proof. We reduce checking realisability to solving a reachability game on a temporal graph that is only polynomially larger. More precisely, given a game $G = (V, E, col)$ consider the game $G' = (V', E', col')$ over vertices $V' \stackrel{\text{def}}{=} V \times C$ that keep track of the maximum colour seen so far. That is, the ownership of vertices and colours are lifted directly as $(s, c) \in V_1' \iff s \in V_1$ and $col'(s, c) \stackrel{\text{def}}{=} col(s)$, and for any $i \in \mathbb{N}$, $s, t, s_0 \in V$, $c, d \in C$, we let (t, d) be an i-successor of (s, c) if, and only if, both t is an i-successor of s and $d = \max\{c, col(t)\}$.

Consider some relation $B \subseteq V \times C$. We have that B is s-realisable if, and only if, Player 1 wins the punctual reachability game on G' from vertex $(s, col(s))$ at time 0, towards target vertices $B \subseteq V'$ at target time K. Indeed, any winning Player 1 strategy in this game witnesses that B is s-realisable and vice versa. By Theorem 2, the existence of such a winning strategy can be verified in polynomial space by backwards-computing the winning region.   □

The following defines a small, and **PSPACE**-verifiable certificate for Player 1 to win the parity game on a periodic temporal graph.

Definition 5 (Certificates). *Given temporal parity game (V, E, col) with period K, a certificate for Player 1 winning the game from initial vertex $s_0 \in V$ is a multigraph where the vertex set $V' \subseteq V$ contains s_0, and edges $E' \subseteq V' \times C \times V'$ are labelled by colours, such that*

1. *For every $s \in V'$, the set $Post(s) \stackrel{\text{def}}{=} \{(t, c) \mid (s, c, t) \in E'\}$ is s-realisable.*
2. *The maximal colour on every cycle reachable from s_0 is even.*

Notice that condition 1 implies that no vertex in a certificate is a deadlock. A certificate intuitively allows to derive Player 1 strategies based on those witnessing the realisability condition.

Example 3. Consider the game from Example 1 played on the temporal graph with period 15. A certificate for Player 1 winning from state v at time 0 is depicted in Fig. 3. Indeed, the Player 1 strategy mentioned in Example 1 (aim

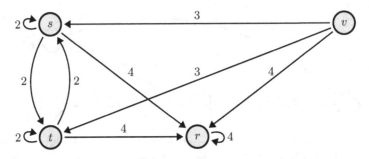

Fig. 3: A certificate that Player 1 wins the game in Example 1 from state v at time 0.

to alternate between s and t) witnesses that $Post(v) = \{(s,3),(t,3),(r,4)\}$ is v-realisable because it allows Player 1 to enforce that after $K = 15$ steps from v, the game ends up in one of those states via paths whose colour is dominated by $col(v) = 3$ or $col(r) = 4$.

Lemma 4. *Player 1 wins the parity game on G from vertex s_0 if, and only if, there exists a certificate.*

Proof. For the backward implication we argue that a certificate C allows to derive a winning strategy for Player 1 in the parity game G. By the realisability assumption (1), for each vertex $s \in V$ there must exist a Player 1 strategy σ_s with $R_s^{\sigma_s} = Post(s)$ that tells her how to play in G for K rounds if the starting time is a multiple of K. Moreover, suppose she plays according to σ_s for K rounds and let t and c be the vertex reached and maximal colour seen on the way. Then by definition of the summaries, $(t,c) \in R_s^{\sigma_s} = Post(s)$ and so in the certificate C there must be some edge $s \xrightarrow{c} t$.

Suppose Player 1 continues to play in G like this forever: From time $i \cdot K$ to $(i+1) \cdot K$ she plays according to some strategy σ_{s_i} determined by the vertex s_i reached at time $i \cdot K$. Any consistent infinite play ρ in G, chosen by her opponent, describes an infinite walk ρ' in C such that the colour seen in any step $i \in \mathbb{N}$ of ρ' is precisely the dominant colour on ρ between rounds iK and $(i+1)K$. Therefore the dominant colours seen infinitely often on ρ and ρ' are the same and, by certificate condition (2) on the colouring of cycles, even. We conclude that the constructed strategy for Player 1 is winning.

For the forward implication, assume that Player 1 wins the game on G from vertex s at time 0. Since the game G is played on a temporal graph with period K, its expansion up to time $K - 1$ is an ordinary parity game on a static graph with vertices $V \times \{0, 1, \ldots, K-1\}$ where the second component indicates the time modulo K. Therefore, by positional determinacy of parity games (Proposition 1), we can assume that Player 1 wins in G using a strategy σ that is itself periodic. That is, $\sigma(hv) = \sigma(h'v)$ for any two histories h, h' of lengths $|h| \equiv |h'| \mod K$. Moreover, we can safely assume that σ is uniform, meaning that it is winning from any vertex $(s, 0)$ for which a winning strategy exists. Such a strategy induces a multigraph $C = (V, E')$ where the edge relation is defined by $(s, c, t) \in E' \iff (t, c) \in R_s^\sigma$. It remains to show the second condition for C to be a certificate, namely that any cycle in C, reachable from the initial vertex s_0, has an even maximal colour. Suppose otherwise, that C contains a reachable cycle whose maximal colour is odd. Then there must be play in G that is consistent with σ and which sees the same (odd) colour infinitely often. But this contradicts the assumption that σ was winning in G in the first place.   □

Our main theorem is now an easy consequence of the existence of small certificates.

Theorem 3. *Solving parity games on periodic temporal graphs is* PSPACE-*complete.*

Proof. Hardness already holds for reachability games Lemma 2. For the upper bound we show membership in NPSPACE and use Savitch's theorem. By Lemma 4 it suffices to guess and verify a candidate certificate C. These are by definition polynomial in the number of vertices and colours in the given temporal parity game. Verifying the cycle condition (2) is trivial in polynomial time and verifying the realisability condition (1) is in PSPACE by Lemma 3.   □

Remark 1. The PSPACE upper bound in Theorem 3 can easily be extended to games on temporal graphs that are *ultimately* periodic, meaning that there exist $T, K \in \mathbb{N}$ so that for all $n \geq T$, $s \xrightarrow{n} t$ implies $s \xrightarrow{n+K} t$. Such games can be solved by first considering the periodic suffix according to Theorem 3 thereby computing the winning region for Player 1 at time exactly T, and then solving the temporal reachability game with horizon T.

5   Monotonicity

In this section, we consider the effects of monotonicity assumptions on the edge relation with respect to time on the complexity of solving reachability games. We first show that reachability games remain PSPACE-hard even if the edge relation is decreasing (or increasing) with time. We then give a fragment for which the problem becomes solvable in polynomial time.

Increasing and Decreasing temporal graphs: Let the edge between vertices $u, v \in V$ of a temporal graph be referred to as *decreasing* if $u \xrightarrow{i+1} v$ implies $u \xrightarrow{i} v$ for all $i \in \mathbb{N}$, i.e. edges can only disappear over time. Similarly, call the edge *increasing* if for all $i \in \mathbb{N}$ we have that $u \xrightarrow{i} v$ implies $u \xrightarrow{i+1} v$; i.e. an edge available at current time continues to be available in the future. A temporal graph is decreasing (increasing) if all its edges are. We assume that the times at which edge availability changes are given in binary. More specifically, every edge is given as inequality constraint of the form $\Phi_{u,v}(x) \stackrel{\text{def}}{=} x \leq n$ (respectively $x \geq n$) for some $n \in \mathbb{N}$.

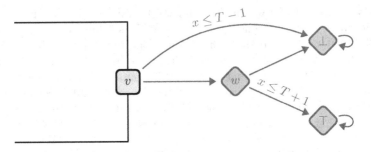

Fig. 4: Reduction from a punctual reachability game to a reachability game on a temporal graph that is finite and decreasing, see Theorem 4. Components added are shown in red.

Although both restrictions imply that the graph is ultimately static, we observe that solving reachability games on such monotonically increasing or decreasing temporal graphs remains PSPACE-complete.

Theorem 4. *Solving reachability and Parity games on decreasing (respectively increasing) temporal graphs is PSPACE-complete.*

Proof. The upper bound holds for parity games as the description of the temporal graph explicitly includes a maximal time T from which the graph becomes static. One can therefore solve the Parity game for the static suffix graph (in NP) and then apply the PSPACE procedure (Theorem 2) to solve for temporal reachability towards the winning region at time T. Alternatively, the same upper bound also follows from Theorem 3 and Remark 1.

For the lower bound we reduce from punctual reachability games which are PSPACE-hard by Lemma 2. Consider a (static) graph G and a target time $T \in \mathbb{N}$ given in binary. Without loss of generality, assume that the target vertex v has no outgoing edges. We convert G into a temporal graph G' with $V' = V \cup \{w, \top, \bot\}$, $V_1' = (V_1 \setminus \{v\}) \cup \{w\}$, $V_2' = V' \setminus V_1'$ and new target $\top$. The vertex $\bot$ is a sink state and the original target vertex v is now controlled by Player 2. Edge availabilities are $v \xrightarrow{x} \bot$ if $x \leq T - 1$, $v \xrightarrow{x} w$ if $x \leq T + 1$, $w \xrightarrow{x} \top$ if $x \leq T + 1$, and all other edges disappear after time $T + 1$. The constructed temporal graph is finite and decreasing. See Fig. 4. The construction ensures that the only way to reach $\top$ is to reach v at time T, w at time $T + 1$ and take the edge from w to $\top$ at time $T + 1$. Player 1 wins in G' if and only if she wins the punctual reachability game on G.

A similar reduction works in the case of increasing temporal graphs by switching the ownership of vertices v and w. The vertex v, now controlled by Player 1 has the edge $v \xrightarrow{x} w$ at times $x \geq T$ and the edge $v \longrightarrow \bot$ at all times. The vertex w now controlled by Player 2 has the edge $w \longrightarrow \top$ available at all times but the edge $w \xrightarrow{x} \bot$ becomes available at time $x \geq T + 2$. □

Declining and improving temporal games: We now consider the restriction where all edges controlled by one player are increasing and those of the over player are decreasing. Taking the perspective of the system Player 1, we call a game on a temporal graph *declining* if all edges $u \longrightarrow v$ with $u \in V_1$ are decreasing and all edges $u \longrightarrow v$ with $u \in V_2$ are increasing. Note that *declining* is a property of the game and not the graph as the definition requires a distinction based on ownership of vertices, which is specified by the game and not the underlying graph. From now on, we refer to such games as declining temporal reachability (or parity) games. Notice that Player 1 has fewer, and Player 2 has more choices to move at later times. Analogously, call the game *improving* if the conditions are reversed, i.e., all edges $u \longrightarrow v$ with $u \in V_1$ are increasing and all edges $u \longrightarrow v$ with $u \in V_2$ are decreasing.

We show that declining (and improving) temporal reachability games can be solved in polynomial time.

Theorem 5. *Solving declining (respectively improving) temporal reachability games is in* P.

Proof. We first give the proof for declining games. Consider the reachability game on the expansion with vertices $V \times \mathbb{N}$ such that the target set is $F \times \mathbb{N}$. For $k \in \mathbb{N}$ let $W_k \subseteq V$ be the set of those vertices u such that Player 1 has a winning strategy from (u, k). We first show that

$$W_{i+1} \subseteq W_i \tag{1}$$

For sake of contradiction, suppose there exists $u \in W_{i+1} \setminus W_i$. Let σ_{i+1}^1 be a (positional) winning strategy from $(u, i+1)$ for Player 1 in the expansion. Since $u \notin W_i$, by positional determinacy of reachability games (Proposition 1), Player 2 has a winning strategy σ_i^2 from (u, i). Consider a strategy σ_i^1 for Player 1, such that for all $v \in V_1$, $\sigma_i^1(v, k) \stackrel{\text{def}}{=} \sigma_{i+1}^1(v, k+1)$, for all $k \geq i$. Similarly, let σ_{i+1}^2 be the strategy for Player 2, such that for all $v \in V_2$, $\sigma_{i+1}^2(v, k+1) = \sigma_i^2(v, k)$, for all $k \geq i$, Note that this is well defined because by definition of declining games, i.e, $v \xrightarrow{k+1} u$ implies $v \xrightarrow{k} u$ for all $v \in V_1$, and $v \xrightarrow{k} u$ implies $v \xrightarrow{k+1} u$, for all $v \in V_2$. Starting from the vertex $(u, i+1)$, the pair of strategies $(\sigma_{i+1}^1, \sigma_{i+1}^2)$ defines a unique play π_{i+1}, which is winning for Player 1. Similarly, the pair of strategies (σ_i^1, σ_i^2) define a play π_i which is winning for Player 2 starting from (u, i). However, the two plays visit the same set of states, particularly, (v, k) is visited in π_i if and only if $(v, k+1)$ is visited in π_{i+1}. Therefore, either both are winning for Player 1 or both are losing for Player 2, which is a contradiction. Let $N \subseteq \mathbb{N}$ be the set of times at which the graph changes, i.e.

$$N = \{c \mid \exists \Phi_{u,v}(x) = x \triangleleft c, \text{ where } \triangleleft \in \{\leq, \geq\}\}\}$$

Let $m \stackrel{\text{def}}{=} \max(N)$ be the latest time any edge availability changes. We show that $W_m = W_k$ for all $k \geq m$. To see this, note that W_m is equal to the winning region for Player 1 in the (static) reachability game played on $G_m = (V, E_m)$,

---

Algorithm 1 Algorithm for declining games with set of change times N and $m = \max(N)$

$W \leftarrow Solve(G_m)$ $\triangleright$ Computes Player 1 winning region in G_m
while $N \neq \emptyset$ **do**
    $n \leftarrow max(N)$
    if $(Pre_1(W \times \{n\}) = W$ **then**
       $N \leftarrow N \setminus n$ $\triangleright$ Accelerate to next change time
    else
       $W \leftarrow Pre_1(W)$
       $N \leftarrow N \cup \{n-1\} \setminus \{n\}$
    end if
end while

---

where $E_m = \{(u,v) \mid u \xrightarrow{m} v\}$. Consider a (positional) winning strategy σ_m for Player 1 in G_m and define a positional strategy $\sigma(v,k) = \sigma_m(v)$, for $k \geq m$. Since the graph is static after time m, this is well defined. Starting from a vertex (u,k), a vertex $(v, k+k')$ is visited on a σ-consistent path if and only if there is a σ_m-consistent path $u \longrightarrow_{k'} v$. Therefore, σ is a winning strategy from any vertex (v,k) such that $k \geq m$ and $v \in W_m$. Moreover, the set W_m can be computed in time $\mathcal{O}(|V|^2)$ by solving the reachability game on G_m [13, Theorem 12].

To solve reachability on declining temporal games, we can first compute the winning region W_m in the stabilised game G_m. This means $W_m \times [m, \infty)$ is winning for Player 1. To win the declining temporal reachability game, Player 1 can play the punctual reachability game with target set W_m at target time m. The winning region for Player 1 at time 0 can therefore be computed as $Pre_1^m(W_m \times \{m\})$ as outlined in the proof of Theorem 2. Note that naïvely this only gives a PSPACE upper bound as in the worst case, we would compute Pre_1 an exponential (m) times.

To overcome this, note that in the expansion graph $Pre_1^i(W_m \times \{m\}) = W_{m-i} \times \{m-i\}$. According to Eq. (1), $W_{m-i} \subseteq W_{m-i'}$ for $i' > i$. Let i, i' be such that $m-i$ and $m-i'$ are both consecutive change points, i.e, $m-i, m-i' \in N$ and $\forall \ell \in N.\ell < m - i' \vee \ell > m - i$. Since the edge availability of the graph does not change between time $m - i'$ and $m - i$, we have $W_{m-i-1} = W_{m-i}$ implies $W_{m-i'} = W_{m-i}$. Therefore, we can accelerate the Pre_1 computation and directly move to the time step $m - i'$, i.e, the i'th iteration in the computation. This case is illustrated at time $n' = m - i'$ in Fig. 5.

With this change, our algorithm runs the Pre_1 computation at most $|V| + |N|$, as each Pre_1 computation either corresponds to a step a time in N when the graph changes, or a step in which the winning region grows such as at time n in Fig. 5. Since each Pre_1 computation can be done in polynomial time, we get a PTIME algorithm in this case, shown in Algorithm 1.

The case for improving temporal reachability games can be solved similarly. Instead of computing the winning region for Player 1 in G_m, we start with computing the winning region W_m^2 for Player 2 in G_m and switch the roles of Player 1 and Player 2, i.e, Player 2 has the punctual reachability objective with

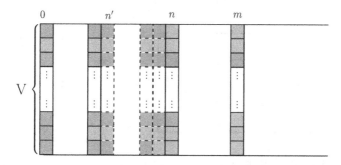

Fig. 5: Illustration of Algorithm 1. The blue vertices at time i denote the winning region W_i for Player 1. The times $n, n' \in N$ and Pre_1 computation at change point n increases the winning region but is stable at time n'.

target set W_m^2 and target time m, which can be solved as above. This gives us an algorithm to compute the winning region for Player 2 and by determinacy of reachability games on infinite graphs, we can compute the winning region for Player 1 at time 0 as well.   □

Remark 2. Algorithm 1 also works for parity objectives by changing step 1, where Solve(G_m) would amount to solving the parity game on the static graph G_m. This can be done in quasi-polynomial time and therefore gives a quasi-polynomial time algorithm to solve declining (improving) temporal parity games and in particular, gives membership in the complexity class NP ∩ coNP.

Since the declining (improving) restriction on games on temporal graphs allow for improved algorithms, a natural question is to try to lift this approach to a larger class of games on temporal graphs. Note that the above restrictions are a special case of eventually periodic temporal graphs with a prefix of time m followed by a periodic graph with period 1. Now, we consider temporal graphs of period $K > 1$ such that the game arena is declining (improving) within each period. Formally, a game on a temporal graph G is *periodically declining* (improving) if there exists a period K such that for all $k \in \mathbb{N}$, $k \in E(u,v)$ if and only if $k+K \in E(u,v)$; and the game on the finite temporal graph resulting from G by making the graph constant from time K onwards, is declining (improving). We prove that this case is PSPACE-hard, even with reachability objectives.

Theorem 6. *Solving periodically declining (improving) temporal reachability games is PSPACE-complete.*

Proof. The upper bound follows from the general case of parity games on periodic temporal graphs in Theorem 3. The lower bound is by reduction from punctual reachability games. See Fig. 6. Given a (static) graph G with target state v and target time T, we obtain a periodically declining game G' with period $K = T+1$, vertices $V \cup \{w, \bot, \top\}$, new target $\top$, such that $V_1' = V_1 \cup \{w, \bot, \top\}$ and $V_2' = V_2$.

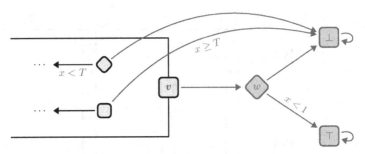

Fig. 6: Reduction from a punctual reachability game to a reachability game on a temporal graphs that is periodic and declining, see Theorem 6. Parts added are shown in red.

We assume without loss of generality that the original target v is a Player 1 vertex, i.e, $v \in V_1$.

We describe the edge availability in G' up to the period $K = T + 1$. For all edges (s, t) of the original graph G, such that $s \in V_1$, the edge $s \xrightarrow{x} t$ is available if and only if $x < T$. Moreover for all $s \in V_1 \setminus \{v\}$, there is a new edge $s \xrightarrow{x} \perp$ available at all times $x \leq T$. For all $s \in V_2$, there is an edge $s \xrightarrow{x} t$ is available at all times (until end of period) and $s \xrightarrow{x} \perp$ is available after time $x \geq T$. These edges ensure that if a play in the original punctual reachability game ends in a vertex of the game other than v at time T, then Player 2 can force the play to reach the sink state $\perp$ and win.

From the original target v, there is an edge to the new state w at all times. From the state w, there are edges $w \longrightarrow \perp$ at all times and $w \xrightarrow{x} \top$ if $x = 0$. If the state w is reached at time k such that $1 < k < T+1$, then the play is forced to go to $\perp$. The only winning strategy for Player 1 is to reach v at time T, w at time $T + 1$ at which the time is reset due to periodicity. The edge $w \xrightarrow{T+1} \top$ is now available for Player 1 and they can reach the new target $\top$.

The lower bound for the case of periodically increasing temporal reachability games follows by the same construction and using the duality between improving and declining games on temporal graphs. Given a punctual reachability game G with vertices $V = V_1 \uplus V_2$ with target set F, we obtain the dual punctual reachability game $\hat{G}$ with same target time by first switch the ownership of vertices, i.e, $\hat{V}_i = V_{3-i}$, $i \in \{1, 2\}$ and make the new target as $V \setminus F$. It is easy to see that Player 1 wins G if and only if Player 2 wins $\hat{G}$.

Applying the same construction as shown in Fig. 6 to $\hat{G}$, we obtain a periodically declining temporal reachability game $\hat{G}'$, preserving the winner. Now switching the ownership of vertices in $\hat{G}'$ yields a periodically improving temporal reachability game G' which is winning for Player 1 if and only if Player 1 wins G. $\square$

6   Conclusion

In this work we showed that parity games on ultimately periodic temporal graphs are solvable in polynomial space. The lower bound already holds for the very special case of punctual reachability games, and the PSPACE upper bound, which improves on the naïve exponential-space algorithm on the unfolded graph, is achieved by proving the existence of small, PSPACE-verifiable certificates.

We stress again that all constructions are effective no matter how the temporal graphs are defined, as long as checking edge availability for binary encoded times is no obstacle. In the paper we use edge constraints given in the existential fragment of Presburger arithmetic but alternate representations, for example using compressed binary strings of length $h(G)$ given as Straight-Line Programs [5, Section 3] would equally work. Checking existence of edge at time i would correspond to querying whether the ith bit is 1 or not which is P-complete [27, Theorem 1].

The games considered here are somewhat orthogonal to parity games played on the configuration graphs of timed automata, where time is continuous, and constraints are *quantifier-free* formulae involving possibly more than one variable (clocks). Solving parity games on timed automata with two clocks is complete for EXP but is in P if there is at most one one clock [38] [16, Contribution 3(d)]. Games on temporal graphs with quantifier-free constraints corresponds to a subclass of timed automata games with two-clocks, with intermediate complexity of PSPACE. This is because translating a temporal graph game to a timed automata game requires two clocks: one to hold the global time used to check the edge predicate and one to ensure that time progresses one unit per step.

An interesting continuation of the work presented here would be to consider mean-payoff games [11] played on temporal graphs, possibly with dynamic step-rewards depending on the time. If rewards are constant but the edge availability is dynamic, then our arguments for improved algorithms on declining/improving graphs would easily transfer. However, the PSPACE upper bound using summaries seems trickier, particularly checking realisability of suitable certificates.

Acknowledgements This work was supported by the Engineering and Physical Sciences Research Council (EPSRC), grant EP/V025848/1. We thank Viktor Zamaraev and Sven Schewe for fruitful discussions and constructive feedback.

References

1. Automata Logics, and Infinite Games: A Guide to Current Research. Springer-Verlag (2002). https://doi.org/10.1007/3-540-36387-4
2. Akrida, E.C., Mertzios, G.B., Spirakis, P.G., Zamaraev, V.: Temporal vertex cover with a sliding time window. Journal of Computer and System Sciences **107**, 108–123 (2020). https://doi.org/https://doi.org/10.1016/j.jcss.2019.08.002
3. Alur, R., Dill, D.L.: A theory of timed automata. Theor. Comput. Sci. **126**(2), 183 – 235 (1994). https://doi.org/10.1016/0304-3975(94)90010-8

4. Avni, G., Ghorpade, P., Guha, S.: A Game of Pawns. In: International Conference on Concurrency Theory. Leibniz International Proceedings in Informatics (LIPIcs), vol. 279, pp. 16:1–16:17. Schloss Dagstuhl – Leibniz-Zentrum für Informatik (2023). https://doi.org/10.4230/LIPIcs.CONCUR.2023.16

5. Babai, L., Szemeredi, E.: On the complexity of matrix group problems i. In: Annual Symposium on Foundations of Computer Science. pp. 229–240 (1984). https://doi.org/10.1109/SFCS.1984.715919

6. Calude, C.S., Jain, S., Khoussainov, B., Li, W., Stephan, F.: Deciding parity games in quasipolynomial time. In: Symposium on Theory of Computing. pp. 252–263 (2017). https://doi.org/10.1145/3055399.3055409

7. Chandra, A.K., Kozen, D.C., Stockmeyer, L.J.: Alternation. Journal of the ACM (JACM) 28(1), 114–133 (1981). https://doi.org/10.1145/322234.322243

8. Chatterjee, K., Henzinger, T.A., Prabhu, V.S.: Timed Parity Games: Complexity and Robustness. Logical Methods in Computer Science Volume 7, Issue 4 (Dec 2011). https://doi.org/10.2168/LMCS-7(4:8)2011

9. Colcombet, T., Fijalkow, N.: Universal Graphs and Good for Games Automata: New Tools for Infinite Duration Games. In: International Conference on Foundations of Software Science and Computational Structures. LNCS, vol. 11425, pp. 1–26. Springer (2019). https://doi.org/10.1007/978-3-030-17127-8_1

10. De Carufel, J.L., Flocchini, P., Santoro, N., Simard, F.: Cops & robber on periodic temporal graphs: Characterization and improved bounds. In: Structural Information and Communication Complexity. pp. 386–405. Springer Nature Switzerland (2023). https://doi.org/10.1007/978-3-031-32733-9_17

11. Ehrenfeucht, A., Mycielski, J.: Positional strategies for mean payoff games. International Journal of Game Theory 8(2), 109–113 (Jun 1979). https://doi.org/10.1007/BF01768705

12. Erlebach, T., Hoffmann, M., Kammer, F.: On temporal graph exploration. Journal of Computer and System Sciences 119, 1–18 (2021). https://doi.org/https://doi.org/10.1016/j.jcss.2021.01.005

13. Fijalkow, N., Bertrand, N., Bouyer-Decitre, P., Brenguier, R., Carayol, A., Fearnley, J., Gimbert, H., Horn, F., Ibsen-Jensen, R., Markey, N., Monmege, B., Novotný, P., Randour, M., Sankur, O., Schmitz, S., Serre, O., Skomra, M.: Games on graphs (2023). https://doi.org/10.48550/arXiv.2305.10546

14. Flocchini, P., Mans, B., Santoro, N.: Exploration of periodically varying graphs. In: Algorithms and Computation. pp. 534–543. Springer Berlin Heidelberg (2009). https://doi.org/10.1007/978-3-642-10631-6_55

15. Haase, C.: A survival guide to presburger arithmetic. SIGLOG News 5(3), 67–82 (2018). https://doi.org/10.1145/3242953.3242964

16. Hansen, T.D., Ibsen-Jensen, R., Miltersen, P.B.: A faster algorithm for solving one-clock priced timed games (2013). https://doi.org/10.1007/978-3-642-40184-8_37

17. Holme, P., Saramäki, J.: Temporal Network Theory (01 2019). https://doi.org/10.1007/978-3-030-23495-9

18. Holzer, M.: On emptiness and counting for alternating finite automata. In: International Conference on Developments in Language Theory. pp. 88–97 (1995)

19. Jančar, P., Sawa, Z.: A note on emptiness for alternating finite automata with a one-letter alphabet. Inf. Process. Lett. 104(5), 164–167 (2007). https://doi.org/10.1016/j.ipl.2007.06.006

20. Jiang, T., Ravikumar, B.: A note on the space complexity of some decision problems for finite automata. Inf. Process. Lett. 40(1), 25–31 (1991). https://doi.org/https://doi.org/10.1016/S0020-0190(05)80006-7

21. Jurdziński, M., Trivedi, A.: Reachability-time games on timed automata. In: International Colloquium on Automata, Languages and Programming. pp. 838–849. Springer Berlin Heidelberg (2007). https://doi.org/10.1007/978-3-540-73420-8_72

22. Jurdziński, M.: Deciding the winner in parity games is in up ∩ co-up. Inf. Process. Lett. **68**(3), 119–124 (1998). https://doi.org/https://doi.org/10.1016/S0020-0190(98)00150-1

23. Jurdziński, M., Lazić, R.: Succinct Progress Measures for Solving Parity Games. In: Annual IEEE Symposium on Logic in Computer Science. pp. 1–9. IEEE Computer Society (2017). https://doi.org/10.1109/LICS.2017.8005092

24. Kuhn, F., Lynch, N., Oshman, R.: Distributed computation in dynamic networks. In: Symposium on Theory of Computing. p. 513–522. STOC '10, Association for Computing Machinery (2010). https://doi.org/10.1145/1806689.1806760

25. Lehtinen, K., Parys, P., Schewe, S., Wojtczak, D.: A Recursive Approach to Solving Parity Games in Quasipolynomial Time. Logical Methods in Computer Science **18**(1), 8:1–18 (2022). https://doi.org/10.46298/lmcs-18(1:8)2022

26. Lehtinen, K., Boker, U.: Register Games. Logical Methods in Computer Science **16**(2), 6:1–6:25 (2020). https://doi.org/10.23638/LMCS-16(2:6)2020

27. Lifshits, Y., Lohrey, M.: Querying and embedding compressed texts. In: International Symposium on Mathematical Foundations of Computer Science. pp. 681–692. Springer Berlin Heidelberg (2006). https://doi.org/10.1007/11821069_59

28. Maler, O., Pnueli, A., Sifakis, J.: On the synthesis of discrete controllers for timed systems. In: International Symposium on Theoretical Aspects of Computer Science. pp. 229–242. Springer Berlin Heidelberg (1995). https://doi.org/10.1007/3-540-59042-0_76

29. Mertzios, G.B., Molter, H., Niedermeier, R., Zamaraev, V., Zschoche, P.: Computing maximum matchings in temporal graphs. Journal of Computer and System Sciences **137**, 1–19 (2023). https://doi.org/https://doi.org/10.1016/j.jcss.2023.04.005

30. Mertzios, G.B., Molter, H., Zamaraev, V.: Sliding window temporal graph coloring. Journal of Computer and System Sciences **120**, 97–115 (2021). https://doi.org/https://doi.org/10.1016/j.jcss.2021.03.005

31. Michail, O.: An Introduction to Temporal Graphs: An Algorithmic Perspective, pp. 308–343. Springer International Publishing (2015). https://doi.org/10.1007/978-3-319-24024-4_18

32. Michail, O., Chatzigiannakis, I., Spirakis, P.G.: Causality, influence, and computation in possibly disconnected synchronous dynamic networks. Journal of Parallel and Distributed Computing **74**(1), 2016–2026 (2014). https://doi.org/10.1016/j.jpdc.2013.07.007

33. Michail, O., Spirakis, P.G.: Traveling salesman problems in temporal graphs. In: International Symposium on Mathematical Foundations of Computer Science. pp. 553–564. Springer Berlin Heidelberg (2014). https://doi.org/10.1016/j.tcs.2016.04.006

34. Pnueli, A., Rosner, R.: On the synthesis of a reactive module. In: Annual Symposium on Principles of Programming Languages. p. 179–190. POPL '89, Association for Computing Machinery (1989). https://doi.org/10.1145/75277.75293

35. Pnueli, A.: The temporal logic of programs. In: Annual Symposium on Foundations of Computer Science. p. 46–57. SFCS '77, IEEE Computer Society (1977). https://doi.org/10.1109/SFCS.1977.32

36. Ravi, R.: Rapid rumor ramification: Approximating the minimum broadcast time. In: Proceedings 35th Annual Symposium on Foundations of Computer Science. pp. 202–213 (1994). https://doi.org/10.1109/SFCS.1994.365693
37. Scarpellini, B.: Complexity of subcases of presburger arithmetic. Transactions of the American Mathematical Society **284**, 203–218 (1984). https://doi.org/10.1090/s0002-9947-1984-0742421-9
38. Trivedi, A.: Competitive optimisation on timed automata. Ph.D. thesis, University of Warwick (April 2009)

Categorical Semantics

Drawing from an Urn is Isometric

Bart Jacobs[(✉)]

Institute for Computing and Information Sciences,
Radboud University, Nijmegen, The Netherlands
bart@cs.ru.nl

Abstract. Drawing (a multiset of) coloured balls from an urn is one of the most basic models in discrete probability theory. Three modes of drawing are commonly distinguished: multinomial (draw-replace), hypergeometric (draw-delete), and Pólya (draw-add). These drawing operations are represented as maps from urns to distributions over multisets of draws. The set of urns is a metric space via the Wasserstein distance. The set of distributions over draws is also a metric space, using Wasserstein-over-Wasserstein. The main result of this paper is that the three draw operations are all isometries, that is, they preserve the Wasserstein distances.

Keywords: probability · urn drawing · Wasserstein distance.

1 Introduction

We start with an illustration of the topic of this paper. We consider a situation with a set $C = \{R, G, B\}$ of three colours: red, green, blue. Assume that we have two urns υ_1, υ_2 with 10 coloured balls each. We describe these urns as multisets of the form:

$$\upsilon_1 = 8|G\rangle + 2|B\rangle \qquad \text{and} \qquad \upsilon_2 = 5|R\rangle + 4|G\rangle + 1|B\rangle.$$

Recall that a multiset is like a set, except that elements may occur multiple times. Here we describe urns as multisets using 'ket' notation $|-\rangle$. It separates multiplicities of elements (before the ket) from the elements in the multiset (inside the ket). Thus, urn υ_1 contains 8 green balls and 2 blue balls (and no red ones). Similarly, urn υ_2 contains 5 red, 4 green, and 1 blue ball(s).

Below, we shall describe the Wasserstein distance between multisets (of the same size). How this works does not matter for now; we simply posit that the Wasserstein distance $d(\upsilon_1, \upsilon_2)$ between these two urns is $\frac{1}{2}$ — where we assume the discrete distance on the set C of colours.

We turn to draws from these two urns, in this introductory example of size two. These draws are also described as multisets, with elements from the set $C = \{R, G, B\}$ of colours. There are six multisets (draws) of size 2, namely:

$$2|R\rangle \quad 1|R\rangle + 1|G\rangle \quad 2|G\rangle \quad 1|R\rangle + 1|B\rangle \quad 2|B\rangle \quad 1|G\rangle + 1|B\rangle. \qquad (1)$$

As we see, there are three draws with 2 balls of the same colour, and three draws with balls of different colours.

N. Kobayashi and J. Worrell (Eds.): FoSSaCS 2024, LNCS 14574, pp. 101–120, 2024.
https://doi.org/10.1007/978-3-031-57228-9_6

We consider the hypergeometric probabilities associated with these draws, from the two urns. Let's illustrate this for the draw $1|G\rangle + 1|B\rangle$ of one green ball and one blue ball from the urn υ_1. The probability of drawing $1|G\rangle + 1|B\rangle$ is $\frac{16}{45}$; it is obtained as sum of:

- first drawing-and-deleting a green ball from $\upsilon_1 = 8|G\rangle + 2|B\rangle$, with probability $\frac{8}{10}$. It leaves an urn $7|G\rangle + 2|B\rangle$, from which we can draw a blue ball with probability $\frac{2}{9}$. Thus drawing "first green then blue" happens with probability $\frac{8}{10} \cdot \frac{2}{9} = \frac{8}{45}$.
- Similarly, the probability of drawing "first blue then green" is $\frac{2}{10} \cdot \frac{8}{9} = \frac{8}{45}$.

We can similarly compute the probabilities for each of the above six draws (1) from urn υ_1. This gives the hypergeometric distribution, which we write using kets-over-kets as:

$$hg[2](\upsilon_1) = \frac{28}{45}\left|2|G\rangle\right\rangle + \frac{16}{45}\left|1|G\rangle + 1|B\rangle\right\rangle + \frac{1}{45}\left|2|B\rangle\right\rangle.$$

The fraction written before a big ket is the probability of drawing the multiset (of size 2), written inside that big ket, from the urn υ_1.

Drawing from the second urn υ_2 gives a different distribution over these multisets (1). Since urn υ_2 contains red balls, they additionally appear in the draws.

$$hg[2](\upsilon_2) = \frac{2}{9}\left|2|R\rangle\right\rangle + \frac{4}{9}\left|1|R\rangle + 1|G\rangle\right\rangle + \frac{2}{15}\left|2|G\rangle\right\rangle$$
$$+ \frac{1}{9}\left|1|R\rangle + 1|B\rangle\right\rangle + \frac{4}{45}\left|1|G\rangle + 1|B\rangle\right\rangle.$$

We can also compute the distance between these two hypergeometric distributions over multisets. It involves a Wasserstein distance, over the space of multisets (of size 2) with their own Wasserstein distance. Again, details of the calculation are skipped at this stage. The distance between the above two hypergeometric draw-distributions is:

$$d\Big(hg[2](\upsilon_1), hg[2](\upsilon_2)\Big) = \frac{1}{2} = d(\upsilon_1, \upsilon_2).$$

This coincidence of distances is non-trivial. It holds, in general, for arbitrary urns (of the same size) over arbitrary metric spaces of colours, for draws of arbitrary sizes. Moreover, the same coincidence of distances holds for the multinomial and Pólya modes of drawing. These coincidences are the main result of this paper, see Theorems 1, 2, and 3 below.

In order to formulate and obtain these results, we describe multinomial, hypergeometric and Pólya distributions in the form of (Kleisli) maps:

$$\mathcal{D}(X) \xrightarrow{\;mn[K]\;} \mathcal{D}\big(\mathcal{M}[K](X)\big) \xleftarrow[\;pl[K]\;]{\;hg[K]\;} \mathcal{M}[L](X) \qquad (2)$$

They all produce distributions (indicated by $\mathcal{D}$), in the middle of this diagram, on multisets (draws) of size K, indicated by $\mathcal{M}[K]$, over a set X of colours. Details will be provided below. Using the maps in (2), the coincidence of distances that we saw above can be described as a preservation property, in terms of distance preserving maps — called isometries. At this stage we wish to emphasise that the representation of these different drawing operations as maps in (2) has a categorical background. It makes it possible to formulate and prove basic properties of drawing from an urn, such as naturality in the set X of colours. Also, as shown in [8] for the multinomial and hypergeometric case, drawing forms a monoidal transformation (with 'zipping' for multisets as coherence map). This paper demonstrates that the three draw maps (2) are even more well-behaved: they are all isometries, that is, they preserve Wasserstein distances. This is a new and amazing fact.

This paper concentrates on the mathematics behind these isometry results, and not on interpretations or applications. We do like to refer to interpretations in machine learning [14] where the distance that we consider on colours in an urn is called the *ground distance*. Actual distances between colours are used there, based on experiments in psychophysics, using perceived differences [16].

The Wasserstein — or Wasserstein-Kantorovich, or Monge-Kantorovich — distance is the standard distance on distributions and on multisets, going back to [12]. After some preliminaries on multisets and distributions, and on distances in general, Sections 4 and 5 of this paper recall the Wasserstein distance on distributions and on multisets, together with some basic results. The three subsequent Sections 6 – 8 demonstrate that multinomial, hypergeometric and Pólya drawing are all isometric. Distances occur on multiple levels: on colours, on urns (as multisets or distributions) and on draw-distributions. This may be confusing, but many illustrations are included.

2 Preliminaries on multisets and distributions

A *multiset* over a set X is a finite formal sum of the form $\sum_i n_i | x_i \rangle$, for elements $x_i \in X$ and natural numbers $n_i \in \mathbb{N}$ describing the multiplicities of these elements x_i. We shall write $\mathcal{M}(X)$ for the set of such multisets over X. A multiset $\varphi \in \mathcal{M}(X)$ may equivalently be described in functional form, as a function $\varphi \colon X \to \mathbb{N}$ with finite support: $supp(\varphi) := \{ x \in X \mid \varphi(x) \neq 0 \}$. Such a function $\varphi \colon X \to \mathbb{N}$ can be written in ket form as $\sum_{x \in X} \varphi(x) | x \rangle$. We switch back-and-forth between ket form and functional form and use the formulation that best suits a particular situation.

For a multiset $\varphi \in \mathcal{M}(X)$ we write $\|\varphi\| \in \mathbb{N}$ for the *size* of the multiset. It is the total number of elements, including multiplicities:

$$\|\varphi\| := \sum_{x \in X} \varphi(x).$$

For a number $K \in \mathbb{N}$ we write $\mathcal{M}[K](X) \subseteq \mathcal{M}(X)$ for the subset of multisets of size K. There are 'accumulation' maps $acc \colon X^K \to \mathcal{M}[K](X)$ turning lists into multisets via $acc(x_1, \ldots, x_K) := 1|x_1\rangle + \cdots + 1|x_K\rangle$. For instance $acc(c, b, a, c, a, c) = 2|a\rangle + 1|b\rangle + 3|c\rangle$. A standard result (see [10]) is that for a multiset $\varphi \in \mathcal{M}[K](X)$ there are $(\varphi) := \frac{K!}{\varphi\!\:}$ many sequences $x \in X^K$ with $acc(x) = \varphi$, where $\varphi\!\: = \prod_x \varphi(x)!$.

Multisets $\varphi, \psi \in \mathcal{M}(X)$ can be added and compared elementwise, so that $(\varphi + \psi)(x) = \varphi(x) + \psi(x)$ and $\varphi \leq \psi$ means $\varphi(x) \leq \psi(x)$ for all $x \in X$. In the latter case, when $\varphi \leq \psi$, we can also subtract $\psi - \varphi$ elementwise.

The mapping $X \mapsto \mathcal{M}(X)$ is functorial: for a function $f \colon X \to Y$ we have $\mathcal{M}(f) \colon \mathcal{M}(X) \to \mathcal{M}(Y)$ given by $\mathcal{M}(f)(\varphi)(y) = \sum_{x \in f^{-1}(y)} \varphi(x)$. This map $\mathcal{M}(f)$ preserves sums and size.

For a multiset $\tau \in \mathcal{M}(X \times Y)$ on a product set we can take its two marginals $\mathcal{M}(\pi_1)(\tau) \in \mathcal{M}(X)$ and $\mathcal{M}(\pi_2)(\tau) \in \mathcal{M}(Y)$ via functoriality, using the two projection functions $\pi_1 \colon X \times Y \to X$ and $\pi_2 \colon X \times Y \to Y$. Starting from $\varphi \in \mathcal{M}(X)$ and $\psi \in \mathcal{M}(Y)$, we say that $\tau \in \mathcal{M}(X \times Y)$ is a *coupling* of φ, ψ if φ and ψ are the two marginals of τ. We define the *decoupling* map:

$$\mathcal{M}(X \times Y) \xrightarrow{\; dcpl := \langle \mathcal{M}(\pi_1), \mathcal{M}(\pi_2) \rangle \;} \mathcal{M}(X) \times \mathcal{M}(Y) \tag{3}$$

The inverse image $dcpl^{-1}(\varphi, \psi) \subseteq \mathcal{M}(X \times Y)$ is thus the subset of couplings of φ, ψ.

A *distribution* is a finite formal sum of the form $\sum_i r_i | x_i \rangle$ with multiplicities $r_i \in [0,1]$ satisfying $\sum_i r_i = 1$. Such a distribution can equivalently be described as a function $\omega \colon X \to [0,1]$ with finite support, satisfying $\sum_x \omega(x) = 1$. We write $\mathcal{D}(X)$ for the set of distributions on X. This $\mathcal{D}$ is functorial, in the same way as $\mathcal{M}$. Both $\mathcal{D}$ and $\mathcal{M}$ are monads on the category **Sets** of sets and functions, but we only use this for $\mathcal{D}$. The unit and multiplication / flatten maps $unit \colon X \to \mathcal{D}(X)$ and $flat \colon \mathcal{D}^2(X) \to \mathcal{D}(X)$ are given by:

$$unit(x) := 1 | x \rangle \qquad flat(\Omega) := \sum_{x \in X} \left(\sum_{\omega \in \mathcal{D}(X)} \Omega(\omega) \cdot \omega(x) \right) | x \rangle. \tag{4}$$

Kleisli maps $c \colon X \to \mathcal{D}(Y)$ are also called channels and written as $c \colon X \rightsquigarrow Y$. Kleisli extension $c \gg (-) \colon \mathcal{D}(X) \to \mathcal{D}(Y)$ for such a channel, is defined on $\omega \in \mathcal{D}(X)$ as:

$$c \gg \omega := flat\big(\mathcal{D}(c)(\omega)\big) = \sum_{y \in Y} \left(\sum_{x \in X} \omega(x) \cdot c(x)(y) \right) | y \rangle.$$

Channels $c \colon X \rightsquigarrow Y$ and $d \colon Y \rightsquigarrow Z$ can be composed to $d \odot c \colon X \rightsquigarrow Z$ via $(d \odot c)(x) := d \gg c(x)$. Each function $f \colon X \to Y$ gives rise to a deterministic channel $\langle f \rangle := unit \circ f \colon X \rightsquigarrow Y$, that is, via $\langle f \rangle(x) = 1 | f(x) \rangle$.

An example of a channel is arrangement $arr \colon \mathcal{M}[K](X) \to \mathcal{D}(X^K)$. It maps a multiset $\varphi \in \mathcal{M}[K](X)$ to the uniform distribution of sequences that accumulate to φ.

$$arr(\varphi) := \sum_{x \in acc^{-1}(\varphi)} \frac{1}{(\varphi)} | x \rangle = \sum_{x \in acc^{-1}(\varphi)} \frac{\varphi!}{K!} | x \rangle. \tag{5}$$

One can show that $\langle acc \rangle \odot arr = \mathcal{D}(acc) \circ arr = unit \colon \mathcal{M}[K](X) \to \mathcal{D}(\mathcal{M}[K](X))$. The composite in the other direction produces the uniform distribution of all permutations of a sequence:

$$arr \odot \langle acc \rangle = arr \circ acc = prm \qquad \text{where} \qquad prm(\boldsymbol{x}) := \sum_{t \colon K \overset{\cong}{\to} K} \frac{1}{K!} | t(\boldsymbol{x}) \rangle, \tag{6}$$

in which $\underline{t}(x_1, \ldots, x_K) := (x_{t(1)}, \ldots, x_{t(K)})$. In writing $t\colon K \xrightarrow{\cong} K$ we implicitly identify the number K with the set $\{1, \ldots, K\}$.

Each multiset $\varphi \in \mathcal{M}(X)$ of non-zero size can be turned into a distribution via normalisation. This operation is called frequentist learning, since it involves learning a distribution from a multiset of data, via counting. Explicitly:

$$Flrn(\varphi) := \sum_{x \in X} \frac{\varphi(x)}{\|\varphi\|} \,|\, x \,\rangle.$$

For instance, if we learn from an urn with three red, two green and five blue balls, we get the probability distribution for drawing a ball of a particular colour from the urn:

$$Flrn\Big(3|\,R\,\rangle + 2|\,G\,\rangle + 5|\,B\,\rangle\Big) = \tfrac{3}{10}|\,R\,\rangle + \tfrac{1}{5}|\,G\,\rangle + \tfrac{1}{2}|\,B\,\rangle.$$

This map $Flrn$ is a natural transformation (but not a map of monads).

Given two distributions $\omega \in \mathcal{D}(X)$ and $\rho \in \mathcal{D}(Y)$, we can form their parallel product $\omega \otimes \rho \in \mathcal{D}(X \times Y)$, given in functional form as:

$$\big(\omega \otimes \rho\big)(x, y) := \omega(x) \cdot \rho(y).$$

Like for multisets, we call a joint distribution $\tau \in \mathcal{D}(X \times Y)$ a *coupling* of $\omega \in \mathcal{D}(X)$ and $\rho \in \mathcal{D}(Y)$ if ω, ρ are the two marginals of τ, that is if, $\mathcal{D}(\pi_1)(\tau) = \omega$ and $\mathcal{D}(\pi_2)(\tau) = \rho$. We can express this also via a decouple map $dcpl = \langle \mathcal{D}(\pi_1), \mathcal{D}(\pi_2)\rangle$ as in (3).

An *observation* on a set X is a function of the form $p\colon X \to \mathbb{R}$. Such a map p, together with a distribution $\omega \in \mathcal{D}(X)$, is called a random variable — but confusingly, the distribution is often left implicit. The map $p\colon X \to \mathbb{R}$ will be called a *factor* if it restricts to non-negative reals $X \to \mathbb{R}_{\geq 0}$. Each element $x \in X$ gives rise to a point observation $\mathbf{1}_x\colon X \to \mathbb{R}$, with $\mathbf{1}_x(x') = 1$ if $x = x'$ and $\mathbf{1}_x(x') = 0$ if $x \neq x'$. For a distribution $\omega \in \mathcal{D}(X)$ and an observation $p\colon X \to \mathbb{R}$ on the same set X we write $\omega \models p$ for the validity (expected value) of p in ω, defined as (finite) sum: $\sum_{x \in X} \omega(x) \cdot p(x)$. We shall write $Obs(X) = \mathbb{R}^X$ and $Fact(X) = (\mathbb{R}_{\geq 0})^X$ for the sets of observations and factors on X.

3 Preliminaries on metric spaces

A metric space will be written as a pair (X, d_X), where X is a set and $d_X\colon X \times X \to \mathbb{R}_{\geq 0}$ is a distance function, also called metric. This metric satisfies:

- $d_X(x, x') = 0$ iff $x = x'$;
- symmetry: $d_X(x, x') = d_X(x', x)$;
- triangular inequality: $d_X(x, x'') \leq d_X(x, x') + d_X(x', x'')$.

Often, we drop the subscript X in d_X if it is clear from the context. We use the standard distance $d(x, y) = |x - y|$ on real and natural numbers.

Definition 1. *Let (X, d_X), (Y, d_Y) be two metric spaces.*

1. A function $f: X \to Y$ *is called* short *(or also* non-expansive*) if:*

$$d_Y\bigl(f(x), f(x')\bigr) \leq d_X\bigl(x, x'\bigr), \qquad \text{for all } x, x' \in X.$$

Such a map is called an isometry *or an* isometric embedding *if the above inequality* $\leq$ *is an actual equality* $=$. *This implies that the function* f *is injective, and thus an 'embedding'.*

We write $\mathbf{Met}_S$ *for the category of metric spaces with short maps between them.*

2. A function $f: X \to Y$ *is* Lipschitz *or* M-Lipschitz, *if there is a number* $M \in \mathbb{R}_{>0}$ *such that:*

$$d_Y\bigl(f(x), f(x')\bigr) \leq M \cdot d_X\bigl(x, x'\bigr), \qquad \text{for all } x, x' \in X.$$

The number M *is sometimes called the Lipschitz constant. Thus, a short function is Lipschitz, with constant 1. We write* $\mathbf{Met}_L$ *for the category of metric spaces with Lipschitz maps between them (with arbitrary Lipschitz constants).*

Lemma 1. *For two metric spaces* (X_1, d_1) *and* (X_2, d_2) *we equip the cartesian product* $X_1 \times X_2$ *of sets with the sum of the two metrics:*

$$d\bigl((x_1, x_2), (x_1', x_2')\bigr) := d_{X_1}(x_1, x_1') + d_{X_2}(x_2, x_2'). \tag{7}$$

With the usual projections and tuples this forms a product in the category $\mathbf{Met}_L$. □

The product $\times$ also exists in the category $\mathbf{Met}_S$ of metric spaces with short maps. There, it forms a *monoidal* product (a tensor $\otimes$) since there are no diagonals. In the setting of $[0, 1]$-bounded metrics (with short maps) one uses the maximum instead of the sum (7) in order to form products (possibly infinite). In the category $\mathbf{Met}_L$ the products $X_1 \times X_2$ with maximum and with sum of distances are isomorphic, via the identity maps. This works since for $r, s \in \mathbb{R}_{\geq 0}$ one as $\max(r, s) \leq r + s$ and $r + s \leq 2 \cdot \max(r, s)$.

4 The Wasserstein distance between distributions

This section introduces the Wasserstein distance between probability distributions and recalls some basic results. There are several equivalent formulations for this distance. We express it in terms of validity and couplings, see also *e.g.* [1,3,6,4].

Definition 2. *Let* (X, d_X) *be a metric space. The* Wasserstein *metric* $d: \mathcal{D}(X) \times \mathcal{D}(X) \to \mathbb{R}_{\geq 0}$ *is defined by any of the three equivalent formulas:*

$$
\begin{aligned}
d\bigl(\omega, \omega'\bigr) &:= \bigwedge_{\tau \in dcpl^{-1}(\omega, \omega')} \tau \models d_X \\
&= \bigvee_{p,\, p' \in Obs(X),\, p \oplus p' \leq d_X} \omega \models p + \omega' \models p' \\
&= \bigvee_{q \in Facts(X)} \bigl| \omega \models q - \omega' \models q \bigr|.
\end{aligned}
\tag{8}
$$

This turns $\mathcal{D}(X)$ *into a metric space. The operation* $\oplus$ *in the second formulation is defined as* $(p \oplus p')(x, x') = p(x) + p'(x')$. *The set* $\mathrm{Fact}_S(X) \subseteq \mathrm{Fact}(X)$ *in the third formulation is the subset of short factors* $X \to \mathbb{R}_{\geq 0}$. *To be precise, we should write* $\mathrm{Fact}_S(X, d_X)$ *since the distance* d_X *on* X *is a parameter, but we leave it implicit for convenience. The meet* $\bigwedge$ *and joins* $\bigvee$ *in (8) are actually reached, by what are called the* optimal *coupling and the* optimal *observations / factor.*

In this definition it is assumed that X is a metric space. This includes the case where X is simply a set, with the discrete metric (where different elements have distance 1). The above Wasserstein distance can then be formulated as what is often called the *total variation distance*. For distributions $\omega, \omega' \in \mathcal{D}(X)$ it is:

$$d(\omega, \omega') = \tfrac{1}{2} \sum_{x \in X} |\omega(x) - \omega'(x)|.$$

This discrete case is quite common, see *e.g.* [11] and the references given there.

The equivalence of the first and second formulation in (8) is an instance of strong duality in linear programming, which can be obtained via Farkas' Lemma, see *e.g.* [13]. The second formulation is commonly associated with Monge. The single factor q in the third formulation can be obtained from the two observations p, p' in the second formulation, and vice-versa. What we call the Wasserstein distance is also called the Monge-Kantorovich distance.

We do not prove the equivalence of the three formulations for the Wasserstein distance $d(\omega, \omega')$ between two distributions ω, ω' in (8), one with a meet $\bigwedge$ and two with a join $\bigvee$. This is standard and can be found in the literature, see *e.g.* [15]. These three formulations do not immediately suggest how to calculate distances. What helps is that the minimum and maxima are actually reached and can be computed. This is done via linear programming, originally introduced by Kantorovich, see [13,15,3]. In the sequel, we shall see several examples of distances between distributions. They are obtained via our own Python implementation of the linear optimisation, which also produces the optimal coupling, observations or factor. This implementation is used only for illustrations.

Example 1. Consider the set X containing the first eight natural numbers, so $X = \{0, 1, \ldots, 7\} \subseteq \mathbb{N}$, with the usual distance, written as d_X, between natural numbers: $d_X(n, m) = |n - m|$. We look at the following two distributions on X.

$$\omega = \tfrac{1}{2}|0\rangle + \tfrac{1}{2}|4\rangle \qquad \omega' = \tfrac{1}{8}|2\rangle + \tfrac{1}{8}|3\rangle + \tfrac{1}{8}|6\rangle + \tfrac{5}{8}|7\rangle.$$

We claim that the Wasserstein distance $d(\omega, \omega')$ is $\frac{15}{4}$. This will be illustrated for each of the three formulations in Definition 2.

– The optimal coupling $\tau \in \mathcal{D}(X \times X)$ of ω, ω' is:

$$\tau = \tfrac{1}{8}|0, 2\rangle + \tfrac{1}{8}|0, 3\rangle + \tfrac{1}{8}|0, 6\rangle + \tfrac{1}{8}|0, 7\rangle + \tfrac{1}{2}|4, 7\rangle.$$

It is not hard to see that τ's first marginal is ω, and its second marginal is ω'. We compute the distances as:

$$d(\omega, \omega') = \tau \models d_X$$
$$= \tfrac{1}{8} \cdot d_X(0,2) + \tfrac{1}{8} \cdot d_X(0,3) + \tfrac{1}{8} \cdot d_X(0,6) + \tfrac{1}{8} \cdot d_X(0,7) + \tfrac{1}{2} \cdot d_X(4,7)$$
$$= \tfrac{2}{8} + \tfrac{3}{8} + \tfrac{6}{8} + \tfrac{7}{8} + \tfrac{3}{2} = \tfrac{18}{8} + \tfrac{3}{2} = \tfrac{9}{4} + \tfrac{6}{4} = \tfrac{15}{4}.$$

- There are the following two optimal observations $p, p' : X \to \mathbb{R}$, described as sums of weighted point predicates:

$$p = -1 \cdot 1_1 - 2 \cdot 1_2 - 3 \cdot 1_3 - 4 \cdot 1_4 - 5 \cdot 1_5 - 6 \cdot 1_6 - 7 \cdot 1_7$$
$$p' = 1 \cdot 1_1 + 2 \cdot 1_2 + 3 \cdot 1_3 + 4 \cdot 1_4 + 5 \cdot 1_5 + 6 \cdot 1_6 + 7 \cdot 1_7.$$

It is not hard to see that $(p \oplus p')(i,j) := p(i) + p'(j) \le d_X(i,j)$ holds for all $i, j \in X$. Using the second formulation in (8) we get:

$$\left(\omega \models p\right) + \left(\omega' \models p'\right)$$
$$= \tfrac{1}{2} \cdot p(0) + \tfrac{1}{2} \cdot p(4) + \tfrac{1}{8} \cdot p'(2) + \tfrac{1}{8} \cdot p'(3) + \tfrac{1}{8} \cdot p'(6) + \tfrac{5}{8} \cdot p'(7)$$
$$= \tfrac{-4}{2} + \tfrac{2}{8} + \tfrac{3}{8} + \tfrac{6}{8} + \tfrac{35}{8} = -2 + \tfrac{46}{8} = \tfrac{30}{8} = \tfrac{15}{4}.$$

- Finally, there is a (single) short factor $q : X \to \mathbb{R}_{\ge 0}$ given by:

$$q = 7 \cdot 1_0 + 6 \cdot 1_1 + 5 \cdot 1_2 + 4 \cdot 1_3 + 3 \cdot 1_4 + 2 \cdot 1_5 + 1 \cdot 1_6.$$

Then:

$$\left(\omega \models q\right) - \left(\omega' \models q\right)$$
$$= \tfrac{1}{2} \cdot q(0) + \tfrac{1}{2} \cdot q(4) - \left(\tfrac{1}{8} \cdot q(2) + \tfrac{1}{8} \cdot q(3) + \tfrac{1}{8} \cdot q(6) + \tfrac{5}{8} \cdot q(7)\right)$$
$$= \tfrac{7}{2} + \tfrac{3}{2} - \left(\tfrac{5}{8} + \tfrac{4}{8} + \tfrac{1}{8}\right) = \tfrac{10}{2} - \tfrac{10}{8} = \tfrac{20}{4} - \tfrac{5}{4} = \tfrac{15}{4}.$$

From the fact that the coupling τ, the two observations p, p', and the single factor q produce the same distance one can deduce that they are optimal, using the formula (8).

We proceed with several standard properties of the Wasserstein distance on distributions.

Lemma 2. *In the context of Definition 2, the following properties hold.*

1. *For an M-Lipschitz function $f : X \to Y$, the pushforward map $\mathcal{D}(f) : \mathcal{D}(X) \to \mathcal{D}(Y)$ is also M-Lipschitz; as a result, $\mathcal{D}$ lifts to a functor $\mathcal{D} : \mathbf{Met}_L \to \mathbf{Met}_L$, and also to $\mathcal{D} : \mathbf{Met}_S \to \mathbf{Met}_S$.*
2. *If $f : X \to Y$ is an isometry, then so is $\mathcal{D}(f) : \mathcal{D}(X) \to \mathcal{D}(Y)$.*
3. *For an M-Lipschitz factor $q : X \to \mathbb{R}_{\ge 0}$, the validity-of-$q$ factor $(-) \models q : \mathcal{D}(X) \to \mathbb{R}_{\ge 0}$ is also M-Lipschitz.*
4. *For each element $x \in X$ and distribution $\omega \in \mathcal{D}(X)$ one has: $d\left(1 | x\right), \omega\right) = \omega \models d_X(x, -)$; especially, $d\left(1 | x\right), 1 | x'\right)\right) = d_X(x, x')$, making the map unit$: X \to \mathcal{D}(X)$ an isometry.*

5. *The monad multiplication* flat: $\mathcal{D}^2(X) \to \mathcal{D}(X)$ *is short, so that* $\mathcal{D}$ *lifts from a monad on* **Sets** *to a monad on* **Met**$_S$ *and on* **Met**$_L$.
6. *If a channel* $c: X \to \mathcal{D}(Y)$ *is M-Lipschitz, then so is its Kleisli extension* $c \gg= (-) := $ flat $\circ \, \mathcal{D}(c): \mathcal{D}(X) \to \mathcal{D}(Y)$.
7. *If channel* $c: X \nrightarrow Y$ *is M-Lipschitz and channel* $d: Y \nrightarrow Z$ *is K-Lipschitz, then their (channel) composite* $d \circ c: X \nrightarrow Z$ *is* $(M \cdot K)$-*Lipschitz*.
8. *For distributions* $\omega_i, \omega_i' \in \mathcal{D}(X)$ *and numbers* $r_i \in [0,1]$ *with* $\sum_i r_i = 1$ *one has:*

$$d\left(\sum_i r_i \cdot \omega_i, \sum_i r_i \cdot \omega_i' \right) \leq \sum_i r_i \cdot d(\omega_i, \omega_i').$$

9. *The permutation channel* prm: $X^K \to \mathcal{D}(X^K)$ *from (6) is short.*

Proof. We skip the first two points since they are standard.

3. Let $q: X \to \mathbb{R}_{\geq 0}$ be M-Lipschitz, then $\frac{1}{M} \cdot q: X \to \mathbb{R}_{\geq 0}$ is short. The function $(-) \models q: \mathcal{D}(X) \to \mathbb{R}_{\geq 0}$ is then also M-Lipschitz, since for $\omega, \omega' \in \mathcal{D}(X)$,

$$\begin{aligned} \left| \omega \models q - \omega' \models q \right| &= M \cdot \left| \omega \models \tfrac{1}{M} \cdot q - \omega' \models \tfrac{1}{M} \cdot q \right| \\ &\leq M \cdot \bigvee_{p \in Facts(X)} \left| \omega \models p - \omega' \models p \right| \\ &= M \cdot d(\omega, \omega'). \end{aligned}$$

4. The only coupling of $1|x\rangle, \omega \in \mathcal{D}(X)$ is $1|x\rangle \otimes \omega \in \mathcal{D}(X \times X)$. Hence:

$$d(1|x\rangle, \omega) = 1|x\rangle \otimes \omega \models d_X = \sum_{x' \in X} \omega(x') \cdot d_X(x, x') = \omega \models d_X(x, -).$$

5. We first note that for a distribution of distributions $\Omega \in \mathcal{D}^2(X)$ and a short factor $p: X \to \mathbb{R}_{\geq 0}$ the validity in Ω of the short validity factor $(-) \models p: \mathcal{D}(X) \to \mathbb{R}_{\geq 0}$ from item 3 satisfies:

$$\begin{aligned} \Omega \models \big((-) \models p \big) &= \sum_{\omega \in \mathcal{D}(X)} \Omega(\omega) \cdot \big(\omega \models p \big) \\ &= \sum_{\omega \in \mathcal{D}(X)} \sum_{x \in X} \Omega(\omega) \cdot \omega(x) \cdot p(x) \\ &\overset{(4)}{=} \sum_{x \in X} \text{flat}(\Omega)(x) \cdot p(x) \\ &= \text{flat}(\Omega) \models p. \end{aligned}$$

Thus for $\Omega, \Omega' \in \mathcal{D}^2(X)$,

$$\begin{aligned} d_X &\left(\text{flat}(\Omega), \text{flat}(\Omega') \right) \\ &= \bigvee_{p \in Facts(X)} \left| \text{flat}(\Omega) \models p - \text{flat}(\Omega') \models p \right| \\ &= \bigvee_{p \in Facts(X)} \left| \Omega \models \big((-) \models p \big) - \Omega' \models \big((-) \models p \big) \right| \quad \text{as just shown} \\ &\leq \bigvee_{Q \in Facts(\mathcal{D}(X))} \left| \Omega \models Q - \Omega' \models Q \right| \quad \text{by item 3} \\ &= d_{\mathcal{D}(X)}(\Omega, \Omega'). \end{aligned}$$

6. Directly by points (1) and (5).

7. The channel composite $d \circ c = \mathit{flat} \circ \mathcal{D}(d) \circ c$ consists of a functional composite of M-Lipschitz, K-Lipschitz, and 1-Lipschitz maps, and is thus $(M \cdot K \cdot 1)$-Lipschitz. This uses items 1 and (5).

8. If we have couplings τ_i for ω_i, ω_i', then $\sum_i r_i \cdot \tau_i$ is a coupling of $\sum_i r_i \cdot \omega_i$ and $\sum_i r_i \cdot \omega_i'$. Moreover:

$$d\left(\sum_i r_i \cdot \omega_i, \sum_i r_i \cdot \omega_i'\right) \leq \left(\sum_i r_i \cdot \tau_i\right) \models d_X = \sum_i r_i \cdot \left(\tau_i \models d_X\right).$$

Since this holds for all τ_i, we get: $d\left(\sum_i r_i \cdot \omega_i, \sum_i r_i \cdot \omega_i'\right) \leq \sum_i r_i \cdot d(\omega_i, \omega_i')$.

9. We unfold the definition of the prm map from (6) and use the previous item in the first step below. We also use that the distance between two sequences is invariant under permutation (of both).

$$
\begin{aligned}
d_{\mathcal{D}(X^K)}\big(\mathit{prm}(\boldsymbol{x}), \mathit{prm}(\boldsymbol{y})\big) &\leq \sum_{t:\, K \xrightarrow{\cong} K} \frac{1}{K!} \cdot d_{\mathcal{D}(X^K)}\big(1\big|\underline{t}(\boldsymbol{x})\big\rangle, 1\big|\underline{t}(\boldsymbol{y})\big\rangle\big) \\
&= \sum_{t:\, K \xrightarrow{\cong} K} \frac{1}{K!} \cdot d_{X^K}\big(\underline{t}(\boldsymbol{x}), \underline{t}(\boldsymbol{y})\big) \qquad \text{by item 4} \\
&= \sum_{t:\, K \xrightarrow{\cong} K} \frac{1}{K!} \cdot d_{X^K}\big(\boldsymbol{x}, \boldsymbol{y}\big) = d_{X^K}\big(\boldsymbol{x}, \boldsymbol{y}\big). \qquad \square
\end{aligned}
$$

Later on we need the following facts about tensors of distributions.

Proposition 1. *Let X, Y be metric spaces, and K be a positive natural number.*

1. *The tensor map $\otimes \colon \mathcal{D}(X) \times \mathcal{D}(Y) \to \mathcal{D}(X \times Y)$ is an isometry.*
2. *The K-fold tensor map $\mathit{iid}[K] \colon \mathcal{D}(X) \to \mathcal{D}(X^K)$, given by $\mathit{iid}[K](\omega) := \omega^K = \omega \otimes \cdots \otimes \omega$, is K-Lipschitz. Actually, there is an equality: $d(\omega^K, \rho^K) = K \cdot d(\omega, \rho)$.*

Proof. 1. Let distributions $\omega, \omega' \in \mathcal{D}(X)$ and $\rho, \rho' \in \mathcal{D}(Y)$ be given. For the inequality $d_{\mathcal{D}(X) \times \mathcal{D}(Y)}\big((\omega, \rho), (\omega', \rho')\big) \leq d_{\mathcal{D}(X \times Y)}\big(\omega \otimes \rho, \omega' \otimes \rho'\big)$ one uses that a coupling $\tau \in \mathcal{D}\big((X \times Y) \times (X \times Y)\big)$ of $\omega \otimes \rho, \omega' \otimes \rho' \in \mathcal{D}(X \times Y)$ can be turned into two couplings τ_1, τ_2 of ω, ω' and of ρ, ρ', namely as $\tau_i := \mathcal{D}(\pi_i \times \pi_i)(\tau)$. For the reverse inequality one turns two couplings τ_1, τ_2 of ω, ω' and ρ, ρ' into a coupling τ of $\omega \otimes \rho, \omega' \otimes \rho'$ via $\tau := \mathcal{D}\big(\langle \pi_1 \times \pi_1, \pi_2 \times \pi_2 \rangle\big)(\tau_1 \otimes \tau_2)$.

2. For $\omega, \rho \in \mathcal{D}(X)$ and $K \in \mathbb{N}$, using the previous item, we get:

$$d_{\mathcal{D}(X^K)}\big(\omega^K, \rho^K\big) \overset{1}{=} d_{\mathcal{D}(X)^K}\big((\omega, \ldots, \omega), (\rho, \ldots, \rho)\big) \overset{(7)}{=} K \cdot d_{\mathcal{D}(X)}\big(\omega, \rho\big). \qquad \square$$

5   The Wasserstein distance between multisets

There is also a Wasserstein distance between multisets of the same size. This section recalls the definition and the main results.

Definition 3. *Let* (X, d_X) *be a metric space and* $K \in \mathbb{N}$ *a natural number. We can turn the metric* $d_X \colon X \times X \to \mathbb{R}_{\geq 0}$ *into the* Wasserstein *metric* $d \colon \mathcal{M}[K](X) \times \mathcal{M}[K](X) \to \mathbb{R}_{\geq 0}$ *on multisets (of the same size), via:*

$$
\begin{aligned}
d(\varphi, \varphi') &:= \bigwedge_{\tau \in dcpl^{-1}(\varphi, \varphi')} Flrn(\tau) \models d_X \\
&= \frac{1}{K} \cdot \bigwedge_{x \in acc^{-1}(\varphi),\, x' \in acc^{-1}(\varphi')} d_{X^K}(x, x') \\
&\stackrel{(7)}{=} \frac{1}{K} \cdot \bigwedge_{x \in acc^{-1}(\varphi),\, x' \in acc^{-1}(\varphi')} \sum_{0 \leq i < K} d_X(x_i, x'_i).
\end{aligned}
\tag{9}
$$

All meets in (9) are finite and can be computed via enumeration. Alternatively, one can use linear optimisation. We give an illustration below. The equality of the first two formulations is standard, like in Definition 2, and is used here without proof. There is an alternative formulation of the above distance between multisets that uses bistochastic matrices, see *e.g.* [2,6], but we do not need it here.

Example 2. Consider the following two multisets of size 4 on the set $X = \{1, 2, 3\} \subseteq \mathbb{N}$, with standard distance between natural numbers.

$$\varphi = 3|1\rangle + 1|2\rangle \qquad\qquad \varphi' = 2|1\rangle + 1|2\rangle + 1|3\rangle.$$

The optimal coupling $\tau \in \mathcal{M}[4](X \times X)$ is:

$$\tau = 2|1,1\rangle + 1|1,2\rangle + 1|2,3\rangle.$$

The resulting Wasserstein distance $d(\varphi, \varphi')$ is:

$$Flrn(\tau) \models d_X = \tfrac{1}{2} \cdot d_X(1,1) + \tfrac{1}{4} \cdot d_X(1,2) + \tfrac{1}{4} \cdot d_X(2,3) = \tfrac{1}{4} \cdot 1 + \tfrac{1}{4} \cdot 1 = \tfrac{1}{2}.$$

Alternatively, we may proceed as follows. There are $(\varphi) = \frac{4!}{3! \cdot 1!} = 4$ lists that accumulate to φ, and $(\varphi') = \frac{4!}{2! \cdot 1! \cdot 1!} = 12$ lists that accumulate to φ'. We can align them all and compute the minimal distance. It is achieved for instance at:

$$\tfrac{1}{4} \cdot d_{X^4}\big((1,1,1,2), (1,1,2,3)\big) \stackrel{(7)}{=} \tfrac{1}{4} \cdot (0 + 0 + 1 + 1) = \tfrac{2}{4} = \tfrac{1}{2}.$$

Lemma 3. *We consider the situation in Definition 3.*

1. *Frequentist learning* $Flrn \colon \mathcal{M}[K](X) \to \mathcal{D}(X)$ *is an isometry, for* $K > 0$.
2. *For numbers* $K, n \geq 1$ *the scalar multiplication function* $n \cdot (-) \colon \mathcal{M}[K](X) \to \mathcal{M}[n \cdot K](X)$ *is an isometry.*
3. *The sum of distributions* $+ \colon \mathcal{M}[K](X) \times \mathcal{M}[L](X) \to \mathcal{M}[K+L](X)$ *is short.*
4. *If* $f \colon X \to Y$ *is* M-Lipschitz, then $\mathcal{M}[K](f) \colon \mathcal{M}[K](X) \to \mathcal{M}[K](Y)$ *is* M-Lipschitz *too. Thus, the fixed size multiset functor* $\mathcal{M}[K]$ *lifts to categories of metric spaces* $\mathbf{Met}_S$ *and* $\mathbf{Met}_L$.
5. *For* $K > 0$ *the accumulation map* $acc \colon X^K \to \mathcal{M}[K](X)$ *is* $\frac{1}{K}$-Lipschitz, and *thus short.*

6. *The arrangement channel* $\mathrm{arr}\colon \mathcal{M}[K](X) \rightarrowtail X^K$ *is K-Lipschitz; in fact there is an equality* $d\big(\mathrm{arr}(\varphi), \mathrm{arr}(\varphi')\big) = K \cdot d(\varphi, \varphi')$.

Proof. 1. Via naturality of frequentist learning: if $\tau \in \mathcal{M}[K](X \times X)$ is a coupling of $\varphi, \varphi' \in \mathcal{M}[K](X)$, then $\mathit{Flrn}(\tau) \in \mathcal{D}(X \times X)$ is a coupling of $\mathit{Flrn}(\varphi), \mathit{Flrn}(\varphi') \in \mathcal{D}(X)$. This gives $d(\varphi, \varphi') \leq d\big(\mathit{Flrn}(\varphi), \mathit{Flrn}(\varphi')\big)$. The reverse inequality is a bit more subtle. Let $\sigma \in \mathcal{D}(X \times X)$ be an optimal coupling of $\mathit{Flrn}(\varphi), \mathit{Flrn}(\varphi')$. Then, since any coupling $\tau \in \mathcal{M}[K](X \times X)$ of φ, φ' gives, as we have just seen, a coupling $\mathit{Flrn}(\tau) \in \mathcal{D}(X \times X)$ of $\mathit{Flrn}(\varphi), \mathit{Flrn}(\varphi')$, we obtain, by optimality:

$$d\big(\mathit{Flrn}(\varphi), \mathit{Flrn}(\varphi')\big) = \sigma \models d_X \leq \mathit{Flrn}(\tau) \models d_X.$$

Since this holds for any coupling τ, we get $d\big(\mathit{Flrn}(\varphi), \mathit{Flrn}(\varphi')\big) \leq d(\varphi, \varphi')$.

2. For multisets $\varphi, \varphi' \in \mathcal{M}[K](X)$, by the previous item:

$$\begin{aligned}
d_{\mathcal{M}[K](X)}(\varphi, \varphi') &= d_{\mathcal{D}(X)}\big(\mathit{Flrn}(\varphi), \mathit{Flrn}(\varphi')\big) \\
&= d_{\mathcal{D}(X)}\big(\mathit{Flrn}(n \cdot \varphi), \mathit{Flrn}(n \cdot \varphi')\big) \\
&= d_{\mathcal{M}[n \cdot K](X)}(n \cdot \varphi, n \cdot \varphi').
\end{aligned}$$

3. For multisets $\varphi, \varphi' \in \mathcal{M}[K](X)$ and $\psi, \psi' \in \mathcal{M}[L](X)$, using Lemma 2 (8),

$$\begin{aligned}
d\big(\varphi + \psi, &\,\varphi' + \psi'\big) \\
&\overset{1}{=} d\big(\mathit{Flrn}(\varphi + \psi), \mathit{Flrn}(\varphi' + \psi')\big) \\
&= d\big(\tfrac{K}{K+L} \cdot \mathit{Flrn}(\varphi) + \tfrac{L}{K+L} \cdot \mathit{Flrn}(\psi), \tfrac{K}{K+L} \cdot \mathit{Flrn}(\varphi') + \tfrac{L}{K+L} \cdot \mathit{Flrn}(\psi'),\big) \\
&\leq \tfrac{K}{K+L} \cdot d\big(\mathit{Flrn}(\varphi), \mathit{Flrn}(\varphi')\big) + \tfrac{L}{K+L} \cdot d\big(\mathit{Flrn}(\psi), \mathit{Flrn}(\psi')\big) \\
&\overset{1}{=} \tfrac{K}{K+L} \cdot d(\varphi, \varphi') + \tfrac{L}{K+L} \cdot d(\psi, \psi') \\
&\leq d(\varphi, \varphi') + d(\psi, \psi') \\
&\overset{(7)}{=} d\big((\varphi, \psi), (\varphi', \psi')\big).
\end{aligned}$$

4. Let $f\colon X \to Y$ be M-Lipschitz. We use that frequentist learning Flrn is an isometry and a natural transformation $\mathcal{M}[K] \Rightarrow \mathcal{D}$. For multisets $\varphi, \varphi' \in \mathcal{M}[K](X)$,

$$\begin{aligned}
d_{\mathcal{M}[K](Y)}\big(\mathcal{M}(f)(\varphi), \mathcal{M}(f)(\varphi')\big) \\
\overset{1}{=} d_{\mathcal{D}(Y)}\big(\mathit{Flrn}\big(\mathcal{M}(f)(\varphi)\big), \mathit{Flrn}\big(\mathcal{M}(f)(\varphi')\big)\big) \\
= d_{\mathcal{D}(Y)}\big(\mathcal{D}(f)\big(\mathit{Flrn}(\varphi)\big), \mathcal{D}(f)\big(\mathit{Flrn}(\varphi')\big)\big) \qquad &\text{by naturality of } \mathit{Flrn} \\
\leq M \cdot d_{\mathcal{D}(X)}\big(\mathit{Flrn}(\varphi), \mathit{Flrn}(\varphi')\big) \qquad &\text{by Lemma 2 (1)} \\
\overset{1}{=} d_{\mathcal{M}[K](X)}(\varphi, \varphi').
\end{aligned}$$

5. The map $\mathrm{acc}\colon X^K \to \mathcal{M}[K](X)$ is $\frac{1}{K}$-Lipschitz since for $\boldsymbol{y}, \boldsymbol{y}' \in X^K$,

$$\begin{aligned}
d\big(\mathrm{acc}(\boldsymbol{y}), \mathrm{acc}(\boldsymbol{y}')\big) &= \frac{1}{K} \cdot \bigwedge_{\boldsymbol{x} \in \mathrm{acc}^{-1}(\mathrm{acc}(\boldsymbol{y})), \, \boldsymbol{x}' \in \mathrm{acc}^{-1}(\mathrm{acc}(\boldsymbol{y}'))} d_{X^K}(\boldsymbol{x}, \boldsymbol{x}') \\
&\leq \frac{1}{K} \cdot d_{X^K}(\boldsymbol{y}, \boldsymbol{y}').
\end{aligned}$$

6. For fixed $\varphi, \varphi' \in \mathcal{M}[K](X)$, take arbitrary $\boldsymbol{x} \in \mathrm{acc}^{-1}(\varphi)$ and $\boldsymbol{x}' \in \mathrm{acc}^{-1}(\varphi')$. Then:

$$
\begin{aligned}
d_{\mathcal{D}(X^K)}\big(\mathrm{arr}(\varphi), \mathrm{arr}(\varphi')\big) &= d_{\mathcal{D}(X^K)}\big(\mathrm{arr}(\mathrm{acc}(\boldsymbol{x})), \mathrm{arr}(\mathrm{acc}(\boldsymbol{x}'))\big) \\
&\stackrel{(6)}{=} d_{\mathcal{D}(X^K)}\big(\mathrm{prm}(\boldsymbol{x}), \mathrm{prm}(\boldsymbol{x}')\big) \\
&\leq d_{X^K}\big(\boldsymbol{x}, \boldsymbol{x}'\big) \qquad \text{by Lemma 2 (9).}
\end{aligned}
$$

Since this holds for all $\boldsymbol{x} \in \mathrm{acc}^{-1}(\varphi)$, $\boldsymbol{x}' \in \mathrm{acc}^{-1}(\varphi')$ we get an inequaltiy $d_{\mathcal{D}(X^K)}\big(\mathrm{arr}(\varphi), \mathrm{arr}(\varphi')\big) \leq K \cdot d_{\mathcal{M}[K](X)}(\varphi, \varphi')$, see Definition 3. This inequality is an actual equality since acc, and thus $\mathcal{D}(\mathrm{acc})$, is $\frac{1}{K}$-Lipschitz:

$$
\begin{aligned}
d_{\mathcal{M}[K](X)}\big(\varphi, \varphi'\big) &= d_{\mathcal{D}(\mathcal{M}[K](X))}\big(1|\varphi\rangle, 1|\varphi'\rangle\big) \\
&= d_{\mathcal{D}(\mathcal{M}[K](X))}\Big(\mathcal{D}(\mathrm{acc})\big(\mathrm{arr}(\varphi)\big), \mathcal{D}(\mathrm{acc})\big(\mathrm{arr}(\varphi')\big)\Big) \\
&\leq \tfrac{1}{K} \cdot d_{\mathcal{D}(X^K)}\big(\mathrm{arr}(\varphi), \mathrm{arr}(\varphi')\big) \qquad \qquad \square
\end{aligned}
$$

6 Multinomial drawing is isometric

Multinomial draws are of the draw-and-replace kind. This means that a drawn ball is returned to the urn, so that the urn remains unchanged. Thus we may use a distribution $\omega \in \mathcal{D}(X)$ as urn. For a draw size number $K \in \mathbb{N}$, the multinomial distribution $mn[K](\omega) \in \mathcal{D}(\mathcal{M}[K](X))$ on multisets / draws of size K can be defined via accumulated sequences of draws:

$$
\begin{aligned}
mn[K](\omega) &:= \mathcal{D}(\mathrm{acc})\big(\omega^K\big) \\
&= \mathcal{D}(\mathrm{acc})\big(iid[K](\omega)\big) \\
&= \sum_{\varphi \in \mathcal{M}[K](X)} (\varphi) \cdot \prod_{x \in X} \omega(x)^{\varphi(x)} \, |\varphi\rangle.
\end{aligned}
\tag{10}
$$

We recall that $(\varphi) = \frac{K!}{\prod_x \varphi(x)!}$ is the number of sequences that accumulate to a multiset / draw $\varphi \in \mathcal{M}[K](X)$. A basic result from [8, Prop. 3] is that applying frequentist learning to the draws yields the original urn:

$$
Flrn \gg mn[K](\omega) = \omega.
\tag{11}
$$

We can now formulate and prove our first isometry result.

Theorem 1. *Let X be an arbitrary metric space (of colours), and $K > 0$ be a positive natural (draw size) number. The multinomial channel*

$$
\mathcal{D}(X) \xrightarrow{\quad mn[K] \quad} \mathcal{D}\big(\mathcal{M}[K](X)\big)
$$

is an isometry. This involves the Wasserstein metric (8) for distributions over X on the domain $\mathcal{D}(X)$, and the Wasserstein metric for distributions over multisets of size K, with their Wasserstein metric (9), on the codomain $\mathcal{D}\big(\mathcal{M}[K](X)\big)$.

Proof. Let distributions $\omega, \omega' \in \mathcal{D}(X)$ be given. The map $mn[K]$ is short since:

$$d_{\mathcal{D}(\mathcal{M}[K](X))}\Big(mn[K](\omega),\ mn[K](\omega')\Big)$$
$$\overset{(10)}{=} d_{\mathcal{D}(\mathcal{M}[K](X))}\Big(\mathcal{D}(acc)(iid[K](\omega)),\ \mathcal{D}(acc)(iid[K](\omega'))\Big)$$
$$\leq \tfrac{1}{K}\cdot d_{\mathcal{D}(X^K)}\Big(iid[K](\omega),\ iid[K](\omega')\Big) \qquad \text{by Lemma 3 (5)}$$
$$= \tfrac{1}{K}\cdot K \cdot d_{\mathcal{D}(X)}(\omega,\ \omega') \qquad\qquad\ \text{by Proposition 1 (2)}$$
$$= d_{\mathcal{D}(X)}(\omega,\ \omega').$$

There is also an inequality in the other direction, via:

$$d_{\mathcal{D}(X)}(\omega,\ \omega') \overset{(11)}{=} d_{\mathcal{D}(X)}\Big(Flrn \gg= mn[K](\omega),\ Flrn \gg= mn[K](\omega')\Big)$$
$$\leq d_{\mathcal{D}(\mathcal{M}[K](X))}\Big(mn[K](\omega),\ mn[K](\omega')\Big).$$

The latter inequality follows from the fact that frequentist learning *Flrn* is short, see Lemma 3 (1), and that Kleisli extension *Flrn* $\gg= (-)$ is thus short too, see Lemma 2 (6). ❏

Example 3. Consider the following two distributions $\omega, \omega' \in \mathcal{D}(\mathbb{N})$.

$$\omega = \tfrac{1}{3}|0\rangle + \tfrac{2}{3}|2\rangle \quad \text{and} \quad \omega' = \tfrac{1}{2}|1\rangle + \tfrac{1}{2}|2\rangle \quad \text{with} \quad d(\omega,\omega') = \tfrac{1}{2}.$$

This distance $d(\omega,\omega')$ involves the standard distance on $\mathbb{N}$, using the optimal coupling $\tfrac{1}{3}|0,1\rangle + \tfrac{1}{6}|2,1\rangle + \tfrac{1}{2}|2,2\rangle \in \mathcal{D}(\mathbb{N} \times \mathbb{N})$.

We take draws of size $K = 3$. There are 10 multisets of size 3 over $\{0,1,2\}$:

$$\varphi_1 = 3|0\rangle \qquad \varphi_2 = 2|0\rangle + 1|1\rangle \qquad \varphi_3 = 1|0\rangle + 2|1\rangle \qquad \varphi_4 = 3|1\rangle$$
$$\varphi_5 = 2|0\rangle + 1|2\rangle \qquad \varphi_6 = 1|0\rangle + 1|1\rangle + 1|2\rangle \qquad \varphi_7 = 2|1\rangle + 1|2\rangle$$
$$\varphi_8 = 1|0\rangle + 2|2\rangle \qquad \varphi_9 = 1|1\rangle + 2|2\rangle \qquad \varphi_{10} = 3|2\rangle.$$

These multisets occur in the following multinomial distributions of draws of size 3.

$$mn[3](\omega) = \tfrac{1}{27}|\varphi_1\rangle + \tfrac{2}{9}|\varphi_5\rangle + \tfrac{4}{9}|\varphi_8\rangle + \tfrac{8}{27}|\varphi_{10}\rangle$$
$$mn[3](\omega') = \tfrac{1}{8}|\varphi_4\rangle + \tfrac{3}{8}|\varphi_7\rangle + \tfrac{3}{8}|\varphi_9\rangle + \tfrac{1}{8}|\varphi_{10}\rangle.$$

The optimal coupling $\tau \in \mathcal{D}\big(\mathcal{M}[3](\mathbb{N}) \times \mathcal{M}[3](\mathbb{N})\big)$ between these two multinomial distributions is:

$$\tau = \tfrac{1}{27}\Big|\varphi_1,\varphi_4\Big\rangle + \tfrac{19}{216}\Big|\varphi_5,\varphi_4\Big\rangle + \tfrac{1}{8}\Big|\varphi_{10},\varphi_{10}\Big\rangle + \tfrac{29}{216}\Big|\varphi_5,\varphi_7\Big\rangle$$
$$+ \tfrac{5}{72}\Big|\varphi_8,\varphi_7\Big\rangle + \tfrac{3}{8}\Big|\varphi_8,\varphi_9\Big\rangle + \tfrac{37}{216}\Big|\varphi_{10},\varphi_7\Big\rangle.$$

We compute the distance between the multinomial distributions, using $d_{\mathcal{M}} = d_{\mathcal{M}[3](\mathbb{N})}$.

$$d\big(mn[3](\omega), mn[3](\omega')\big) = \tau \models d_{\mathcal{M}}$$
$$= \tfrac{1}{27}\cdot d_{\mathcal{M}}(\varphi_1,\varphi_4) + \tfrac{19}{216}\cdot d_{\mathcal{M}}(\varphi_5,\varphi_4) + \tfrac{1}{8}\cdot d_{\mathcal{M}}(\varphi_{10},\varphi_{10}) + \tfrac{29}{216}\cdot d_{\mathcal{M}}(\varphi_5,\varphi_7)$$
$$+ \tfrac{5}{72}\cdot d_{\mathcal{M}}(\varphi_8,\varphi_7) + \tfrac{3}{8}\cdot d_{\mathcal{M}}(\varphi_8,\varphi_9) + \tfrac{37}{216}\cdot d_{\mathcal{M}}(\varphi_{10},\varphi_7)$$
$$= \tfrac{1}{27}\cdot 1 + \tfrac{19}{216}\cdot 1 + \tfrac{1}{8}\cdot 0 + \tfrac{29}{216}\cdot\tfrac{2}{3} + \tfrac{5}{72}\cdot\tfrac{2}{3} + \tfrac{3}{8}\cdot\tfrac{1}{3} + \tfrac{37}{216}\cdot\tfrac{2}{3} = \tfrac{1}{2}.$$

As predicted in Theorem 1, this distance coincides with the distance $d(\omega, \omega') = \frac{1}{2}$ between the original urn distributions. One sees that the computation of the distance between the draw distributions is more complex, involving 'Wasserstein over Wasserstein'.

7 Hypergeometric drawing is isometric

We start with some preparatory observations on probabilistic projection and drawing.

Lemma 4. *For a metric space X and a number K, consider the probabilistic projection-delete PD and probabilistic draw-delete DD channels.*

$$X^{K+1} \xrightarrow{\ PD\ } \mathcal{D}(X^K) \qquad\qquad \mathcal{M}[K+1](X) \xrightarrow{\ DD\ } \mathcal{D}(\mathcal{M}[K](X))$$

They are defined via deletion of elements from sequences and from multisets:

$$PD(x_1, \ldots, x_{K+1}) := \sum_{1 \leq i \leq K+1} \frac{1}{K+1} \big| x_1, \ldots, x_{i-1}, x_{i+1}, \ldots, x_{K+1} \big\rangle$$

$$DD(\psi) := \sum_{x \in \text{supp}(\psi)} \frac{\psi(x)}{K+1} \big| \psi - 1 | x \rangle \big\rangle$$

$$= \sum_{x \in \text{supp}(\psi)} \text{Flrn}(\psi)(x) \big| \psi - 1 | x \rangle \big\rangle.$$

Then:

1. ⟨acc⟩ ⊚ PD = DD ⊚ ⟨acc⟩;
2. Flrn ⫐ DD(ψ) = Flrn(ψ);
3. PD is $\frac{K}{K+1}$-Lipschitz, and thus short;
4. DD is an isometry.

Proof. The first point is easy and the second one is [8, Lem. 5 (ii)].

3. For $x, y \in X^{K+1}$, via Lemma 2 (8) and (4),

$$d\big(PD(x), PD(y)\big) = d\left(\sum_{1 \leq i \leq K+1} \frac{1}{K+1} \big| x_1, \ldots, x_{i-1}, x_{i+1}, \ldots, x_{K+1} \big\rangle, \right.$$
$$\left. \sum_{1 \leq i \leq K+1} \frac{1}{K+1} \big| y_1, \ldots, y_{i-1}, y_{i+1}, \ldots, y_{K+1} \big\rangle \right)$$

$$\leq \sum_{1 \leq i \leq K+1} \frac{1}{K+1} \cdot d\Big(1 \big| x_1, \ldots, x_{i-1}, x_{i+1}, \ldots, x_{K+1} \big\rangle, $$
$$1 \big| y_1, \ldots, y_{i-1}, y_{i+1}, \ldots, y_{K+1} \big\rangle \Big)$$

$$= \sum_{1 \leq i \leq K+1} \frac{1}{K+1} \cdot d_{X^K}\Big((x_1, \ldots, x_{i-1}, x_{i+1}, \ldots, x_{K+1}), $$
$$(y_1, \ldots, y_{i-1}, y_{i+1}, \ldots, y_{K+1}) \Big)$$

$$= \sum_{1 \leq i \leq K+1} \frac{1}{K+1} \cdot K \cdot d_X(x_i, y_i)$$

$$\overset{(7)}{=} \frac{K}{K+1} \cdot d_{X^{K+1}}(x, y).$$

4. Via item 1 we get:

$$\langle acc \rangle \circ PD \circ arr \;=\; DD \circ \langle acc \rangle \circ arr \;=\; DD \circ unit \;=\; DD. \qquad (*)$$

Now we can show that DD is short: for $\psi, \psi' \in \mathcal{M}[K+1](X)$

$$
\begin{aligned}
d_{\mathcal{D}(\mathcal{M}[K](X))}&\big(DD(\psi),\, DD(\psi')\big)\\
&\overset{(*)}{=}\; d_{\mathcal{D}(\mathcal{M}[K](X))}\Big(\mathcal{D}(acc)\big(PD \gg\!\!= arr(\psi)\big),\, \mathcal{D}(acc)\big(PD \gg\!\!= arr(\psi')\big)\Big)\\
&\leq\; \tfrac{1}{K} \cdot d_{\mathcal{D}(X^K)}\Big(PD \gg\!\!= arr(\psi),\, PD \gg\!\!= arr(\psi')\Big)\\
&\leq\; \tfrac{1}{K} \cdot \tfrac{K}{K+1} \cdot d_{\mathcal{D}(X^{K+1})}\big(arr(\psi),\, arr(\psi')\big)\\
&=\; \tfrac{1}{K+1} \cdot (K+1) \cdot d_{\mathcal{M}[K+1](X))}\big(\psi,\, \psi'\big)\\
&=\; d_{\mathcal{M}[K+1](X))}\big(\psi,\, \psi'\big).
\end{aligned}
$$

For the reverse inequality we use item 2 and the fact that $Flrn$ is a short:

$$
\begin{aligned}
d_{\mathcal{D}(\mathcal{M}[K](X))}&\big(DD(\psi), DD(\psi')\big)\\
&\geq\; d_{\mathcal{D}(\mathcal{M}[K](X))}\Big(Flrn \gg\!\!= DD(\psi),\, Flrn \gg\!\!= DD(\psi')\Big)\\
&=\; d_{\mathcal{D}(X)}\big(Flrn(\psi),\, Flrn(\psi')\big)\\
&=\; d_{\mathcal{M}[K+1](X)}\big(\psi, \psi'\big). \qquad\qquad \square
\end{aligned}
$$

The hypergeometric channel $hg[K]\colon \mathcal{M}[L](X) \to \mathcal{D}\big(\mathcal{M}[K](X)\big)$, for urn size $L \geq K$, where K is the draw size, is an iteration of draw-delete's, see [8, Thm. 6]:

$$hg[K](\upsilon) := \underbrace{DD \circ \cdots \circ DD}_{L-K \text{ times}} = \sum_{\varphi \in \mathcal{M}[K](X),\, \varphi \leq \upsilon} \frac{\binom{\upsilon}{\varphi}}{\binom{L}{K}}\, |\varphi\rangle, \qquad (12)$$

where $\binom{\upsilon}{\varphi} := \prod_{x \in X} \binom{\upsilon(x)}{\varphi(x)}$.

Theorem 2. *The hypergeometric channel* $hg[K]\colon \mathcal{M}[L](X) \to \mathcal{D}\big(\mathcal{M}[K](X)\big)$ *defined in* (12), *for* $L \geq K$, *is an isometry.*

Proof. We see in (12) that $hg[K]$ is a (channel) iteration of isometries DD, and thus of short maps; hence it it short itself. Via iterated use of Lemma 4 (2) we get $Flrn \gg\!\!= hg[K](\psi) = Flrn(\psi)$. This gives the inequality in the other direction, like in the proof of Lemma 4 (2):

$$
\begin{aligned}
d_{\mathcal{M}[K+1](X)}\big(\psi, \psi'\big) &=\; d_{\mathcal{D}(X)}\big(Flrn(\psi), Flrn(\psi')\big)\\
&=\; d_{\mathcal{D}(\mathcal{M}[K](X))}\Big(Flrn \gg\!\!= hg[K](\psi),\, Flrn \gg\!\!= hg[K](\psi')\Big)\\
&\leq\; d_{\mathcal{D}(\mathcal{M}[K](X))}\big(hg[K](\psi), hg[K](\psi')\big). \qquad\qquad \square
\end{aligned}
$$

The very beginning of this paper contains an illustration of this result, for urns over the set of colours $C = \{R, G, B\}$, considered as a discrete metric space.

8   Pólya drawing is isometric

Hypergeometric distributions use the draw-delete mode: a drawn ball is removed from the urn. The less well-known Pólya draws [7] use the draw-add mode. This means that a drawn ball is returned to the urn, together with another ball of the same colour (as the drawn ball). Thus, with hypergeometric draws the urn decreases in size, so that only finitely many draws are possible, whereas with Pólya draws the urn grows in size, and the drawing may be repeated arbitrarily many times. As a result, for Pólya distributions we do not need to impose restrictions on the size K of draws. We do have to restrict draws from urn v to multisets $\varphi \in \mathcal{M}[K](X)$ with $supp(\varphi) \subseteq supp(v)$ since we can only draw balls of colours that are in the urn. Pólya distributions are formulated in terms of multi-choose binomials $\left(\!\binom{n}{m}\!\right) := \binom{n+m-1}{m} = \frac{(n+m-1)!}{m! \cdot (n-1)!}$, for $n > 0$. This multi-choose number $\left(\!\binom{n}{m}\!\right)$ is the number of multisets of size m over a set with n elements, see [9,10] for details.

$$pl[K](v) := \sum_{\varphi \in \mathcal{M}[K](X),\, supp(\varphi) \subseteq supp(v)} \frac{\left(\!\binom{v}{\varphi}\!\right)}{\left(\!\binom{L}{K}\!\right)} \,|\,\varphi\,\rangle, \tag{13}$$

where $\left(\!\binom{v}{\varphi}\!\right) := \prod_{x \in supp(v)} \left(\!\binom{v(x)}{\varphi(x)}\!\right)$.

Theorem 3. *Each Pólya channel* $pl[K] \colon \mathcal{M}[L](X) \to \mathcal{D}\big(\mathcal{M}[K](X)\big)$, *for urn and draw sizes* $L > 0, K > 0$, *is an isometry.*

Proof. One inequality follows by exploiting the equation $Flrn \gg= pl[K](\psi) = Flrn(\psi)$ like in previous sections. The reverse inequality, for shortness, involves a draw-store-add channel of the form:

$$\mathcal{M}[L](X) \times \mathcal{M}[N](X) \xrightarrow{\quad DSA \quad} \mathcal{D}\big(\mathcal{M}[L](X) \times \mathcal{M}[N+1](X)\big)$$

defined as:

$$DSA(v,\varphi) := \sum_{x \in supp(v+\varphi)} Flrn(v+\varphi)(x) \,\big|\, v, \varphi + 1 | x \,\rangle$$
$$= 1|v\rangle \otimes \left(\sum_{x \in supp(v+\varphi)} Flrn(v+\varphi)(x)|\varphi + 1|x\,\rangle \right).$$

With some effort one shows that this channel DSA is short and that the Pólya channel can be expressed via iterated draw-store-add's, namely as:

$$pl[K](v) = \mathcal{D}(\pi_2)\Big(\big(\underbrace{DSA \circ \cdots \circ DSA}_{K \text{ times}} \big)(v, 0) \Big),$$

where $0 \in \mathcal{M}[0](X)$ is the empty multiset. This makes the Pólya channel $pl[K]$ short, and thus an isometry. $\square$

We illustrate that the Pólya channel is an isometry.

Example 4. We take as space of colours $X = \{0, 10, 50\} \subseteq \mathbb{N}$ with two urns:

$$v_1 = 3|0\rangle + 1|10\rangle \qquad v_2 = 1|0\rangle + 2|10\rangle + 1|50\rangle.$$

The distance between these urns is 15, via the optimal coupling $1|0,0\rangle + 2|0,10\rangle + 1|10,50\rangle$, yielding $\frac{1}{4} \cdot (0-0) + \frac{1}{2} \cdot (10-0) + \frac{1}{4} \cdot (50-10) = 5 + 10 = 15$.

We look at Pólya draws of size $K = 2$. This gives distributions:

$$pl[2](v_1) = \frac{3}{5}\left|2|0\rangle\right\rangle + \frac{3}{10}\left|1|0\rangle + 1|10\rangle\right\rangle + \frac{1}{10}\left|2|10\rangle\right\rangle$$

$$pl[2](v_2) = \frac{1}{10}\left|2|0\rangle\right\rangle + \frac{1}{5}\left|1|0\rangle + 1|10\rangle\right\rangle + \frac{3}{10}\left|2|10\rangle\right\rangle + \frac{1}{10}\left|1|0\rangle + 1|50\rangle\right\rangle$$
$$+ \frac{1}{5}\left|1|10\rangle + 1|50\rangle\right\rangle + \frac{1}{10}\left|2|50\rangle\right\rangle$$

We compute the distance between these two distributions via the last formulation in (8), using the optimal short factor $p \colon \mathcal{M}[2](X) \to \mathbb{R}_{\geq 0}$ given by:

$$p(2|0\rangle) = 0 \qquad p(1|0\rangle + 1|10\rangle) = 5 \qquad p(2|10\rangle) = 10$$
$$p(1|0\rangle + 1|50\rangle) = 25 \qquad p(1|10\rangle + 1|50\rangle) = 30 \qquad p(2|50\rangle) = 50.$$

Then:

$$pl[2](v_1) \models p = \frac{3}{5} \cdot 0 + \frac{3}{10} \cdot 5 + \frac{1}{10} \cdot 10 = \frac{5}{2}$$
$$pl[2](v_2) \models p = \frac{1}{10} \cdot 0 + \frac{1}{5} \cdot 5 + \frac{3}{10} \cdot 10 + \frac{1}{10} \cdot 25 + \frac{1}{5} \cdot 30 + \frac{1}{10} \cdot 50 = \frac{35}{2}.$$

As predicted by Theorem 3, the distance between the Pólya distributions then coincides with the distance between the urns:

$$d\Big(pl[2](v_1), pl[2](v_2)\Big) = \left| pl[2](v_1) \models p - pl[2](v_2) \models p \right|$$
$$= \frac{35}{2} - \frac{5}{2} = 15 = d(v_1, v_2).$$

9   Conclusions

Category theory provides a fresh look at the area of probability theory, see *e.g.* [5] or [10] for an overview. Its perspective allows one to formulate and prove new results. This paper demonstrates that draw operations, viewed as (Kleisli) maps, are incredibly well-behaved: they preserve Wasserstein distances. Such distances on urns filled with coloured balls are relatively simple, starting from a 'ground' metric on the set of colours. But on draw distributions, the distances involve Wasserstein-over-Wasserstein. This paper concentrates on drawing from an urn. A natural question is whether other probabilistic operations, as Kleisli maps, preserve distance. This is a topic for further investigation.

Acknowledgments

Thanks are due to the anonymous reviewers for their detailed comments that improved an earlier version of this work.

References

1. F. van Breugel. An introduction to metric semantics: operational and denotational models for programming and specification languages. *Theor. Comp. Sci.*, 258(1-2):1–98, 2001. `doi: 10.1016/S0304-3975(00)00403-5`.

2. H. Brezis. Remarks on the Monge-Kantorovich problem in the discrete setting. *Comptes Rendus Mathematique*, 356(2):207–213, 2018. `doi:10.1016/j.crma.2017.12. 008`.

3. Y. Deng and W. Du. The Kantorovich metric in computer science: A brief survey. In C. Baier and A. di Pierro, editors, *Quantitative Aspects of Programming Languages*, number 253(3) in Elect. Notes in Theor. Comp. Sci., pages 73–82. Elsevier, Amsterdam, 2009. `doi:10. 1016/j.entcs.2009.10.006`.

4. J. Desharnais, V. Gupta, R. Jagadeesan, and P. Panangaden. Metrics for labelled Markov processes. *Theor. Comp. Sci.*, 318:232–354, 2004.

5. T. Fritz. A synthetic approach to Markov kernels, conditional independence, and theorems on sufficient statistics. *Advances in Math.*, 370:107239, 2020. `doi:10.1016/J.AIM. 2020.107239`.

6. T. Fritz and P. Perrone. A probability monad as the colimit of spaces of finite samples. *Theory and Appl. of Categories*, 34(7):170–220, 2019. `doi:10.48550/arXiv.1712.05363`.

7. F. Hoppe. Pólya-like urns and the Ewens' sampling formula. *Journ. Math. Biology*, 20:91–94, 1984. `doi:10.1007/BF00275863`.

8. B. Jacobs. From multisets over distributions to distributions over multisets. In *Logic in Computer Science*. IEEE, Computer Science Press, 2021. `doi:10.1109/lics52264. 2021.9470678`.

9. B. Jacobs. Urns & tubes. *Compositionality*, 4(4), 2022. `doi:10.32408/ compositionality-4-4`.

10. B. Jacobs. Structured probabilitistic reasoning. Book, in preparation, see `http://www. cs.ru.nl/B.Jacobs/PAPERS/ProbabilisticReasoning.pdf`, 2023.

11. B. Jacobs and A. Westerbaan. Distances between states and between predicates. *Logical Methods in Comp. Sci.*, 16(1), 2020. See `https://lmcs.episciences.org/6154`.

12. L. Kantorovich and G. Rubinshtein. On a space of totally additive functions. *Vestnik Leningrad Univ.*, 13:52–59, 1958.

13. J. Matoušek and B. Gärtner. *Understanding and Using Linear Programming*. Springer Verlag, Berlin, 2006. `doi:10.1007/978-3-540-30717-4`.

14. Y. Rubner, C. Tomasi, and L. Guibas. The Earth Mover's Distance as a metric for image retrieval. *Int. Journ. of Computer Vision*, 40:99–121, 2000. `doi:10.1023/A: 1026543900054`.

15. C. Villani. *Optimal Transport — Old and New*. Springer, Berlin Heidelberg, 2009. `doi: 10.1007/978-3-540-71050-9`.

16. G. Wyszecki and W. Stiles. *Color Science: Concepts and Methods, Quantitative Data and Formulae*. Wiley, 1982.

Enriching Diagrams with Algebraic Operations

Alejandro Villoria$^{(\boxtimes)}$ ⓘ , Henning Basold ⓘ ,
and Alfons Laarman ⓘ

Leiden Institute of Advanced Computer Science, Leiden, The Netherlands
{a.d.villoria.gonzalez,h.basold,a.w.laarman}@liacs.leidenuniv.nl

Abstract. In this paper, we extend diagrammatic reasoning in monoidal
categories with algebraic operations and equations. We achieve this by
considering monoidal categories that are enriched in the category of
Eilenberg-Moore algebras for a monad. Under the condition that this
monad is monoidal and there is an adjunction between the free algebra
functor and the underlying category functor, we construct an adjunction
between symmetric monoidal categories and symmetric monoidal categories enriched over algebras for the monad. This allows us to devise
an extension, and its semantics, of the ZX-calculus with probabilistic
choices by freely enriching over convex algebras, which are the algebras
of the finite distribution monad. We show how this construction can be
used for diagrammatic reasoning of noise in quantum systems.

1 Introduction

Monoidal categories are one way of generalizing algebraic reasoning and they can
be used to draw intuitive diagrams that encapsulate this reasoning graphically.
That monoidal categories are a powerful abstraction has been demonstrated
in countless areas, such as linear logic [19] or quantum mechanics [1], just to
name a few, and are amenable to graphical reasoning [47] with diagrammatic
languages such as the ZX-calculus [15]. Another abstraction of algebraic reasoning are monads [3,37,43] and their algebras, or representations thereof [21,36],
which are distinct from monoidal categories in that identities (like associativity) always hold strictly and they allow rather arbitrary algebraic operations. In
this paper, we set out to combine these two approaches into one framework, in
which monoidal category diagrams can be composed not only sequentially and
in parallel with a tensor product but also with additional algebraic operations.

One such operation is the formation of convex combinations, which can be
used to create a probabilistic mix of two or more diagrams. This occurs, for
instance, when reasoning about the behaviour of noise in quantum circuits.
Figure 1 shows on the left two quantum logic gates, one called G and one
called E that, respectively, model the wanted behaviour and a possible error.
These two gates are mixed, where G gets a probability of 0.9 and E of 0.1.
The trapezoids delimit the combination of the gates, and A and B are the input and output types of the gates[1]. In monoidal categories, the gates in the

---

[1] We read diagrams from top to bottom.

N. Kobayashi and J. Worrell (Eds.): FoSSaCS 2024, LNCS 14574, pp. 121–143, 2024.
https://doi.org/10.1007/978-3-031-57228-9_7

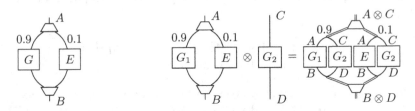

Fig. 1: **Left**: Probabilistic mix of a gate G with an error E. **Right**: Interaction of tensor and convex sum, where double wires visually indicate a tensor product

picture represent morphisms $G, E \colon A \to B$ and our aim is to interpret the trapezoid block as a convex sum $G +_{0.9} E$ of these morphisms, where we define $G +_p E = pG + (1 - p)E$. Such sums should also nicely interact with the tensor product. For instance, if $G_1 \colon A \to B$ and $G_2 \colon C \to D$ are gates, then an identity such as $(G_1 +_{0.9} E) \otimes G_2 = (G_1 \otimes G_2) +_{0.9} (E \otimes G_2)$ should hold for these morphisms of type $A \otimes C \to B \otimes D$, see Figure 1 on the right. Having an operation to form convex combinations together with intuitive identities enables reasoning about, for example, probabilistic combination and noise in quantum circuits.

The difficulty lies in combining monoidal diagrams with algebraic operations such that the algebraic identities and the monoidal identities interact coherently. We will handle this difficulty by using enriched monoidal categories, where the enrichment yields the algebraic operations and the monoidal structure the parallel composition. More precisely, we will assume that the algebraic theory is given by a monad T and that the monoidal categories are enriched over the Eilenberg-Moore category $\mathbf{Alg}^T$ of algebras for this monad. Our aim in this paper is to construct for an arbitrary monoidal category $\mathbf{C}$ an $\mathbf{Alg}^T$-enriched monoidal category $F\mathbf{C}$ that is free in the sense that there is an inclusion $\iota_\mathbf{C} \colon \mathbf{C} \to (F\mathbf{C})_0$ into the underlying category of $F\mathbf{C}$ and for every $\mathbf{Alg}^T$-enriched monoidal category $\underline{\mathbf{D}}$ and monoidal functor $G \colon \mathbf{C} \to \underline{\mathbf{D}}_0$, there is a unique $\mathbf{Alg}^T$-enriched monoidal functor $\bar{G}_0 \colon F\mathbf{C} \to \underline{\mathbf{D}}$ that makes the following diagram commute.

$$
\begin{array}{ccc}
(F\mathbf{C})_0 & \xrightarrow{\bar{G}_0} & \underline{\mathbf{D}}_0 \\
{\scriptstyle \iota_\mathbf{C}} \uparrow & \nearrow {\scriptstyle G} & \\
\mathbf{C} & &
\end{array}
$$

This free construction does not work for all monads, but we show that the free enrichment always exists for monoidal **Set**-monads whose free T-algebra functor is left adjoint to the underlying category functor $\mathbf{Alg}^T(I, -) \colon \mathbf{Alg}^T \to \mathbf{Set}$ for I the monoidal unit of $\mathbf{Alg}^T$.

Contributions
Specifically, we contribute a construction for free enrichment over algebras for some monoidal monads in Theorem 1 and Corollary 1. We also show how the enrichment preserves symmetric monoidal structure in Theorem 3 and Corollary 2.

Given this construction, we demonstrate how a graphical language for reasoning in monoidal categories can be enriched with the free algebras for these monads, which enables diagrammatic reasoning of the interaction between the sequential and parallel compositions with the algebraic structure. We show how the theory can be applied to obtain convex combinations of ZX-diagrams and what the resulting identities of diagrams are. By exploiting the mapping property of the free enrichment, we automatically obtain sound interpretations of these operations and identities. Lastly, we describe how we can use the enrichment of ZX-diagrams to reason about noise in quantum systems.

Related Work

ZX-diagrams are *universal* in the sense that they can in principle represent any linear map between Hilbert spaces of dimension $\mathbb{C}^{2^n}$ [15]. Indeed, sums and linear/convex combinations [25,49] of ZX-diagrams can be encoded within the language, but in practice these representations oftentimes lead to either very large diagrams or to diagrams that do not reveal upon visual inspection the (linear/convex) structure that the diagram is representing. This, in return, diminishes the advantages gained by reasoning in terms of abstract graphical representations. Our perspective of using enrichment keeps the abstraction barrier and thus makes reasoning about convex combinations of diagrams tractable. In general, our theory also covers the recently developed linear combinations of ZX-diagrams such as [50,39] and other, so far unexplored, algebraic operations such as those of join-semilattices. Moreover, the identities that have to be crafted carefully by hand and proven to be sound fall automatically out of our theory. Other related work is that of *Sheet diagrams* [16] and *Tape diagrams* [5], recently developed graphical languages for rig categories, which are categories with two monoidal structures – one for addition and one for multiplication.

Outline

The paper is organised as follows. We start by introducing notation and recalling some background of enriched and monoidal categories in Section 2. In Section 3, we establish the necessary theory to define categories enriched over Eilenberg-Moore algebras and we construct a free enrichment over those algebras. Our next step in Section 4 is to extend these definitions and the free construction to also include monoidal structures on categories, which ensures that these enrichment and monoidal structure coherently interact. Section 5 is devoted to applying our theory to enrich ZX-diagrams with convex sums to reason about probabilistic processes such as quantum noise. We conclude the paper with directions for future work in Section 6.

2 Background

In this section, we recall some terminology from category theory [8,34,35,45] and introduce some notation. We denote the collection of objects of a category $\mathbf{C}$ as $|\mathbf{C}|$, and the morphisms from object A to B as $\mathbf{C}(A, B)$. A *monoidal category*

$(\mathbf{C}, \otimes, I)$ is a category $\mathbf{C}$ together with a functor $\otimes : \mathbf{C} \times \mathbf{C} \to \mathbf{C}$ called the *tensor product* and an object $I \in |\mathbf{C}|$ called the *tensor unit* subject to some conditions [28]. We will often refer to a monoidal category $(\mathbf{C}, \otimes, I)$ as just $\mathbf{C}$. A monoidal category is a *symmetric monoidal* category (SMC) when it also has a *braiding* $\sigma_{A,B} : A \otimes B \to B \otimes A$ such that $\sigma_{B,A} \circ \sigma_{A,B} = \mathrm{Id}_{A \otimes B}$ that is also subject to coherence conditions [28].

Given a monoidal category $(\mathbf{V}, \times, *)$, a $\mathbf{V}$-*(enriched) category* $\underline{\mathbf{C}}$ consists of

- a class $|\underline{\mathbf{C}}|$ of objects,
- for each pair $A, B \in |\underline{\mathbf{C}}|$, an object $\underline{\mathbf{C}}(A, B) \in |\mathbf{V}|$ that we refer to as the *hom-object*,
- for objects $A, B, C \in |\underline{\mathbf{C}}|$, a *composition* morphism $\underline{\circ} : \underline{\mathbf{C}}(B, C) \times \underline{\mathbf{C}}(A, B) \to \underline{\mathbf{C}}(A, C)$ in $\mathbf{V}$, and
- for all $A \in |\underline{\mathbf{C}}|$, an *identity element* $j_A : * \to \underline{\mathbf{C}}(A, A)$

subject to associativity and unit axioms [28]. We say that $\mathbf{V}$ is the *base of enrichment* for $\underline{\mathbf{C}}$. A way to look at the above definition is that we construct a $\mathbf{V}$-enriched category $\underline{\mathbf{C}}$ by identifying morphisms of some category $\mathbf{C}$ as objects from $\mathbf{V}$, which we are able to compose by using the tensor product of $\mathbf{V}$. The most well-known example is that of *locally small* categories, in which the morphisms between two objects form a set, and thus we can see them as objects in the monoidal category $(\mathbf{Set}, \times, *)$ for $\times$ the Cartesian product and $*$ the singleton set.

With a suitable definition of $\mathbf{V}$-functors and $\mathbf{V}$-natural transformations, $\mathbf{V}$-categories organise themselves into a 2-category [28], denoted by $\mathbf{V}$-$\mathbf{Cat}$. For an SMC $(\mathbf{V}, \times, *)$, $\mathbf{V}$-$\mathbf{Cat}$ is also an SMC as follows. We define for $\mathbf{V}$-categories $\underline{\mathbf{C}}$ and $\underline{\mathbf{D}}$ a $\mathbf{V}$-category $\underline{\mathbf{C}} \otimes \underline{\mathbf{D}}$ with objects $|\underline{\mathbf{C}} \otimes \underline{\mathbf{D}}|$ being the categorical product and hom-objects $(\underline{\mathbf{C}} \otimes \underline{\mathbf{D}})((A, B), (C, D)) = \underline{\mathbf{C}}(A, C) \times \underline{\mathbf{D}}(B, D)$. The unit is given by $j_{(A,B)} = * \cong * \times * \xrightarrow{u_A \times v_B} \underline{\mathbf{C}}(A, A) \times \underline{\mathbf{D}}(B, B)$ in terms of the units u of $\underline{\mathbf{C}}$ and v of $\underline{\mathbf{D}}$. Similarly, one also defines the composition for $\underline{\mathbf{C}} \otimes \underline{\mathbf{D}}$ in terms of the composition morphisms of $\underline{\mathbf{C}}$ and $\underline{\mathbf{D}}$, appealing to the symmetry in $\mathbf{V}$ [28, Sec. 1.4]. The tensor product also extends to $\mathbf{V}$-functors and $\mathbf{V}$-natural transformations, which makes it a 2-functor. Finally, one defines I to be the unit $\mathbf{V}$-category with one object 0 and $I(0, 0) = *$ and we thus obtain, with suitable definitions of associators etc., a symmetric monoidal 2-category $(\mathbf{V}$-$\mathbf{Cat}, \otimes, I)$.

Most of the categories we are interested in are also *dagger-compact* categories ($\dagger$-CC). These are SMCs $(\mathbf{C}, \otimes, I)$ with some additional structure. First, they are equipped with an endofunctor $\dagger : \mathbf{C}^{op} \to \mathbf{C}$, that satisfies $(\mathrm{Id}_A)^\dagger = \mathrm{Id}_A$, $(g \circ f)^\dagger = f^\dagger \circ g^\dagger$, $(f^\dagger)^\dagger = f$, and $(f \otimes g)^\dagger = f^\dagger \otimes g^\dagger$. And secondly, for every object A there exists a *dual* A^* such that there exists *unit* $\eta_A : I \to A \otimes A^*$ and *counit* $\epsilon_A : A^* \otimes A \to I$ morphisms subject to some conditions [20].

We are interested in categories that let us reason about quantum mechanics. One of them is **FdHilb**, the category of finite dimensional Hilbert spaces of the form $\mathbb{C}^n$ and linear maps as morphisms. The category **Qubit** is the (full) subcategory of **FdHilb** with objects Hilbert spaces of the form $\mathbb{C}^{2^n}$ and linear maps. Similarly, the category **CPM(Qubit)** [48] has objects $\mathbb{C}^{2^n}$ and morphisms completely positive linear maps between them [13]. We usually work in **Qubit** when reasoning about pure quantum evolutions and in **CPM(Qubit)** when impure quantum evolutions (such as noise) can take place. All of these categories are †-CC, with the monoidal structure $\otimes$ given by the usual Kronecker product of vector spaces and the dagger † being the conjugate transpose.

3 Algebraic Enrichment

In this section, we are going to recall the concept of monoidal and affine monads, and discuss some properties of the *Eilenberg-Moore* category of a monad. We also start applying the *Distribution monad* and the *Multiset monad* to running examples that will be of interest in later sections.

The Distribution monad $(\mathcal{D}, \mu, \eta)^2$ contains the functor $\mathcal{D} : \mathbf{Set} \to \mathbf{Set}$ that maps a set A to the set $\mathcal{D}(A)$ of (finitely supported) probability distributions over elements of A. We write probability distributions as formal convex sums: $\sum_a p_a[a] \in \mathcal{D}(A)$ such that $a \in A, p_a \in [0,1]$, and $\sum_a p_a = 1$. $\mathcal{D}$ acts on a morphism f by simply applying f to the underlying set: $(\mathcal{D}f)(\sum_a p_a[a]) = \sum_a p_a[f(a)]$. The unit of the monad is the map $\eta : A \to \mathcal{D}(A) : a \mapsto 1[a]$ (the Dirac distribution), and the multiplication μ "flattens" a distribution of distributions by multiplying the probabilities together: $\mu : \mathcal{D}(\mathcal{D}(A)) \to \mathcal{D}(A) : \sum_q p_q[\sum_a q_a[a]] \mapsto \sum_a r_a[a]$ where $r_a = \sum_q p_q q_a$ [23].

The functor $\mathcal{D}$ is also a monoidal functor, which makes $(\mathcal{D}, \mu, \eta)$ a *monoidal monad*. In particular, this means that there exists a map:

$$\nabla : \mathcal{D}(A) \times \mathcal{D}(B) \to \mathcal{D}(A \times B) : \left(\sum_a p_a[a], \sum_b p_b[b] \right) \mapsto \sum_{a,b} p_a p_b[(a,b)]$$

for every $A, B \in |\mathbf{Set}|$.

A monad $T : \mathbf{C} \to \mathbf{C}$ is *affine* if there is an isomorphism $T(*) \cong *$ for $*$ the terminal object of $\mathbf{C}$ [24]. This is the case for $\mathcal{D}$.

If $\mathcal{D}$ is a monad for expressing convex combinations of elements of a set, the Multiset monad $\mathcal{M}$ is its analogue for linear combinations with coefficients over some semiring.

We recall that given any monad T in $\mathbf{C}$ we can construct its Eilenberg-Moore category, with objects T-*algebras* of the form (A, α_A) for $A \in |\mathbf{C}|$ and T-*action* $\alpha_A : T(A) \to A$ such that $\alpha_A \circ T(\alpha_A) = \alpha_A \circ \mu_A$ and $\alpha_A \circ \eta_A = \mathrm{Id}_A$. Algebra homomorphisms $f : (A, \alpha_A) \to (B, \alpha_B)$ are morphisms of the underlying objects $f : A \to B$ that commute with the action: $f \circ \alpha_A = \alpha_B \circ T(f)$. The identity and composition follow from the ones for the underlying objects [35].

For a monad T on $\mathbf{C}$, we have that $\mathbf{Alg}^T$ is complete whenever $\mathbf{C}$ is complete. Cocompleteness is not as immediate, but if $\mathbf{C} = \mathbf{Set}$ then $\mathbf{Alg}^T$ is also cocomplete [3,22]. This makes $\mathbf{Alg}^T$ over monads on $\mathbf{Set}$ a complete and cocomplete category and in particular, $\mathbf{Alg}^T$ has reflexive coequalizers, which we use to define the tensor product of algebras.

---

2 We will often refer to a monad (T, μ, η) as T.

When T is a monoidal monad on a monoidal category $(\mathbf{C}, \otimes, I)$, the tensor product of T algebras $(A, a), (B, b)$, denoted $(A, a) \otimes^T (B, b)$, is (if it exists) defined as the coequalizer diagram [46,10]

$$F(T(A) \otimes T(B)) \xrightarrow[\;F(a \otimes b)\;]{\overset{\mu \cdot F(\nabla)}{\longrightarrow}} F(A \otimes B) \xrightarrow{\;q\;} (A, a) \otimes^T (B, b), \qquad (1)$$

where $F : \mathbf{C} \to \mathbf{Alg}^T : A \mapsto (T(A), \mu)$ is the left adjoint to the *forgetful* functor $U : \mathbf{Alg}^T \to \mathbf{C} : (A, \alpha_A) \mapsto A$ that maps objects to their free algebras over T. Given that we need $\mathbf{Alg}^T$ to be monoidal in order to use it as a base of enrichment, diagram (1) above is a convenient representation of the tensor product of algebras. The rest of the structure to make $\mathbf{Alg}^T$ a (symmetric) monoidal category follows under certain conditions, in particular when $(\mathbf{C}, \otimes, I)$ is a closed (S)MC and the coequalizer (1) exists for all algebras $(A, a), (B, b)$ [46].

We can define (symmetric) monoidal structure in the category of *free algebras* as follows. Using (1) for the category of free algebras over a monoidal monad, we have that the following diagram forms a coequalizer.

$$F(TT(A) \otimes TT(B)) \xrightarrow[\;F(\mu \otimes \mu)\;]{\overset{\mu \cdot F(\nabla)}{\longrightarrow}} F(T(A) \otimes T(B)) \xrightarrow{\;\mu \cdot F(\nabla)\;} F(A \otimes B) \qquad (2)$$

Therefore, $(T(A), \mu) \otimes^T (T(B), \mu) \cong F(A \otimes B)$ [46, Prop. 2.5.2]. The monoidal unit is $I^T = (T(I), \mu)$, while the associator, unitor, and symmetry (if present) are the images of the ones in $(\mathbf{C}, \otimes, I)$ under F. We then have that $(\mathbf{Alg}^T, \otimes^T, I^T)$ is the (symmetric) monoidal category of free T-algebras.

A functor $F : (\mathbf{V}_1, \otimes_{\mathbf{V}_1}, I_{\mathbf{V}_1}) \to (\mathbf{V}_2, \otimes_{\mathbf{V}_2}, I_{\mathbf{V}_2})$ between two monoidal categories can be lifted to a 2-functor $F_* : \mathbf{V}_1\text{-Cat} \to \mathbf{V}_2\text{-Cat}$ [7]. This is called a *change of enriching*, where we turn a $\mathbf{V}_1$-category into a $\mathbf{V}_2$-category. Indeed, given a $\mathbf{V}_1$-category $\underline{\mathbf{C}}$, we can construct the $\mathbf{V}_2$-category $F_*\underline{\mathbf{C}}$ by defining $|F_*\underline{\mathbf{C}}| := |\underline{\mathbf{C}}|$ and, for every $A, B \in |F_*\underline{\mathbf{C}}|$, the hom-objects are $F_*\underline{\mathbf{C}}(A, B) := F(\underline{\mathbf{C}}(A, B))$ with composition and identity element following from the ones in $\underline{\mathbf{C}}$ under F. For a symmetric monoidal category $(\mathbf{V}, \otimes, I)$, an important instance is the functor $\mathbf{V}(I, -)_* : \mathbf{V}\text{-Cat} \to \mathbf{Cat}$ called the *underlying category* functor and it is denoted by $(-)_0$.

The following lemma states explicitly the case when one of the enriching categories is $\mathbf{Set}$.

Lemma 1 ([7, Prop. 6.4.7]). *Let $(\mathbf{V}, \otimes, I)$ be a closed symmetric monoidal category with coproducts. Then the hom-functor $\mathbf{V}(I, -) : \mathbf{V} \to \mathbf{Set}$ has a left adjoint F that sends a set X to $F(X) = \coprod_X I$, the X-th fold copower of I. Moreover, F is a strong morphism of symmetric monoidal categories and the induced 2-functor F_* is left-adjoint to the underlying category functor $(-)_0$.*

Theorem 1. *A monoidal monad T on $(\mathbf{Set}, \times, *)$ endows the category of T-algebras $\mathbf{Alg}^T$ with a bicomplete (complete and cocomplete) closed SMC structure. This allows to lift the free-forgetful adjunction of Lemma 1 as a change of enriching between $\mathbf{Alg}^T$-categories and $\mathbf{Set}$-categories for a monoidal T.*

Proof. The proof follows from Lemma 1 and previous arguments. Given that $\mathbf{Alg}^T$ for T a monad on $\mathbf{Set}$ is bicomplete, then coequalizer (1) exists and we can define tensor products of algebras. We can then make $\mathbf{Alg}^T$ a symmetric monoidal category given that $\mathbf{Set}$ is closed symmetric monoidal. Finally, $\mathbf{Alg}^T$ can be made into a closed category following [32] given that $\mathbf{Set}$ has equalizers. We can then use Lemma 1 to create a change of enriching between $\mathbf{Alg}^T$-categories and $\mathbf{Set}$-categories.

Corollary 1. *Let T be a monoidal monad on $\mathbf{Set}$ defined by a free-forgetful adjunction $U : \mathbf{Alg}^T \rightleftarrows \mathbf{Set} : L$. If L is naturally isomorphic to the functor F from Lemma 1, that is $(T(-), \mu) \cong \coprod_{(-)}(TI, \mu)$, then the induced 2-functor L_* is left adjoint to the underlying category functor $(-)_0$. This lets us use Theorem 1 to enrich locally small categories with the free algebras over T. The condition $L \cong F$ holds, in particular, when T is an affine monad.*

Proof. Whenever we have that $L \cong F$, the enrichment over free T-algebras comes simply from substituting F with L in Lemma 1 and Theorem 1. To see that this condition holds when T is affine, we construct hom isomorphisms $\mathbf{Alg}^T(L(X), Y) \cong \mathbf{Set}(X, U(Y)) \cong \mathbf{Set}(X, \mathbf{Alg}^T(I^T, Y)) \cong \mathbf{Alg}^T(F(X), Y)$ for some $X \in |\mathbf{Set}|, Y \in |\mathbf{Alg}^T|$, with the second and last isomorphism coming from their respective adjunctions. The remaining one is due to $T(*) \cong *$, which allows us to get algebra homomorphisms $h : (T(*), \mu) \mapsto (T(Y), \mu)$ from maps $h' : * \mapsto Y$, while the other direction just requires to forget the homomorphism structure.

Let us construct an example for Theorem 1 and relate it to graphical languages. If we have a locally small monoidal category $\mathbf{C}$ with morphisms $f, g : A \to B, h : B \to C$, represented graphically as $\boxed{f}, \boxed{g}, \boxed{h}$, we can freely enrich $\mathbf{C}$ over $\mathbf{Alg}^D$ following the change of enriching category method above. Then, we can realize graphically a probabilistic process involving f and g with probability 0.9 and 0.1 respectively, followed by applying h deterministically (that is, it occurs with probability 1) afterwards as follows.

$$(3)$$

Intuitively, we distinguish between probabilistic and deterministic processes by having the former enclosed within *distribution brackets* (in the same way as we would represent them as a formal sum $0.9[f] + 0.1[g]$), that we choose to depict as trapezoids in this paper. Deterministic processes are depicted without the bracket enclosing mostly as syntactic sugar, otherwise they would simply have a single choice with probability 1. We can see how wires can have weights inside this environment, and how each wire represents a *probabilistic choice*. Intuition also tells us that we could for example rewrite the diagram above by distributing h over the two probabilistic branches. We will discuss in later sections which graphical rules capture the interactions present in these enriched categories.

It is natural to then ask if our enriched category **C** maintained its monoidal structure, and if other desired properties (such as braiding and symmetry, if present) would still hold too. We will address this in the next section.

4   Enriched Monoidal Categories

Recall from Section 2 that a symmetric monoidal category **V** gives rise to a symmetric monoidal 2-category $(\mathbf{V\text{-}Cat}, \otimes, I)$ of **V**-categories. This structure allows us to define an enriched (symmetric) monoidal category to be a (symmetric) pseudo-monoid in **V-Cat** [18], which amounts to the following explicit definition [33,38]. Let us denote by S the symmetry isomorphism of **V-Cat**. A symmetric monoidal **V**-category is a tuple $(\underline{\mathbf{C}}, \odot, U, \alpha, \lambda, \rho, \sigma)$ consisting of:

- a **V**-enriched category $\underline{\mathbf{C}}$
- a **V**-functor $U : I \to \underline{\mathbf{C}}$
- a **V**-functor $\odot : \underline{\mathbf{C}} \otimes \underline{\mathbf{C}} \to \underline{\mathbf{C}}$
- **V**-natural isomorphisms $\alpha : \odot \circ (\odot \otimes \mathrm{Id}_{\underline{\mathbf{C}}}) \to \odot \circ (\mathrm{Id}_{\underline{\mathbf{C}}} \otimes \odot)$ (associator), $\lambda : \odot \circ (U \otimes \mathrm{Id}_{\underline{\mathbf{C}}}) \to \mathrm{Id}_{\underline{\mathbf{C}}}$ (left unitor), $\rho : \odot \circ (\mathrm{Id}_{\underline{\mathbf{C}}} \otimes U) \to \mathrm{Id}_{\underline{\mathbf{C}}}$ (right unitor), and $\sigma : \odot \to \odot \circ S$ (symmetry)

subject to the expected coherence axioms [18]. A (symmetric) monoidal **V**-functor $(\underline{\mathbf{C}}, \odot_1, U_1) \to (\underline{\mathbf{D}}, \odot_2, U_2)$ is a lax (symmetric) pseudo-monoid homomorphism, which means that it consists of a **V**-functor $h : \underline{\mathbf{C}} \to \underline{\mathbf{D}}$ and two **V**-natural transformations $h^0 : U_2 \to h \circ U_1$ and $h^2 : \odot_2 \circ (h \otimes h) \to h \circ \odot_1$ that are coherent with the associators, unitors and symmetries [33]. Together, symmetric monoidal **V**-categories and functors form a category **V-SMCat**.

Our goal is now to lift the adjunction between enriched categories from Theorem 1 to also include enriched monoidal structure. To this end, we introduce *lax monoidal strict 2-functors*, which are tuples (G, G^0, G^2) where $G : \mathbf{V} \to \mathbf{W}$ is a strict functor of 2-categories and $(G, G^0, G^2) : (\mathbf{V}, \otimes, I) \to (\mathbf{W}, \times, *)$ is a lax monoidal functor on the underlying 1-categories.

Theorem 2. *Lax monoidal strict 2-functors* $(G, G^0, G^2) : (\mathbf{V}, \otimes, I) \to (\mathbf{W}, \times, *)$ *induce 2-functors* $\mathbf{PMon}(G) : \mathbf{PMon}(\mathbf{V}, \otimes, I) \to \mathbf{PMon}(\mathbf{W}, \times, *)$ *between 2-categories of pseudo-monoids, lax homomorphisms and 2-cells that are compatible with the homomorphism structures. If a G has a monoidal left adjoint F, then* $\mathbf{PMon}(F)$ *is left adjoint to* $\mathbf{PMon}(G)$*. Finally, if the monoidal categories and functors are symmetric, then the adjunction can be improved to one between symmetric pseudo-monoids.*

Proof. The details and appropriate diagram chases are written in [51, Appendix A], which go through the following steps. We begin by showing that $\mathbf{PMon}(G)$

maps a pseudo-monoid $(\underline{\mathbf{C}}, \odot, U, \alpha, \lambda, \rho, \sigma)$ in $(\mathbf{V}\text{-}\mathbf{Cat}, \otimes, I)$ to a pseudo-monoid $(G\underline{\mathbf{C}}, G(\odot) \circ G^2, G(U) \circ G^0, G\alpha, G\lambda, G\rho, G\sigma)$ in $(\mathbf{W}\text{-}\mathbf{Cat}, \times, *)$ by checking that it fulfills the pseudo-monoid axioms [18].

Similarly, we check that (G, G^0, G^2) maps a pseudo-monoid homomorphism (h, h^0, h^2) to a pseudo-monoid homomorphism $(G(h), G(h^0), G(h^2))$.

If G has a strict left adjoint F, which is also strong monoidal, we show that $(F, F^0, F^2) \dashv (G, G^0, G^2)$ is a *monoidal 2-adjunction* if the mates [27] of G^0 and G^2 are the inverses of F^0 and F^2, respectively, as in the following equations, where $\beta_{A,B}$ is the natural isomorphism $\mathbf{W}(A, GB) \xrightarrow{\cong} \mathbf{V}(FA, B)$ and η the unit of the adjunction:

$$(F^0)^{-1} = \beta(G^0) \quad \text{and} \quad (F^2)^{-1} = \beta(G^2) \circ F(\eta \times \eta).$$

The following theorem, which shows that the change of enrichment extends to monoidal enriched categories, follows from Theorem 2 using that $\mathbf{V}\text{-}\mathbf{SMCat} = \mathbf{PMon}(\mathbf{V}\text{-}\mathbf{Cat}, \otimes, I)$ and that the change of enrichment gives a lax monoidal 2-adjunction [7,17].

Theorem 3. *If* $(G, G^0, G^2) \colon (\mathbf{V}, \otimes, I) \to (\mathbf{W}, \times, *)$ *is a symmetric monoidal functor between symmetric monoidal categories with a monoidal left adjoint* (F, F^0, F^2), *then there are adjunctions that commute with the forgetful functors as in the following diagram.*

$$
\begin{array}{ccc}
\mathbf{V}\text{-}\mathbf{SMCat} & \underset{G_*}{\overset{F_*}{\rightleftarrows}}_{\!\perp} & \mathbf{W}\text{-}\mathbf{SMCat} \\
\downarrow & & \downarrow \\
\mathbf{V}\text{-}\mathbf{Cat} & \underset{G_*}{\overset{F_*}{\rightleftarrows}}_{\!\perp} & \mathbf{W}\text{-}\mathbf{Cat}
\end{array}
$$

From this theorem and combining the results from Section 3 we can derive the following corollary, which is our main tool for building monoidal diagrams that are enriched with algebraic operations.

Corollary 2. *A monoidal monad* T *on* **Set** *with an adjunction between the free* T-*algebra functor and the underlying category functor (see Corollary 1) gives a free-underlying adjunction* $(-)_0 \colon \mathbf{Alg}^T\text{-}\mathbf{Cat} \rightleftarrows \mathbf{Cat} \colon F_*$. *This adjunction lifts to an adjunction between the 2-categories of symmetric monoidal* $\mathbf{Alg}^T$-*enriched and* **Set**-*enriched categories* $(-)_0 \colon \mathbf{Alg}^T\text{-}\mathbf{SMCat} \rightleftarrows \mathbf{SMCat} \colon F_*$. *More explicitly, given such a monad* T *on* **Set** *and a SMC* $(\mathbf{C}, \otimes, I, \alpha, \lambda, \rho, \sigma)$ *we can construct the freely* $\mathbf{Alg}^T$-*enriched SMC* $(\underline{\mathbf{C}}, \underline{\otimes}, \underline{I}, \underline{\alpha}, \underline{\lambda}, \underline{\rho}, \underline{\sigma})$.

Knowing that we keep the symmetric monoidal structure after doing the free enrichment, we can justify drawing parallel composition of probabilistic operations in diagram form. Continuing example (3), let us have another probabilistic process in which f' and g' occur with probabilities 0.7 and 0.3 (respectively) parallelly composed. Then we can draw the following picture.

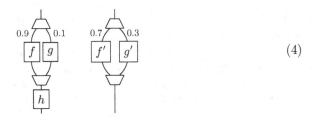

$$(4)$$

5   Applications: ZX-calculus

In this section, we show an example application of the categorical constructions of the previous sections. In particular, we are interested in demonstrating how we can take the Distribution monad and enrich the quantum categories of interest for reasoning about probabilistic processes in quantum systems. Most importantly, we show how the *ZX-calculus*, a graphical calculus for reasoning about quantum processes, can be appropriately extended to accommodate the extra structure on said categories and how additional graphical rewrite rules capture the interaction of probabilistic and deterministic quantum operations. We begin with a general introduction to quantum computing and ZX-calculus, and then follow with the enrichment of our categories of interest, together with the introduction of the extended notation, and we finish by giving an example of how we can use this for diagrammatic reasoning of noise in quantum systems.

5.1   Quantum Computing

When referring to quantum systems and operations, we have to make a distinction whenever we take *impure* operations into account. In the pure states formalisms, quantum states are normalized vectors in a Hilbert space of dimension $\mathbb{C}^{2^n}$, with n the number of *qubits* (quantum bits) of the system. It is common to use Dirac bra-ket notation to represent states, for example, some important single-qubit states are $|0\rangle = \left[\begin{smallmatrix}1\\0\end{smallmatrix}\right], |1\rangle = \left[\begin{smallmatrix}0\\1\end{smallmatrix}\right], |+\rangle = \frac{1}{\sqrt{2}} \cdot (|0\rangle + |1\rangle), |-\rangle = \frac{1}{\sqrt{2}} \cdot (|0\rangle - |1\rangle)$. We operate on qubits by performing unitary transformations U on the quantum states. A multi-qubit quantum system with states $|\psi\rangle$ and $|\phi\rangle$ corresponds to the tensor (Kronecker) product of the quantum states: $|\psi\rangle \otimes |\phi\rangle$. We will represent the n-fold tensor product of a state $|\psi\rangle$ by $|\psi^n\rangle$. Simultaneous (but independent) operations also follow from tensoring unitaries.

When we take into consideration the possibility of applying non-unitary operations we require a more general framework, which is the *density matrix* and *completely positive maps* formalism. In this case, quantum states are positive semidefinite Hermitian matrices ρ of trace one. We write them as $\rho = \sum_i p_i |\psi_i\rangle\langle\psi_i|$ (where $\langle\psi_i| = |\psi_i\rangle^\dagger$, for $\dagger$ the conjugate transpose), that is, a *statistical ensemble* of quantum states $|\psi_i\rangle$ (as density matrices) with probability p_i. Operations on density matrices are completely positive (CP) maps of the form $\Phi : \rho \to \sum_i K_i \rho K_i^\dagger$ with the condition $\sum_i K_i K_i^\dagger \leq 1$ (notice how unitary maps fall inside this description too). When we want to reason about quantum systems in the presence of noise, we then have to use the density matrix and CP map formalism. For more information on quantum computing, we refer the reader to [40], and for a more categorical introduction to [20].

5.2   The ZX-calculus

The ZX-calculus [15] is a graphical language for reasoning about quantum states and processes as diagrams. The language consists of a set of *generators*, which are the green and red[3] *spiders* (also called Z and X spiders), the *Hadamard box*, the *identity wire*, the *swap*, the *cup*, the *cap*, and the *empty diagram*. In Figure 2 we can see the generators of the ZX-calculus and their signature, with input wire(s) coming from the top and outputs going to the bottom. Spiders have a *phase* $\alpha \in [0, 2\pi)$, which as we will see later is omitted when $\alpha = 0$. We can also see how to sequentially compose ($\circ$) arbitrary diagrams by connecting inputs with outputs, and how to parallelly compose diagrams (as a tensor product $\otimes$) by placing them side by side.

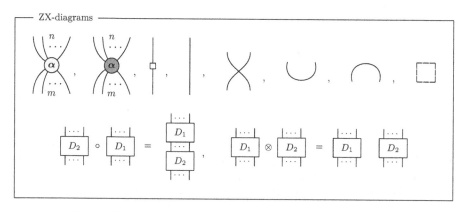

Fig. 2: ZX-diagrams generators and how to compose them.

Each of the generators has a *standard interpretation* $[\![\cdot]\!]$ as a linear map in $\mathbb{C}^{2^n}$ that we can find in Figure 3.

Categorically, ZX-diagrams form the category **ZX** with $|\mathbf{ZX}| = \mathbb{N}$ (where some $n \in \mathbb{N}$ is the number of wires, which we can think of as an n-qubit quantum system) and morphisms being the generators. The standard interpretation is a (monoidal) functor $[\![\cdot]\!] : \mathbf{ZX} \to \mathbf{Qubit}$ that acts on objects as $[\![n]\!] = n$ and on morphisms as defined in Figure 3 [53].

ZX-diagrams come with a set of *rewrite rules* that form the ZX-calculus. These rewrite rules let us transform a diagram into a different one while preserving the semantics (i.e. the interpretation). We have collected the rules in

---

[3] Light and dark in grayscale, respectively.

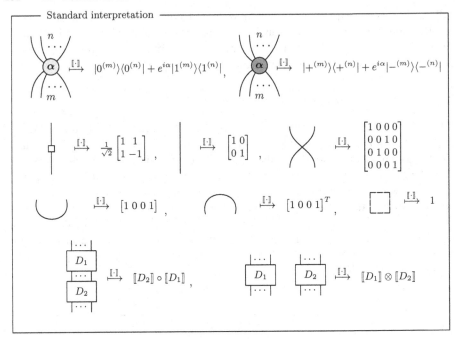

Fig. 3: Standard interpretation of ZX-diagrams.

Figure 4. There is also an important additional rule that can be summarized as the *only connectivity matters* rule, which states that we can deform diagrams at will without changing their meaning, as long as we maintain the connectivity between the generators unchanged. For a thorough explanation of each rule we refer the reader to [15,52].

The ZX-calculus satisfies important properties. ZX-diagrams are *universal*, meaning that any linear map f of the form $f : \mathbb{C}^{2^n} \to \mathbb{C}^{2^m}$ can be represented as a ZX-diagram. The rewrite rules are *sound*, meaning that they do not change the interpretation of the diagram as a linear map. They are also *complete*, which ensures that if two diagrams have the same interpretation, the ruleset is powerful enough to always let us transform one diagram into the other. These properties ensure that the ZX-calculus can be used as a tool for reasoning about quantum computing, as it has been already demonstrated in tasks such as quantum circuit optimisation [30], verification of quantum circuits [41], simulation [31], and as a reasoning tool [4,29].

In (5) we have example one- and two-qubit gates as ZX-diagrams. We also see the computational basis $\{|0\rangle, |1\rangle\}$ and Hadamard basis $\{|+\rangle, |-\rangle\}$ states.

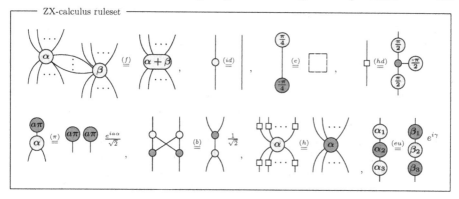

Fig. 4: ZX-calculus ruleset. All rules also hold when swapping the colors of the spiders. In (eu) we omit the calculation of the angles, which can be found in [52].

$$Z = \begin{bmatrix} 1 & 0 \\ 0 & -1 \end{bmatrix} = \pi \quad X = \begin{bmatrix} 0 & 1 \\ 1 & 0 \end{bmatrix} = \pi \quad Y = iXZ = \begin{bmatrix} 0 & -i \\ i & 0 \end{bmatrix} = i\,\pi \atop \pi \tag{5}$$

$$\text{CNOT} = \begin{bmatrix} 1 & 0 & 0 & 0 \\ 0 & 1 & 0 & 0 \\ 0 & 0 & 0 & 1 \\ 0 & 0 & 1 & 0 \end{bmatrix} = \sqrt{2} \qquad \text{CZ} = \begin{bmatrix} 1 & 0 & 0 & 0 \\ 0 & 1 & 0 & 0 \\ 0 & 0 & 1 & 0 \\ 0 & 0 & 0 & -1 \end{bmatrix} = \sqrt{2}$$

$$|0\rangle = \tfrac{1}{\sqrt{2}} \qquad |1\rangle = \tfrac{1}{\sqrt{2}}\,\pi \qquad |+\rangle = \tfrac{1}{\sqrt{2}} \qquad |-\rangle = \tfrac{1}{\sqrt{2}}\,\pi$$

5.3 Enriching the Categories Qubit and CPM(Qubit)

Our motivation is to highlight certain types of relevant physical phenomena (probabilistic processes) that are present in quantum systems within our categories. It is then natural to use the Distribution monad $\mathcal{D}$ together with the construction explained in the previous sections to enrich our categories for quantum reasoning.

Indeed, we take **CPM(Qubit)** and perform a free enrichment over $\mathcal{D}$. What we get is the category $(F_*\mathbf{CPM(Qubit)}, \otimes, I)$ consisting of the same objects as **CPM(Qubit)** and morphisms (incl. identity) for objects A, B the free algebras over $\mathcal{D}$ of the hom-set **CPM(Qubit)**(A, B). Composites of morphisms are the free algebra over the composite in **CPM(Qubit)**, and the SMC structure is preserved thanks to Corollary 2.

We also define the non-freely enriched category $\overline{\mathbf{CPM(Qubit)}}$ so we can interpret probability distributions as CP maps. For this, we define SMC-structure in the non-free $(\mathbf{Alg}^{\mathcal{D}}, \otimes^{\mathcal{D}}, \mathcal{D}(*))$, with a tensor product of algebras defined by the coequalizer (1), with a more detailed description in [51, Appendix B]. The category $(\overline{\mathbf{CPM(Qubit)}}, \odot, U)$ has the same objects as **CPM(Qubit)** and for every pair of objects A, B the hom-object is an algebra $(\mathbf{CPM}(\mathbf{Qubit})(A, B), \alpha)$ with the $\mathcal{D}$-action turning a formal convex sum of linear maps into an actual sum by scalar multiplication and addition. Composition of hom-objects follows from composition in $\mathbf{CPM}(\mathbf{Qubit})$, and for an object A the identity element is $j_A : (*, \alpha) \to (\mathrm{Id}_A, \alpha)$. We define now its symmetric monoidal structure following the definition of enriched SMC from the beginning of Section 4. The tensor product $\odot$ on objects is the same as in $\mathbf{CPM}(\mathbf{Qubit})$, and on hom-objects it is the tensor product in $\mathbf{CPM}(\mathbf{Qubit})$ to the underlying sets: $\odot$: $(\mathbf{CPM}(\mathbf{Qubit})(A, A'), \alpha) \otimes^{\mathcal{D}} (\mathbf{CPM}(\mathbf{Qubit})(B, B'), \alpha) \to (\mathbf{CPM}(\mathbf{Qubit})(A \otimes B, A' \otimes B'), \alpha)$. The unit U is the one in $\mathbf{CPM}(\mathbf{Qubit})$. The associator, unitors, and symmetry all follow from applying the ones in $\mathbf{CPM}(\mathbf{Qubit})$.

In the following sections, we will interpret ZX-diagrams into $F_* \mathbf{CPM}(\mathbf{Qubit})$ as probability distributions of CP maps. From there, to interpret probability distributions as CP maps, we define the functor $\langle\!\langle \cdot \rangle\!\rangle : F_* \mathbf{CPM}(\mathbf{Qubit}) \to \mathbf{CPM}(\mathbf{Qubit})$ that sends objects to themselves and applies the monad algebra to hom-objects i.e. we "evaluate" a probability distribution over CP maps by multiplying the probabilities with the corresponding map and then adding all maps together.

Technically, we can also enrich $\mathbf{Qubit}$ in the same way as we did with $\mathbf{CPM}(\mathbf{Qubit})$, but density matrices and CP maps are the more sensible choices to talk about probabilistic mixtures of operations. On the other hand, enriching $\mathbf{Qubit}$ (or $\mathbf{CPM}(\mathbf{Qubit})$) over algebras of the multiset monad $\mathcal{M}$ leads to an enrichment over commutative monoids that exposes addition of linear maps [20]. This was recently formulated in [39,50] as a way to "split" parameterised Pauli rotation gates in ZX-calculus in such a way that the parameter relocates from its place inside the spider as a phase to a scalar on a wire using the identity $e^{i\alpha P} = \cos \alpha I + i \sin \alpha P$ for P a Pauli matrix (or any matrix satisfying $P^2 = I$).

5.4   Enriched ZX-diagrams and Their Interpretation

In the same manner as the ZX-calculus is a language for reasoning in $\mathbf{Qubit}$, we can create a graphical language with extra structure to reason in our enriched categories. Since we are going to be enriching $\mathbf{CPM}(\mathbf{Qubit})$, we first need to see how to turn the ZX-calculus into a graphical language for CP maps. This is done straightforwardly by adding a *discard* operation $\bar{\overline{\overline{}}}$ to the list of generators of Figure 2 plus additional rewrite rules (that we choose to omit here) stating that isometries can be discarded [13]. The interpretation of a ZX-diagram D as a CP map is then a superoperator $\rho \to [\![D]\!] \rho [\![D]\!]^\dagger$, for $[\![D]\!]$ the standard interpretation of D as in Figure 3 [9].

We then construct an enriched graphical language for $\mathbf{CPM}(\mathbf{Qubit})$ by building on top of the ZX-calculus for CP maps. The notation will be similar to the running examples we have given throughout the text (cf. 1,3,4). The main idea is as follows. We take the generators of the ZX-calculus and allow them to be freely wrapped between opening Δ and closing ∇ *distribution brackets*.

Intuitively, we interpret diagrams that are within distribution brackets as a probabilistic mixture of operations: diagrams placed side by side correspond to different probabilistic choices with some weight attached to the corresponding wires. Within each choice, sequential (and as we will see later, parallel) composition is allowed. The main difference to the usual graphical languages for monoidal categories is that the parallel composition of each choice does not correspond to the tensor product. In a way, we also subsume ZX-diagrams by drawing diagrams that are not enclosed by distribution brackets, which are then interpreted as an operation that occurs with probability 1.

For example, we can represent the single-qubit *depolarizing channel* [40] Φ : $\rho \mapsto (1-p)\rho + \frac{p}{3}(X\rho X + Y\rho Y + Z\rho Z)$ that leaves a quantum state ρ unchanged with probability $(1-p)$ or applies an X, Y or Z error with probability $\frac{p}{3}$ each with the diagram on the left in Figure 5.

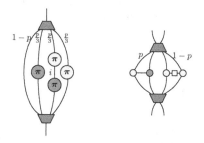

Fig. 5: **Left:** Diagrammatic representation of the depolarizing channel. **Right:** Diagrammatic representation of a mixture of two-qubit gates.

We need to take extra care when handling scalars inside the brackets. Indeed, what we have inside distribution brackets is a formal convex sum of ZX-diagrams (or, in the general case, string diagrams), meaning that the SMC rewriting axioms apply to each summand independently. Since summands are also juxtaposed, it might seem like this notation allows for the transfer of scalars from one summand to another. The crux is that, since what is enclosed by trapezoids is a formal sum, we cannot drag scalars from one summand to another using those same monoidal category axioms. This means that we can consider the probabilities (and any scalar factor if present, such as the imaginary unit in Figure 5) to be bound to the wires themselves, and only interact with the ZX-diagrams (or generally string diagrams) that belong to that summand. An alternative is to encapsulate each summand in "bubbles" for stronger visual separation [50,39].

A probabilistic mixture of operations with multiple inputs or outputs looks similar to the 1-to-1 case, with the caveat that we need to be more careful in the positioning of the wires as to distinguish between tensor product and probabilistic choice.[4] For example, if we want to represent applying the CNOT

---

[4] One could use *scalable notation* [12] to allow wires to be multi-qubit quantum registers. This would help with the distinction when diagrams are larger in practice.

gate and the CZ gate with probabilities p and $1 - p$ respectively we would get the diagram on the right in Figure 5.

We can consider this extra notation as the result of a free enrichment of **ZX** over $\mathbf{Alg}^{\mathcal{D}}$ giving us the category of enriched ZX-diagrams $F_*\mathbf{ZX}$. We can then define the interpretation $[\![\cdot]\!]_{\mathcal{D}}$ of an $\mathbf{Alg}^{\mathcal{D}}$-enriched diagram as a monoidal functor from the category of enriched ZX-diagrams to $\underline{\mathbf{CPM(Qubit)}}$. This functor factors through $F_*\mathbf{CPM(Qubit)}$ as follows:

$$F_*\mathbf{ZX} \xrightarrow{\langle\!\langle\cdot\rangle\!\rangle_*} F_*\mathbf{CPM(Qubit)}$$
$$[\![\cdot]\!]_{\mathcal{D}} \searrow \qquad \downarrow \langle\!\langle\cdot\rangle\!\rangle$$
$$\underline{\mathbf{CPM(Qubit)}}$$

Where $\langle\!\langle\cdot\rangle\!\rangle_*$ interprets an enriched ZX-diagram as a probabilistic mixture of operations which is then evaluated by $\langle\!\langle\cdot\rangle\!\rangle$ as explained in Section 5.3. An example of the interpretation of an arbitrary distribution of ZX-diagrams of arbitrary size can be seen in (6). When using the multiset monad $\mathcal{M}$ instead the interpretation $[\![\cdot]\!]_{\mathcal{M}}$ is similar.

$$\xrightarrow{[\![\cdot]\!]_{\mathcal{D}}} \quad \rho \mapsto \sum_i p_i [\![D_i]\!] \, \rho \, [\![D_i]\!]^\dagger \tag{6}$$

Enriched ZX-diagrams are universal, that is, any morphism in $\underline{\mathbf{CPM(Qubit)}}$ can be represented by an enriched ZX-diagram. Indeed, since $\underline{\mathbf{CPM(Qubit)}}$ is still made of CP maps between Hilbert spaces, we can use universality of the ZX-calculus alone to represent any morphism in $\underline{\mathbf{CPM(Qubit)}}$.

5.5   Additional Rules for Enriched ZX-diagrams

With the new notation we can have new rewrite rules too, some of which were already introduced in [50,39] ((**es**), (**ep**), (**ec**), and (**eδ**)) for the case of linear combinations. We will display them here, including additional rules. The ruleset of the enriched ZX-calculus for the distribution and multiset monads is the same as the one for ZX-calculus plus additional rules that capture the interaction between sums, products, tensor products, and scalars. We can see the additional rules arising from the enrichment in Figure 6, which intuitively state:

- (**es**): The enriched sequential composition rule shows how to sequentially compose distributions. Intuitively this rule follows from products distributing over addition.
- (**ep**): The enriched parallel composition rule is the same as (**es**), but for parallel composition instead of sequential.

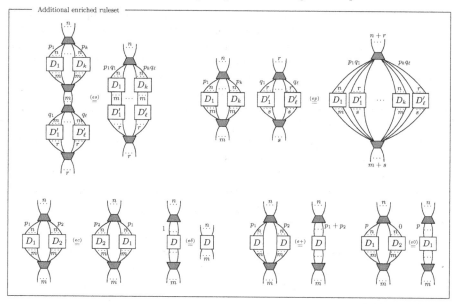

Fig. 6: Additional rules for the enriched ZX-calculus, alongside the ones of Figure 4. Diagrams D, D' are arbitrary ZX-diagrams and weights p, q are probabilities.

- **(ec)**: The enriched commutativity rule shows that bracketed diagrams are invariant under permutation of the branches.
- **(eδ)**: The enriched Dirac delta distribution rule provides a shorthand for the trivial Dirac delta distribution.
- **(e+)**: The enriched addition rule shows that we can remove a branch if it is identical to some other by adding the probabilities.
- **(e0)**: The enriched 0-probability rule allows us to remove branches with 0 probability .

Rules **(es)**,**(ep)**,**(ec)** and **(eδ)** were proven to be sound in [39] but in the context of linear combinations of diagrams interpreted in **Qubit**. We show that these rules still hold as an enrichment in $\mathbf{Alg}^{\mathcal{D}}$ and interpreted in **CPM(Qubit)** in [51, Appendix C]. Finding a complete ruleset (i.e. one that can $\overline{\text{show } D_1 = D_2}$ whenever $[\![D_1]\!]_{\mathcal{D}} = [\![D_2]\!]_{\mathcal{D}}$) for enriched diagrams remains to be done. A possible direction to tackle this problem would be to translate enriched diagrams into ZXW [49] diagrams, which is a complete diagrammatic language with a *W-spider* that can encode addition of phases. Another alternative would be to translate into the controlled form of [25].

We conclude with a demonstration of how we can use this extension of the ZX-calculus to study the effectiveness of Quantum Error Mitigation (QEM) techniques for different noise models. Quantum Error Mitigation [11] are the series of techniques that are used to reduce the effects of noise in near-term quantum systems. One such technique is *Symmetry Verification* [6], which states that given a Hamiltonian (Hermitian operator that determines the evolution of a system) $\hat{H}$, and a symmetry S (an operator that commutes with $\hat{H}$ i.e. $[\hat{H}, S] = \hat{H}S - S\hat{H} = 0$), one can perform measurements of S to verify if the state that (ideally) evolves under $\hat{H}$ was affected by errors. Indeed, under the assumption that the initial state is a $(+1)$ eigenvector of S, then it will stay that way under ideal evolution under $\hat{H}$. This implies that if there is an error E that anti-commutes with S (i.e. $\{E, S\} = ES + SE = 0$) at some point in the computation, we can measure S to detect a change in the eigenvalue. Symmetry verification then proposes to perform a postselection on the result $(+1)$, meaning that we discard computations that give a (-1) outcome when measuring S.

Given a noisy state ρ_{noisy} and a symmetry S, the probability of outcome $(+1)$ when measuring S is given by $p(+1) = \text{tr}(P_{+1}\rho_{noisy})$, for $P_{+1} = \frac{I+S}{2}$ the projector onto the $(+1)$ eigenspace of S and tr the trace operator. This value tells us then with which probability the measurement "accepts" a noisy state, and can be used to compare the effectiveness of different choices of S given a certain noise model [26]. Let us consider $\rho_{\text{noisy}} = \Phi(U\,|0\rangle)$ for Φ the depolarizing noise channel and U some single-qubit unitary – in other words, we have a single layer of depolarizing noise at the end of our computation. For simplicity, let us further assume that our state before the depolarizing channel is the $(+1)$ eigenvector of some Pauli operator e.g. the Pauli X, then we have $U = H$ (the Hadamard gate) and we can draw $p(+1) = \text{tr}(\frac{I+S}{2}\rho_{noisy})$ diagrammatically (up to scalar factor, see [51, Appendix D]) as the following diagram:

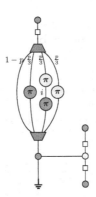

From top to bottom, the diagram represents applying H to the $|0\rangle$ state, followed by a depolarizing noise channel and the verification of X in the form of a CNOT gate controlled on an auxiliary qubit. The auxiliary qubit on the right starts in the $|0\rangle$ state and has a Hadamard gate applied to it before and after the CNOT. It is then postselected into $\langle 0|$, which is the corresponding state for the $(+1)$ outcome. The last operation in the form of $\stackrel{\perp}{=}$ corresponds to the trace. The full diagrammatic calculation is in [51, Appendix D]. With similar diagrams, we can study diagrammatically how well different QEM techniques mitigate certain noise models, and apply them to representations of quantum algorithms that, for example, have one layer of errors for every time step.

6   Discussion and Future Work

In this work, we have shown how to construct freely enriched symmetric monoidal categories over the algebras of a monoidal monad on **Set** that satisfies $F \dashv \mathbf{Alg}^T(I, -)$ for F the free T-algebra functor and I the unit of the monoidal structure, which is the case in particular when T is an affine monad. We have then taken this construction and developed a graphical language that captures the additional algebraic structure of the morphisms for the case of the Distribution monad. We then show how we can use this to study classical probabilistic processes in quantum systems, a highly relevant type of operation for near-term quantum applications. In particular, we extend the ZX-calculus to make it a language for reasoning in an enriched version of **CPM(Qubit)**.

We believe that this work opens several directions for future research. The most evident one is to prove *completeness* of the enriched diagrams, which in turn would facilitate automated implementations for tasks such as simulation of noisy quantum systems, fine-tuned quantum circuit optimization techniques for specific quantum devices, or comparison of the effectiveness of different Quantum Error Mitigation techniques. An interesting venue would be to use enrichment over $\mathcal{M}$ to reason about *quantum circuit pre- and post-processing* techniques, such as *circuit cutting* [44,42], in which quantum circuits are "split" into linear combinations of smaller ones that are executed separately. We also believe that it could be possible to integrate monads that capture quantum behaviours into our construction to represent in enriched ZX-diagrams phenomena such as *superposition of execution orders*, like what is done in the Many-Worlds calculus [14].

Strongly related to completeness is to have presentations of the diagrams in terms of generators and equations. We achieved this by hand in Section 5 by using that the algebras for the distribution monad can be presented as convex algebras with a family of operations $+_p$. The question is then what the analogue of convex monads is when using algebras presented by Lawvere theories or sketches [21,36].

We are also interested in finding other monads that could capture interesting processes outside of the quantum realm. For example, the *non-empty powerset monad* could be used to encode non-deterministic operations and be used for reasoning about a third party operating on a shared quantum system.

Acknowledgements This work was funded by the European Union under Grant Agreement 101080142, EQUALITY project. AV was partly supported by project PRG 946 funded by the Estonian Research Council. The authors would like to thank anonymous reviewers for pointing out reference [2] regarding the tensor product of convex algebras.

References

1. Abramsky, S., Coecke, B.: A categorical semantics of quantum protocols (Mar 2007), http://arxiv.org/abs/quant-ph/0402130, arXiv:quant-ph/0402130

2. Banaschewski, B., Nelson, E.: Tensor products and bimorphisms. Canadian Mathematical Bulletin **19**(4), 385–402 (1976). https://doi.org/10.4153/CMB-1976-060-2
3. Barr, M., Wells, C.: Toposes, Triples and Theories, vol. 278. Springer-Verlag New York (1985)
4. de Beaudrap, N., Horsman, D.: The ZX calculus is a language for surface code lattice surgery. Quantum **4**, 218 (jan 2020). https://doi.org/10.22331/q-2020-01-09-218, https://doi.org/10.22331%2Fq-2020-01-09-218
5. Bonchi, F., Di Giorgio, A., Santamaria, A.: Deconstructing the Calculus of Relations with Tape Diagrams (Oct 2022), http://arxiv.org/abs/2210.09950, arXiv:2210.09950 [cs]
6. Bonet-Monroig, X., Sagastizabal, R., Singh, M., O'Brien, T.E.: Low-cost error mitigation by symmetry verification. Physical Review A **98**(6), 062339 (Dec 2018). https://doi.org/10.1103/PhysRevA.98.062339, http://arxiv.org/abs/1807.10050, arXiv:1807.10050 [quant-ph]
7. Borceux, F.: Handbook of Categorical Algebra, Encyclopedia of Mathematics and its Applications, vol. 2. Cambridge University Press (1994). https://doi.org/10.1017/CBO9780511525865
8. Borceux, F.: Handbook of Categorical Algebra: Volume 1: Basic Category Theory, Encyclopedia of Mathematics and Its Applications, vol. 1. Cambridge University Press, Cambridge (1994). https://doi.org/10.1017/CBO9780511525858
9. Borgna, A., Perdrix, S., Valiron, B.: Hybrid quantum-classical circuit simplification with the ZX-calculus. vol. 13008, pp. 121–139 (2021). https://doi.org/10.1007/978-3-030-89051-3_8, http://arxiv.org/abs/2109.06071, arXiv:2109.06071 [quant-ph]
10. Brandenburg, M.: Tensor categorical foundations of algebraic geometry (Oct 2014), http://arxiv.org/abs/1410.1716, arXiv:1410.1716 [math]
11. Cai, Z., Babbush, R., Benjamin, S.C., Endo, S., Huggins, W.J., Li, Y., McClean, J.R., O'Brien, T.E.: Quantum Error Mitigation (Jun 2023), http://arxiv.org/abs/2210.00921, arXiv:2210.00921 [quant-ph]
12. Carette, T., Horsman, D., Perdrix, S.: SZX-calculus: Scalable Graphical Quantum Reasoning
13. Carette, T., Jeandel, E., Perdrix, S., Vilmart, R.: Completeness of Graphical Languages for Mixed States Quantum Mechanics (Feb 2019), http://arxiv.org/abs/1902.07143, arXiv:1902.07143 [quant-ph]
14. Chardonnet, K., de Visme, M., Valiron, B., Vilmart, R.: The many-worlds calculus (2022)
15. Coecke, B., Duncan, R.: Interacting Quantum Observables: Categorical Algebra and Diagrammatics. New Journal of Physics **13**(4), 043016 (Apr 2011). https://doi.org/10.1088/1367-2630/13/4/043016, http://arxiv.org/abs/0906.4725, arXiv:0906.4725 [quant-ph]
16. Comfort, C., Delpeuch, A., Hedges, J.: Sheet diagrams for bimonoidal categories (Dec 2020), http://arxiv.org/abs/2010.13361, arXiv:2010.13361 [math]
17. Cruttwell, G.: Normed spaces and the change of base for enriched categories (2008)
18. Day, B., Street, R.: Monoidal Bicategories and Hopf Algebroids. Advances in Mathematics **129**(1), 99–157 (Jul 1997). https://doi.org/10.1006/aima.1997.1649, https://www.sciencedirect.com/science/article/pii/S0001870897916492
19. de Paiva, V.: Categorical Semantics of Linear Logic for All. In: Pereira, L.C., Haeusler, E.H., de Paiva, V. (eds.) Advances in Natural Deduction: A Celebration of Dag Prawitz's Work, pp. 181–192. Trends in Logic, Springer Netherlands, Dordrecht (2014). https://doi.org/10.1007/978-94-007-7548-0_9

20. Heunen, C., Vicary, J.: Categories for Quantum Theory: An Introduction. Oxford University Press (11 2019). https://doi.org/10.1093/oso/9780198739623.001.0001, https://doi.org/10.1093/oso/9780198739623.001.0001
21. Hyland, M., Power, J.: The Category Theoretic Understanding of Universal Algebra: Lawvere Theories and Monads. Electronic Notes in Theoretical Computer Science **172**, 437–458 (Apr 2007). https://doi.org/10.1016/j.entcs.2007.02.019
22. Jacobs, B.: Convexity, duality and effects. In: Calude, C.S., Sassone, V. (eds.) Theoretical Computer Science. pp. 1–19. Springer Berlin Heidelberg, Berlin, Heidelberg (2010)
23. Jacobs, B.: New Directions in Categorical Logic, for Classical, Probabilistic and Quantum Logic. Logical Methods in Computer Science **Volume 11, Issue 3**, 1600 (Oct 2015). https://doi.org/10.2168/LMCS-11(3:24)2015, https://lmcs.episciences.org/1600
24. Jacobs, B.: Affine Monads and Side-Effect-Freeness. In: Hasuo, I. (ed.) Coalgebraic Methods in Computer Science. pp. 53–72. Springer International Publishing, Cham (2016)
25. Jeandel, E., Perdrix, S., Veshchezerova, M.: Addition and Differentiation of ZX-diagrams (Mar 2023). https://doi.org/10.48550/arXiv.2202.11386, http://arxiv.org/abs/2202.11386, arXiv:2202.11386 [quant-ph]
26. Kakkar, A., Larson, J., Galda, A., Shaydulin, R.: Characterizing error mitigation by symmetry verification in QAOA. In: 2022 IEEE International Conference on Quantum Computing and Engineering (QCE). IEEE (sep 2022). https://doi.org/10.1109/qce53715.2022.00086, https://doi.org/10.1109%2Fqce53715.2022.00086
27. Kelly, G.M., Street, R.: Review of the elements of 2-categories. In: Kelly, G.M. (ed.) Category Seminar. pp. 75–103. Lecture Notes in Mathematics, Springer, Berlin, Heidelberg (1974). https://doi.org/10.1007/BFb0063101
28. Kelly, M.: Basic Concepts of Enriched Category Theory. No. 64 in Lecture Notes in Mathematics, Cambridge University Press, reprints in theory and applications of categories, no. 10 (2005) edn. (1982), http://www.tac.mta.ca/tac/reprints/articles/10/tr10abs.html
29. Kissinger, A.: Phase-free zx diagrams are css codes (...or how to graphically grok the surface code) (2022)
30. Kissinger, A., van de Wetering, J.: Reducing the number of non-clifford gates in quantum circuits. Physical Review A **102**(2) (aug 2020). https://doi.org/10.1103/physreva.102.022406, https://doi.org/10.1103%2Fphysreva.102.022406
31. Kissinger, A., van de Wetering, J.: Simulating quantum circuits with ZX-calculus reduced stabiliser decompositions. Quantum Science and Technology **7**(4), 044001 (jul 2022). https://doi.org/10.1088/2058-9565/ac5d20, https://doi.org/10.1088%2F2058-9565%2Fac5d20
32. Kock, A.: Closed categories generated by commutative monads. Journal of the Australian Mathematical Society **12**(4), 405–424 (1971). https://doi.org/10.1017/S1446788700010272
33. Kong, L., Yuan, W., Zhang, Z.H., Zheng, H.: Enriched monoidal categories I: Centers (Apr 2021). https://doi.org/10.48550/arXiv.2104.03121, http://arxiv.org/abs/2104.03121
34. Leinster, T.: Basic Category Theory. Cambridge Studies in Advanced Mathematics, Cambridge University Press, Cambridge (2014). https://doi.org/10.1017/CBO9781107360068

35. Mac Lane, S.: Categories for the Working Mathematician, Graduate Texts in Mathematics, vol. 5. Springer, New York, NY (1978). https://doi.org/10.1007/978-1-4757-4721-8, http://link.springer.com/10.1007/978-1-4757-4721-8

36. Manes, E.G.: Algebraic Theories, Graduate Texts in Mathematics, vol. 26. Springer, New York, NY (1976). https://doi.org/10.1007/978-1-4612-9860-1

37. Moggi, E.: Notions of Computation and Monads. Information and Computation **93**(1), 55–92 (1991). https://doi.org/10.1016/0890-5401(91)90052-4

38. Morrison, S., Penneys, D.: Monoidal categories enriched in braided monoidal categories (Jan 2017), http://arxiv.org/abs/1701.00567, arXiv:1701.00567 [math]

39. Muuss, G.: Linear combinations of ZX-diagrams for parameterized quantum circuits (Nov 2022), https://updownup.de/masterthesis.pdf

40. Nielsen, M.A., Chuang, I.L.: Quantum Computation and Quantum Information. Cambridge University Press (2000)

41. Peham, T., Burgholzer, L., Wille, R.: Equivalence checking of quantum circuits with the ZX-calculus. IEEE Journal on Emerging and Selected Topics in Circuits and Systems **12**(3), 662–675 (sep 2022). https://doi.org/10.1109/jetcas.2022.3202204, https://doi.org/10.1109%2Fjetcas.2022.3202204

42. Peng, T., Harrow, A.W., Ozols, M., Wu, X.: Simulating large quantum circuits on a small quantum computer. Physical Review Letters **125**(15) (oct 2020). https://doi.org/10.1103/physrevlett.125.150504, https://doi.org/10.1103%2Fphysrevlett.125.150504

43. Plotkin, G.D., Power, J.: Notions of Computation Determine Monads. In: Nielsen, M., Engberg, U. (eds.) Proceedings of Foundations of Software Science and Computation Structures, 5th International Conference, FOSSACS 2002. Lecture Notes in Computer Science, vol. 2303, pp. 342–356. Springer (2002). https://doi.org/10.1007/3-540-45931-6_24

44. Pérez-Salinas, A., Draškić, R., Tura, J., Dunjko, V.: Reduce&chop: Shallow circuits for deeper problems (May 2023), http://arxiv.org/abs/2212.11862, arXiv:2212.11862 [quant-ph]

45. Riehl, E.: Category Theory in Context. Aurora: Dover Modern Math Originals, Dover Publications (2016), http://www.math.jhu.edu/~eriehl/context/

46. Seal, G.J.: Tensors, monads and actions (Jun 2013), http://arxiv.org/abs/1205.0101, arXiv:1205.0101 [math]

47. Selinger, P.: A Survey of Graphical Languages for Monoidal Categories. In: Coecke, B. (ed.) New Structures for Physics, pp. 289–355. Lecture Notes in Physics, Springer, Berlin, Heidelberg (2011). https://doi.org/10.1007/978-3-642-12821-9_4

48. Selinger, P.: Dagger Compact Closed Categories and Completely Positive Maps: (Extended Abstract). Electronic Notes in Theoretical Computer Science **170**, 139–163 (2007). https://doi.org/https://doi.org/10.1016/j.entcs.2006.12.018, https://www.sciencedirect.com/science/article/pii/S1571066107000606

49. Shaikh, R.A., Wang, Q., Yeung, R.: How to sum and exponentiate Hamiltonians in ZXW calculus (Dec 2022). https://doi.org/10.48550/arXiv.2212.04462, http://arxiv.org/abs/2212.04462, arXiv:2212.04462 [quant-ph]

50. Stollenwerk, T., Hadfield, S.: Diagrammatic Analysis for Parameterized Quantum Circuits (Apr 2022), http://arxiv.org/abs/2204.01307, arXiv:2204.01307 [quant-ph]

51. Villoria, A., Basold, H., Laarman, A.: Enriching diagrams with algebraic operations (2023)

52. Vilmart, R.: A Near-Optimal Axiomatisation of ZX-Calculus for Pure Qubit Quantum Mechanics (Dec 2018), http://arxiv.org/abs/1812.09114, arXiv:1812.09114 [quant-ph]
53. van de Wetering, J.: ZX-calculus for the working quantum computer scientist (Dec 2020), http://arxiv.org/abs/2012.13966, arXiv:2012.13966 [quant-ph]

Monoidal Extended Stone Duality

Fabian Birkmann[1]($\boxtimes$) [ID] *, Henning Urbat[1][ID]*,
and Stefan Milius[1][ID] *

Friedrich-Alexander-Universität Erlangen-Nürnberg, Erlangen, Germany
{fabian.birkmann,stefan.milius,henning.urbat}@fau.de

Abstract. Extensions of Stone-type dualities have a long history in
algebraic logic and have also been instrumental for proving results in
algebraic language theory. We show how to extend abstract categorical
dualities via monoidal adjunctions, subsuming various incarnations of
classical extended Stone and Priestley duality as a special case. Guided by
these categorical foundations, we investigate residuation algebras, which
are algebraic models of language derivatives, and show the subcategory
of derivation algebras to be dually equivalent to the category of profinite
ordered monoids, restricting to a duality between boolean residuation
algebras and profinite monoids. We further extend this duality to capture
relational morphisms of profinite ordered monoids, which dualize to
natural morphisms of residuation algebras.

Keywords: Stone Duality · Profinite Monoids · Regular Languages.

1 Introduction

Marshall H. Stone's representation theorem for boolean algebras, the foundation
for the so called *Stone duality* between boolean algebras and Stone spaces,
manifests a tight connection between logic and topology. It has thus become an
ubiquitous tool in various areas of theoretical computer science, not only in logic,
but also for example in domain theory and automata theory.

From algebraic logic arose the need for extending Stone duality to capture
boolean algebras equipped with additional operators (modelling quantifiers or
modalities). Originating in Jónsson and Tarski's representation theorem for
boolean algebras with operators [21,22], a representation in the spirit of Stone
was proven by Halmos [17]; the general categorical picture of the duality of Kripke
frames and modal algebras is based on an adjunction between operators and
continuous relations developed by Sambin and Vaccaro [31].

In the study of regular languages, the need for extensions of Stone duality
was not discovered until this millenium: while Pippenger [27] has already shown
that the boolean algebra of regular languages on an alphabet Σ corresponds,
under Stone duality, to the Stone space $\widehat{\Sigma^*}$ of profinite words, Gehrke et al. [15]
discovered that, under Goldblatt's form of extended Priestley duality [16], the

---

* All authors are supported by Deutsche Forschungsgemeinschaft (DFG, German
Research Foundation) under project 470467389

© The Author(s) 2024
N. Kobayashi and J. Worrell (Eds.): FoSSaCS 2024, LNCS 14574, pp. 144–165, 2024.
https://doi.org/10.1007/978-3-031-57228-9_8

residuals of language concatenation dualize to multiplication on the space of profinite words. But while categorical frameworks have identified Stone-type dualities to be one of the cornerstones of algebraic language theory [36,30], the correspondence between residuals and multiplication via extended duality has not yet been placed in the categorical big picture. One reason is that, despite some progress in recent years [6,18], extended (Stone) dualities for (co-)algebras are themselves not fully understood as instances of a crisp categorical idea.

Therefore we introduce as our first main contribution a simple, yet powerful framework to extend any categorical duality $\mathbf{C} \simeq^{\mathrm{op}} \hat{\mathbf{C}}$ via *monoidal adjunctions*: For a given adjunction on $\mathbf{C}$ with a strong monoidal right adjoint U we prove a dual equivalence between the category of U-operators on $\mathbf{C}$ to dual operators in the Kleisli category of the monad on $\hat{\mathbf{C}}$ arising from the dual of the given adjunction. We show how to instantiate the abstract extended duality to Priestley duality, which not only recovers Goldblatt's original duality for distributive lattices with operators [16] but also applies more generally to bialgebraic operators with relational morphisms. Guided by our categorical foundations for extended Stone duality we investigate the correspondence between language derivatives and multiplication of profinite words in the setting of *residuation algebras* originally studied by Gehrke [14]. The key observation is that on finite distributive lattices the residuals are equivalent to a *coalgebraic* operator on the lattice, and we show how to lift this correspondence to locally finite structures, i.e. structures built up from finite substructures. By identifying suitable non-full subcategories – derivation algebras and locally finite comonoids, respectively – and an appropriate definition of morphism for residuation algebras, we augment Gehrke's characterization of Stone-topological algebras in terms of residuation algebras to a duality between the categories of derivation algebras and that of profinite ordered monoids:

$$\mathbf{Der} \cong \mathbf{Comon}_{\mathrm{lf}} \simeq^{\mathrm{op}} \mathbf{ProfOrdMon}. \tag{1.1}$$

The above duality clarifies the relation between Gehrke's results and the duality by Rhodes and Steinberg [29] between profinite monoids and counital boolean bialgebras. The extended duality now suggests that the dual equivalence between profinite ordered monoids on one side and locally finite comonoids as well as derivation algebras on the other side extends to a more general duality capturing morphisms of *relational* type of profinite ordered monoids. To this end, we identify a natural notion of relational morphism for residuation algebras and comonoids, and use our abstract extended duality theorem to obtain the dual equivalence

$$\mathbf{RelDer} \cong \mathbf{RelComon}_{\mathrm{lf}} \simeq^{\mathrm{op}} \mathbf{RelProfOrdMon}$$

which extends (1.1) to relational morphisms. To our knowledge, this is the first duality result for relational morphisms of profinite monoids, which have become an ubiquitous tool in algebraic language theory [26] and semigroup theory [29]. Full proof details can be found in the full version [5] of this paper.

Related Work. Duality for (complete) boolean algebras with operators goes back to Jónsson and Tarski [21,22]. This duality was refined by the topological approach via Stone spaces taken by Halmos [17], which allowed to characterize the relations arising as the duals of operators, namely *boolean relations*. Halmos' duality was extended to distributive lattices with (n-ary) operators by Goldblatt [16] and Cignoli [7]. Kupke et al. [24] recognized that boolean relations elegantly describe descriptive frames as coalgebras for the (underlying functor of) the Vietoris monad on Stone spaces; notions of bisimulation for these coalgebras were investigated by Bezhanishvili et al. [2]. Bosangue et al. [6] introduced a framework for dualities over distributive lattices equipped with a theory of operators for a signature, which are dual to certain coalgebras. Hofmann and Nora [18] have taken a categorical approach to extend natural dualities to algebras for a signature equipped with *unary* operators preserving only some of the operations prescribed by the signature; they relate these to coalgebras for (the underlying functor of) a suitable monad T. In their framework T is a parameter required to satisfy certain conditions for the duality to work, while in our work T is already determined by the adjunction. The recent work by Bezhanishvili et al. [1] clarifies the relation between free constructions on distributive lattices and the different versions of the Vietoris monad to derive several dualities between distributive lattices with different types of operators and their corresponding Priestley relations.

Residuated boolean algebras, i.e. boolean algebras with a residuated binary operator,were explicitly considered by Jónsson and Tsinakis [23] to highlight the roles of the residuals in relation algebra. Gehrke et al. [15] discovered the connection between the residuals of the concatenation of regular languages and the multiplication on profinite words and investigated applications to automata theory, most notably a duality-theoretic proof of Eilenberg's variety theorem [8]. The duality theory behind the correspondence of general residuation algebras and Priestley-topological algebras was given via canonical extensions [12,11] and Goldblatt's extended Stone duality [16] by Gehrke [14]. She has also provided conditions under which the dual relations of the residuals is functional; Fussner and Palmigiano [10] have shown that functionality of the dual relation is not equationally definable in the language of residuation algebras.

2   Preliminaries

Readers are assumed to be familiar with basic category theory, such as functors, natural transformations, adjunctions and monoidal categories [25]. We briefly recall the foundations of Stone duality [34] and Priestley duality [28]. By the latter we mean the dual equivalence **DL** $\simeq^{\mathrm{op}}$ **Priest** between the category **DL** of bounded distributive lattices and lattice homomorphisms, and the category **Priest** of Priestley spaces (ordered compact topological spaces in which for every $x \not\leq y$ there exists a clopen up-set containing x but not y) and continuous monotone maps. The duality sends a distributive lattice D to the pointwise-ordered space **DL**$(D, 2)$ of homomorphisms into the two-element lattice (equivalently prime filters, ordered by inclusion), and topologized via pointwise convergence. In the reverse direction, it sends a Priestley space X to the distributive lattice **Priest**$(X, 2)$ of continuous maps into the two-element poset $2 = \{0 \leq 1\}$

with discrete topology (equivalently clopen upsets), with the pointwise lattice structure. Priestley duality restricts to Stone duality **BA** $\simeq^{\mathrm{op}}$ **Stone** between the full subcategories **BA** of boolean algebras and **Stone** of Stone spaces (discretely ordered Priestley spaces). Moreover, it restricts to Birkhoff duality [3] **DL**$_{\mathrm{f}}$ $\simeq^{\mathrm{op}}$ **Pos**$_{\mathrm{f}}$ between finite distributive lattices and finite posets, sending a finite distributive lattice to its poset of join-irreducibles and a poset to its lattice of upsets – note that the pointwise order on homorphisms induces the reverse order on join-irreducibles. For a comprehensive introduction to ordered structures and their dualities, see the first two chapters of the classic textbook by Johnstone [20].

3   Extending Dualities

We present the first contribution of our paper, a general categorical framework for extending Stone-type dualities via monoidal adjunctions, motivated by the extension of Priestley duality to operators due to Goldblatt [16] recovered in Section 4. It serves as the basis for our duality results in the next two sections.

Notation 3.1. (1) For $U\colon \mathbf{C} \to \mathbf{D}$ being right adjoint to $F\colon \mathbf{D} \to \mathbf{C}$ we write $F\colon \mathbf{D} \dashv \mathbf{C} \colon U$ or simply $F \dashv U$. We denote the unit and counit by η and ε and the transposing isomorphisms by

$$(-)^+\colon \mathbf{D}(C, UD) \cong \mathbf{C}(FC, D) \colon (-)^- \quad \text{with} \quad f^+ = \varepsilon \cdot Ff, \quad g^- = Ug \cdot \eta.$$

(2) For dually equivalent categories $\mathbf{C}$ and $\hat{\mathbf{C}}$ we denote the equivalence functors in both directions by $(\hat{-})\colon \mathbf{C} \xrightarrow{\simeq} \hat{\mathbf{C}}$ and $(\hat{-})\colon \hat{\mathbf{C}} \xrightarrow{\simeq} \mathbf{C}$. Moreover, if if $F\colon \mathbf{C} \to \mathbf{D}$ is a functor and $\hat{\mathbf{D}}$ is dual to $\mathbf{D}$, we denote its dual by $\hat{F} = (\hat{-}) \circ F \circ (\hat{-})\colon \hat{\mathbf{C}} \to \hat{\mathbf{D}}$.

(3) The Kleisli category of a monad (T, η, μ) on $\mathbf{C}$ is denoted by $\mathbf{C}_T$. It has the same objects as $\mathbf{C}$ and $\mathbf{C}_T(X, Y) = \mathbf{C}(X, TY)$ with Kleisli composition $g \circ f = \mu \cdot Tg \cdot f$. A morphism $f\colon C \to TD$ of the Kleisli category is *pure* if $f = \eta \cdot f'$ for some $f'\colon C \to D$ in $\mathbf{C}$. (We omit the components of η and μ.)

Assumptions 3.2. We fix monoidal categories $\mathbf{C}, \mathbf{D}$ with dually equivalent categories $\hat{\mathbf{C}}, \hat{\mathbf{D}}$; we regard $\hat{\mathbf{C}}, \hat{\mathbf{D}}$ as monoidal categories with tensor products $\hat{\otimes}$ dual to the tensor products $\otimes$ of $\mathbf{C}, \mathbf{D}$. Moreover, we fix an adjunction $F\colon \mathbf{D} \dashv \mathbf{C} \colon U$ with unit $\eta\colon \mathrm{Id} \to UF$ and counit $\varepsilon\colon FU \to \mathrm{Id}$, and assume that U is a *strong monoidal functor* with associated natural isomorphisms $\lambda\colon UX \otimes UY \cong U(X \otimes Y)$ and $\epsilon\colon I_{\mathbf{D}} \cong UI_{\mathbf{C}}$. One can extend λ to an isomorphism $\lambda\colon \bigotimes_{i=1}^{n} UX_i \cong U(\bigotimes_{i=1}^{n} X_i)$ for all finite n. The dual functor $\hat{U}\colon \hat{\mathbf{C}} \to \hat{\mathbf{D}}$ is a strong monoidal *left* adjoint to $\hat{F}$ and the unit and counit of this dual adjunction are $\hat{\varepsilon}$ and $\hat{\eta}$. We denote the monad dual to the comonad FU by $T = \hat{F}\hat{U}$ with unit $e = \hat{\varepsilon}\colon \mathrm{Id} \to T$ and multiplication $m = \hat{F}\hat{\eta}\hat{U}\colon TT \to T$.

$$
\begin{array}{ccc}
\mathbf{D} & \simeq^{\mathrm{op}} & \hat{\mathbf{D}} \\
F\left(\dashv\right)U & & \hat{F}\left(\vdash\right)\hat{U} \\
\mathbf{C} & \simeq^{\mathrm{op}} & \hat{\mathbf{C}} \circlearrowright T
\end{array}
\tag{3.1}
$$

Remark 3.3. Since $\hat{U}$ is strong monoidal with $\hat{\epsilon}\colon \hat{I}_{\mathbf{D}} \cong \hat{U}\hat{I}_{\mathbf{C}}$ and $\hat{\lambda}\colon \hat{U}X\hat{\otimes}\hat{U}Y \cong \hat{U}(X\hat{\otimes}Y)$ its right adjoint $\hat{F}$ is (lax) monoidal (see e.g. [32, p. 17]) with

$$((\hat{\eta}\hat{\otimes}\hat{\eta})\cdot\hat{\lambda}^{-1})^{-}\colon \hat{F}X\hat{\otimes}\hat{F}Y \to \hat{F}(X\hat{\otimes}Y) \quad \text{and} \quad (\hat{\epsilon}^{-1})^{-}\colon \hat{I}_{\mathbf{C}} \to \hat{F}\hat{I}_{\mathbf{D}}.$$

This makes $\hat{U} \dashv \hat{F}$ a monoidal adjunction, which then induces a monoidal monad $T = \hat{F}\hat{U}$ on $\hat{\mathbf{C}}$. Let $\delta\colon TX\hat{\otimes}TY \to T(X\hat{\otimes}Y)$ denote the witnessing natural transformation, which also extends to any arity. The tensor product $\hat{\otimes}$ of $\hat{\mathbf{C}}$ lifts to the Kleisli category $\hat{\mathbf{C}}_T$; the lifting sends a pair $(f\colon X \to TY, g\colon X' \to TY')$ of $\hat{\mathbf{C}}_T$-morphisms to the $\hat{\mathbf{C}}_T$-morphism $\delta\cdot(f\hat{\otimes}g)\colon X\hat{\otimes}X' \to TY\hat{\otimes}TY' \to T(Y\hat{\otimes}Y')$. This makes $\hat{\mathbf{C}}_T$ itself a monoidal category [33, Prop. 1.2.2] with tensor $\hat{\otimes}$ and the canonical left adjoint $J_T\colon \hat{\mathbf{C}} \to \hat{\mathbf{C}}_T$ a strict monoidal functor.

Definition 3.4. Let $G\colon \mathbf{A} \to \mathbf{B}$ be a functor between monoidal categories, and let $m, n \in \mathbb{N}$. An (m,n)-*ary* G-*operator* consists of an object $A \in \mathbf{A}$ and a morphism $a\colon (GA)^{\otimes m} \to (GA)^{\otimes n}$ of $\mathbf{B}$. An (m,n)-*ary* G-*operator morphism* from (A, b) to (B, b) is a morphism $h\colon GA \to GB$ of $\mathbf{B}$ such that

$$
\begin{array}{ccc}
(GA)^{\otimes m} & \xrightarrow{\;a\;} & (GA)^{\otimes n} \\
{\scriptstyle h^{\otimes m}}\big\downarrow & & \big\downarrow{\scriptstyle h^{\otimes n}} \\
(GB)^{\otimes m} & \xrightarrow{\;b\;} & (GB)^{\otimes n}
\end{array}
$$

commutes. The category of (m, n)-ary G-operators is denoted by $\mathrm{Op}_G^{m,n}(\mathbf{A})$. We call $(m, 1)$-ary G-operators G-*algebras* and $(1, n)$-ary G-operators G-*coalgebras*. If G is strong monoidal we call an operator *pure* if it is of the form $\lambda^{-1}\cdot Ga'\cdot\lambda$, for λ analogous to Assumptions 3.2, and an operator morphism *pure* if it is of the form Gh'.

Note that the full subcategory of $\mathbf{B}$ consisting of the objects in the image of G fully embeds into $\mathrm{Op}_G^{1,1}(\mathbf{A})$ via $GA \mapsto (GA, \mathrm{id}_{GA})$.

Theorem 3.5 (Abstract Extended Duality). *The category of (m, n)-ary U-operators is dually equivalent to the category of (n, m)-ary J_T-operators:*

$$\mathrm{Op}_U^{m,n}(\mathbf{C}) \simeq^{\mathrm{op}} \mathrm{Op}_{J_T}^{n,m}(\hat{\mathbf{C}}).$$

Proof (Sketch). The functor $\mathrm{Op}_{J_T}^{n,m}(\hat{\mathbf{C}}) \to \mathrm{Op}_U^{m,n}(\mathbf{C})$ is defined as follows. An object of $\mathrm{Op}_{J_T}^{n,m}(\hat{\mathbf{C}})$ is an operator $\hat{a}\colon \hat{A}^{\hat{\otimes}n} \to T\hat{A}^{\hat{\otimes}m}$. By dualization, transposition and conjugation with λ it is mapped to

$$\lambda^{-1}\cdot a^{-}\cdot\lambda\colon (UA)^{\otimes m} \cong UA^{\otimes m} \to UA^{\otimes n} \cong (UA)^{\otimes n}.$$

An operator morphism $\hat{f}\colon (\hat{A}, \hat{a}) \to (\hat{B}, \hat{b})$ is mapped to $f^{-}\colon UB \to UA$, the dual of its transpose; a diagram chase shows that this is indeed an operator morphism. One then proves that this yields a dual equivalence $\qquad\square$

An advantage of extending dualities via adjunctions is that adjunctions compose, making the extensions *modular*: let $\mathbf{E}$ be a monoidal category with monoidal adjunctions $F_1\colon \mathbf{E} \dashv \mathbf{C}\colon U_1$ and $F_2\colon \mathbf{D} \dashv \mathbf{E}\colon U_2$ splitting $F \dashv U$, i.e., $F = F_1 F_2$ and $U = U_2 U_1$ and $\lambda = U_2\lambda_1 \cdot \lambda_2 U_1$. Then the following lifting property applies to operators (set $A = B$) as well as operator morphisms (set $m = n = 1$):

Proposition 3.6. *A morphism* $a\colon (UA)^{\otimes m} \to (UB)^{\otimes n}$ *in* $\mathbf{D}$ *lifts to a morphism* $b\colon (U_1A)^{\otimes m} \to (U_1B)^{\otimes n}$ *with* $a = \lambda_2^{-1} \cdot U_2 b \cdot \lambda_2$ *iff the dual of a factors through the canonical monad morphism* $\hat{F}_1\hat{\varepsilon}_2\hat{U}_1\colon T_1 \to T$, *where* $T_1 = \hat{F}_1\hat{U}_1$.

Remark 3.7. (1) A special case of Proposition 3.6 proves that extended dualities preserve purity: splitting $F \dashv U$ into $F_1 = \mathrm{Id} \dashv \mathrm{Id} = U_1$ and $F_2 = F \dashv U = U_2$ we see that a U-operator (or operator morphism) a is pure iff its dual f is pure as a Kleisli morphism, i.e. factors through the unit e of T.

(2) The right adjoint U_2 often is faithful and in this case $\hat{F}_1\hat{\varepsilon}_2\hat{U}_1$ is monic, i.e. T_1 is a submonad of T: faithfulness of U_2 is equivalent to having an epic counit ε_2, hence $\hat{\varepsilon}_2\hat{U}_1$ is mono, and the right adjoint $\hat{F}_1$ preserves monos. In particular, if T is "powerset-like", then $\hat{\mathbf{C}}_T$ is a category of relations, and we think of U-operators (or operator morphisms) of the form $a = \lambda_2^{-1} \cdot U_2 b \cdot \lambda_2$ as dualizing to "more functional" relations. The examples of Section 4.2 illustrate this idea.

4   Example: Extended Priestley Duality

As a first application of our adjoint framework, we investigate the classical Priestley duality (Section 2) and derive a generalized version of Goldblatt's duality [16] between distributive lattices with operators and relational Priestley spaces. We instantiate (3.1) to the following categories and functors, which we will subsequently explain in detail:

$$
\begin{array}{ccccccc}
\mathbf{D} & \simeq^{\mathrm{op}} & \hat{\mathbf{D}} & & \mathbf{JSL} & \simeq^{\mathrm{op}} & \mathbf{StoneJSL} \\
F\left(\dashv\right)U & & \hat{F}\left(\vdash\right)\hat{U} & = & F\left(\dashv\right)U & & \hat{F}\left(\vdash\right)\hat{U} \\
\mathbf{C} & \simeq^{\mathrm{op}} & \hat{\mathbf{C}} \supseteq T & & \mathbf{DL} & \simeq^{\mathrm{op}} & \mathbf{Priest} \circlearrowleft \mathbb{V}_{\downarrow}
\end{array}
$$

Categories The upper duality is Hofman-Mislove-Stralka duality [19] between the category of join-semilattices with bottom and the category of Stone semilattices (i.e. topological join-semilattices with bottom whose underlying topological space is a Stone space) and continuous semilattice homomorphisms. The duality maps a join-semilattice J to the Stone semilattice $\mathbf{JSL}(J,2)$ of semilattice homomorphisms into the two-element semilattice, topologized by pointwise convergence. Equivalently, $\mathbf{JSL}(J,2)$ is the space $\mathrm{Idl}(J)$ of *ideals* (downwards closed and upwards directed subsets) of J, ordered by reverse inclusion, with topology generated by the subbasic open sets $\sigma(j) = \{I \in \mathrm{Idl}(J) \mid j \in I\}$ and their complements for $j \in J$. In the other direction, a Stone semilattice X is mapped to its semilattice $\mathbf{StoneJSL}(X,2)$ of clopen ideals, ordered by inclusion.

Functors The functor $U\colon \mathbf{DL} \to \mathbf{JSL}$ is the obvious forgetful functor. Its left adjoint $F\colon \mathbf{JSL} \to \mathbf{DL}$ maps a join-semilattice to the set $\mathcal{U}^{\circ}_{\mathrm{fg}}(J)$ of finitely generated upsets of J ordered by reverse inclusion. The dual right adjoint $\hat{F}$ of the left adjoint F is the forgetful functor mapping a Stone semilattice to its underlying Priestley space. Indeed, as $U2 = 2$ we compute for the underlying Priestley space $|X|$ of a Stone semilattice X that

$$\hat{F}X = \mathbf{DL}(F(\mathbf{StoneJSL}(X,2)),2) \cong |\mathbf{JSL}(\mathbf{StoneJSL}(X,2),U2)| \cong |X|,$$

and this bijection is a homeomorphism. Its left adjoint $\hat{U}\colon \mathbf{Priest} \to \mathbf{StoneJSL}$ maps a Priestley space X to the space

$$\widehat{U X} = \mathbf{JSL}(U(\mathbf{Priest}(X,2)),2) \cong \mathrm{Idl}(\mathrm{Cl}_\uparrow X) \cong \mathbb{V}_\downarrow X$$

of ideals of clopen upsets of X. This space is isomorphic to the *(downset) Vietoris hyperspace* $\mathbb{V}_\downarrow X$ of X that has as carrier the set of closed downsets of X. The isomorphism $\mathrm{Idl}(\mathrm{Cl}_\uparrow X) \cong \mathbb{V}_\downarrow X$ maps an ideal I to the intersection $\bigcap_{U \in I} X \setminus U$; its inverse sends a closed downset C to the ideal $\{U \in \mathrm{Cl}_\uparrow X \mid C \subseteq X \setminus U\}$ of complements of the basic clopen downsets that contain it. The topology of pointwise convergence on $\mathbf{JSL}(U(\mathbf{Priest}(X,2)),2)$ translates to the *hit-or-miss topology* on $\mathbb{V}_\downarrow X$ generated by the subbasic open sets

$$\{A \subseteq X \text{ closed} \mid A \cap U \neq \emptyset\} \quad \text{for} \quad U \in \mathrm{Cl}_\uparrow X$$

and their complements. For a detailed exposition of these results we refer the reader to the recent work by Bezhanishvili et al. [1]; the free join-semilattice structure on $\mathbb{V}_\downarrow X$ was already observed by Johnstone [20, Sec. 4.8]. The unit $e\colon X \to \mathbb{V}_\downarrow X$ of the Vietoris monad is given by $x \mapsto \downarrow x$ and multiplication is given by union [18]. The monad $\mathbb{V}_\downarrow$ restricts to the full subcategory **Stone** of Stone spaces. We denote the restriction of this monad simply by $\mathbb{V}$.

Remark 4.1 (Continuous Relations). Continuous maps in **Priest** of the form $\rho\colon X \to \mathbb{V}_\downarrow Y$ have a variety of names, we use the term *Priestley relation* as in [7,16] or *Stone relation* if X, Y are Stone spaces. We write $x \rho y$ for $y \in \rho(x)$, and sometimes identify ρ with a subset of $X \times Y$. Let us note that some authors (e.g. [29]) call a relation $R \subseteq X \times Y$ between topological spaces *continuous* if it is closed as a subspace of $X \times Y$. Every Priestley relation is continuous, but a continuous relation between Priestley spaces is generally not a Priestley relation.

Monoidal Structure The category **JSL** of join-semilattices has a tensor product $\otimes$ with the universal property that it extends join-bilinear maps:

$$\mathrm{Bilin}(J \times J', K) \cong \mathbf{JSL}(J \otimes J', K).$$

Join-bilinear maps $J \times J' \to K$ and their corresponding **JSL**-morphisms $J \otimes J' \to K$ are often tacitly identified. The tensor product $\otimes$ makes **JSL** a monoidal category with unit 2, i.e. $2 \otimes J \cong J$. The tensor product has a representation by the generators $\{j \otimes j' \mid j \in J, j' \in J'\}$ and relations

$$j \otimes 0 = 0 \otimes j' = 0, \quad (j_1 \vee j_2) \otimes k = j_1 \otimes k \vee j_2 \otimes k \quad \text{and} \quad j \otimes (k_1 \vee k_2) = j \otimes k_1 \vee j \otimes k_2.$$

We call generating elements $j \otimes j'$ *pure tensors*. If D, D' are bounded distributive lattices then so is $UD \otimes UD'$ [9], with meet given on pure tensors as $(d \otimes d') \wedge (e \otimes e') = (d \wedge e) \otimes (d' \wedge e')$. The lattice $UD \otimes UD'$ moreover is the coproduct of D, D' in **DL**: the coproduct injections are $\iota(d) = d \otimes 1'$ and $\iota'(d') = 1 \otimes d'$ for $d \in D, d' \in D'$, and the copairing of lattice homomorphisms $f \colon D \to E, f' \colon D' \to E$ is given by the extension of the join-bilinar map

$$\wedge \cdot (f \times f') \colon D \times D' \to E, \qquad (d, d') \mapsto f(d) \wedge f(d').$$

Taking coproducts yields a monoidal structure on **DL** and since $U(D + D') = UD \otimes UD'$ the functor U is strong monoidal. The dual monoidal structure on **Priest** takes binary products, and the natural transformation δ of Remark 3.3 is the expected product of sets

$$\delta \colon \mathbb{V}_\downarrow X \times \mathbb{V}_\downarrow Y \to \mathbb{V}_\downarrow (X \times Y), \qquad (C, D) \mapsto C \times D.$$

Spelling out Definition 3.4, the category $\mathrm{Op}_{J_{\mathbb{V}_\downarrow}}^{n,m}(\mathbf{Priest})$ is given as follows:

Definition 4.2. A *$((n, m)$-ary) relational Priestley space* consists of a carrier Priestley space X and a Priestley relation $\rho \colon X^n \to \mathbb{V}_\downarrow X^m$. A *relational morphism* from a relational Priestley space (X, ρ) to (X', ρ') is given by a Priestley relation $\beta \colon X \to \mathbb{V}_\downarrow Y$ such that, for all $\mathbf{x} \in X^n, \mathbf{y} \in X^m, \mathbf{y}' \in X'^m$,

$$\mathbf{x} \, \rho \, \mathbf{y} \wedge (\forall i \colon y_i \, \beta \, y_i') \Rightarrow \exists \mathbf{x}' \colon (\forall i \colon x_i \, \beta \, x_i') \wedge \mathbf{x}' \, \rho' \, \mathbf{y}',$$

and, for all $\mathbf{x} \in X^n, \mathbf{x}' \in X'^n, \mathbf{y}' \in X'^m$,

$$(\forall i \colon x_i \, \beta \, x_i') \wedge \mathbf{x}' \, \rho' \, \mathbf{y}' \Rightarrow \exists \mathbf{y} \colon \mathbf{x} \, \rho \, \mathbf{y} \wedge (\forall i \colon y_i \, \beta \, y_i').$$

We let $\mathrm{Op}_{J_{\mathbb{V}_\downarrow}}^{n,m}(\mathbf{Priest})$ denote the category of (n, m)-ary relational Priestley operators and relational morphisms.

Then Theorem 3.5 instantiates to the following result:

Theorem 4.3 (Extended Priestley duality). *The category of (m, n)-ary U-operators of distributive lattices is dually equivalent to the category of (n, m)-ary relational Priestley spaces and relational morphisms:*

$$\mathrm{Op}_U^{m,n}(\mathbf{DL}) \simeq^{\mathrm{op}} \mathrm{Op}_{J_{\mathbb{V}_\downarrow}}^{n,m}(\mathbf{Priest}).$$

By taking $n = 1$ and restricting the operator morphisms to be pure, we recover Goldblatt's duality [16]. Here, pure relational morphisms are called *bounded morphisms* and n-ary U-algebras $(UD)^{\otimes n} \to UD$ in **JSL** are called n-ary *join-hemimorphisms*.

Corollary 4.4 (Goldblatt, 1989). *The category of distributive lattices with n-ary join-hemimorphisms, and pure morphisms between them, is dually equivalent to the category of $(1, n)$-relational Priestley spaces and bounded morphisms.*

4.1 Deriving Concrete Formulas

We proceed to show how an enriched extension of our adjoint framework can be used to methodically derive concrete (i.e. element-based) formulas for the dual join operator of a continuous relation and vice versa. Let us first observe that all involved categories are *order-enriched*, i.e. the homsets are (pointwise) partially ordered; for **JSL** and **DL** this is clear and relations $X \to \mathbb{V}_{\downarrow}Y$ are ordered by inclusion, as usual. Moreover, from the definitions it is clear that the transposing isomorphisms of the adjunction $F \dashv U$ and the duality $\mathbf{DL} \simeq^{\mathrm{op}} \mathbf{Priest}$ are order-isomorphisms.

Second, in **Priest** we can represent an element $\hat{x}$ of a space $\hat{X}$ as a continuous function $1 \to X$ that we also denote by $\hat{x}$; on the lattice side, elements of a join-semilattice J correspond bijectively to **JSL**-morphisms $2 \to J$.

For the rest of the section we fix a U-algebra $h \colon (UX)^{\otimes n} \to UX$ with dual Priestley relation $\rho \colon \hat{X} \to \mathbb{V}_{\downarrow}\hat{X}^n$. We first show how to express ρ in terms of h. Two elements $\hat{x} \in \hat{X}, \hat{\mathbf{x}} \in \hat{X}^n$ are related by ρ (i.e. $\hat{x} \, \rho \, \hat{\mathbf{x}}$) iff the inequality $e(\hat{\mathbf{x}}) = \mathord{\downarrow}\hat{\mathbf{x}} \leq \rho(\hat{x})$ holds, equivalently, iff the left diagram below commutes laxly:

$$
\begin{array}{ccc}
\hat{X} & \xrightarrow{\ \rho\ } & \mathbb{V}_{\downarrow}\hat{X}^n \\
{\scriptstyle \hat{x}}\big\uparrow & {\scriptstyle \diagdown} & \big\uparrow{\scriptstyle e} \\
1 & \xrightarrow{\ \hat{\mathbf{x}}\ } & \hat{X}^n \xleftarrow{\ \Pi_i \hat{x}_i\ } 1^n
\end{array}
\quad
\begin{array}{ccc}
UX & \xleftarrow{\ h\ } & (UX)^{\otimes n} \\
{\scriptstyle Ux}\big\downarrow & {\scriptstyle \diagdown} & \big\downarrow{\scriptstyle \otimes_i Ux_i} \\
U2 & \xleftarrow{\ \nabla\ } & (U2)^{\otimes n}
\end{array}
$$

The duals of $\hat{x}, \hat{x}_i$ are **DL** morphisms $x, x_i \colon X \to 2$. Under duality and transposition the left diagram corresponds to the right diagram where ∇ is the codiagonal given by n-fold conjunction, i.e. it sends $\bigotimes_{i=1}^n x_i$ to $\bigwedge_{i=1}^n x_i$. Writing $F_z = z^{-1}(1)$ for the prime filter corresponding to a morphism $z \in \mathbf{DL}(X, 2)$ the right diagram yields Goldblatt's formula [16, p. 186] for the dual Priestley relation of an algebra h: we have $\hat{x} \, \rho \, \hat{\mathbf{x}}$ iff $h[\prod_i F_{x_i}] \subseteq F_x$.

To express h in terms of ρ, it suffices to describe $h(\mathbf{x})$ for a pure tensor $\mathbf{x} \in (UX)^{\otimes n}$ by the universal property of the tensor product. We factor $\mathbf{x} = \bigotimes_i x_i \cdot \nabla^{-1} \colon U2 \cong (U2)^{\otimes n} \to (UX)^{\otimes n}$ to see that the element $h(\mathbf{x})$ corresponds to the following morphism representing an element of the join-semilattice UX:

$$
h \cdot \bigotimes_i x_i \cdot \nabla^{-1} \colon U2 \cong (U2)^{\otimes n} \to (UX)^{\otimes n} \to UX.
$$

Its dual is the characteristic function

$$
\hat{X} \xrightarrow{\rho} \mathbb{V}_{\downarrow}\hat{X}^n \xrightarrow{\mathbb{V}_{\downarrow}(\Pi_i C_i)} \mathbb{V}_{\downarrow}(\mathbb{V}_{\downarrow}1)^n \xrightarrow{\mathbb{V}_{\downarrow}\delta} \mathbb{V}_{\downarrow}\mathbb{V}_{\downarrow}1^n \xrightarrow{\mathsf{U}} \mathbb{V}_{\downarrow}1^n \xrightarrow{\mathbb{V}_{\downarrow}\Delta^{-1}} \mathbb{V}_{\downarrow}1 = 2,
$$

where $C_i = \widehat{x_i^+}$ is the clopen upset of $\hat{X}$ dual to

$$
x_i \in \mathbf{JSL}(U2, UX) \cong \mathbf{DL}(FU2, X) \cong \mathbf{Priest}(\hat{X}, \mathbb{V}_{\downarrow}1) \cong \mathbf{Priest}(\hat{X}, 2).
$$

This shows that $h(\mathbf{x}) \in X \cong \mathrm{Cl}_{\uparrow}\hat{X}$ corresponds to the clopen upset

$$
h(\mathbf{x}) = \{a \in \hat{X} \mid \exists (b_1, \ldots, b_n) \in \rho(a) \colon \forall i \colon b_i \in C_i = \widehat{x_i^+}\} \in \mathrm{Cl}_{\uparrow}(\hat{X}),
$$

which is Goldblatt's formula [16, p. 184] for the dual algebra of a relation ρ.

4.2 Partial Functions and Total Relations

As a further application of the adjoint framework we characterize those operators whose dual Priestley relation is a partial function or a total relation, respectively. We achieve this by considering two splittings of the adjunction $F\colon \mathbf{JSL} \dashv \mathbf{DL} :U$ (Proposition 3.6 and Remark 3.7). The tensor on all categories considered is the tensor product of their underlying join-semilattices.

First, split the adjunction into $Q\colon \mathbf{DL_0} \dashv \mathbf{DL} :P$ and $Q'\colon \mathbf{JSL} \dashv \mathbf{DL_0} :P'$, where $\mathbf{DL_0}$ is the category of distributive lattices that are only bounded from below, and P, P' are forgetful functors. The left adjoint Q adds a fresh top element to a lattice in $\mathbf{DL_0}$. The dual submonad $\hat{Q}\hat{P} \hookrightarrow \mathbb{V}_{\downarrow}$ on $\mathbf{Priest}$ is given by

$$\hat{Q}\hat{P}\hat{D} \cong \widehat{QPD} \cong \mathbf{DL}(QPD, 2) \cong \mathbf{DL_0}(PD, P2).$$

Every $f \in \mathbf{DL_0}(PD, P2)$ either satisfies $f(1) = 1$, in which case $f \in \hat{D}$ is prime, or $f(1) = 0$ but then f is the constant zero map $0!\colon PD \to P2$; note that $0!$ is clearly the bottom element in the pointwise ordering of $\mathbf{DL_0}(PD, P2)$, so the monad $\hat{Q}\hat{P}$ just freely adds a bottom element. In particular, the dual category of $\mathbf{DL_0}$ is readily seen to be equivalent to $\mathbf{Priest_0}$, the category of Priestley spaces with a bottom element, and bottom-preserving continuous monotone maps. A continuous relation $\rho\colon X \to \hat{Q}\hat{P}\hat{X}$ is thus simply a *partial continuous function*.

Another splitting of the adjunction $F \dashv U$ is given by $L\colon \mathbf{JSL_1} \dashv \mathbf{DL} :R$ and $L'\colon \mathbf{JSL} \dashv \mathbf{JSL_1} :R'$, where $\mathbf{JSL_1}$ is the category of join-semilattices with both a bottom and top element (which are preserved by homomorphisms). The right adjoints R, R' are forgetful functors. The left adjoint L maps $J \in \mathbf{JSL_1}$ to the distributive lattice $\mathcal{U}^{\circ}_{\mathrm{fg}+}$ of *non-empty* finitely generated upsets of J, ordered by reverse inclusion. The submonad $\hat{L}\hat{R} \hookrightarrow \mathbb{V}_{\downarrow}$ thus maps a Priestley space $\hat{D}$ to

$$\hat{L}\hat{R}\hat{D} \cong \mathbf{DL}(LRD, 2) \cong \mathbf{JSL_1}(RD, R2) \cong \mathbb{V}^{+}_{\downarrow}\hat{D},$$

where $\mathbb{V}^{+}_{\downarrow}$ is the submonad of $\mathbb{V}_{\downarrow}$ taking *non-empty* closed downsets. Morphisms of type $X \to \mathbb{V}^{+}_{\downarrow}Y$ therefore are *total* Priestley relations. Proposition 3.6 thus yields the following result (the unary case is folklore, see e.g. [18, Lemma 4.6]):

Corollary 4.5. *The dual Priestley relation of a U-operator (operator morphism, respectively) is a partial function iff the operator (operator morphism, respectively) preserves non-empty meets, and total iff it preserves $\top$.*

5 Residuation Algebras

The abstract extended duality will now guide us in deriving a categorical duality between *profinite ordered monoids* and a full subcategory of residuation algebras which we call *derivation algebras*. This result is a non-trivial restriction of Gehrke's duality [13,14] between Priestley-topological algebras and residuation algebras. Our result is obtained by combining two ingredients: our framework for extended Stone duality from the previous sections and an isomorphism between residuation algebras and certain lattice coalgebras. The latter is first established for finite algebras via an operator on complete lattices we call *tensor implication*; extending it to locally finite algebras (Definition 5.22) then yields the desired duality with the category of profinite ordered monoids. To this end we introduce the notion of residuation morphism (Definition 5.8). The abstract extended duality then allows us to extend our results to *relational* morphisms of profinite ordered monoids and residuation algebras.

5.1   The Tensor Product of Distributive Lattices Revisited

Notation 5.1. By a *lattice* we always mean a bounded and distributive lattice, i.e. an object of **DL**. We often write de for $d \wedge e$. The dual lattice of D is denoted by D°. The category of meet-semilattices (with a top element) is denoted **MSL**. Analogous to **JSL** it has a tensor product $M \boxtimes M'$ and is dual to the category of Stone meet-semilattices [19]. From now on we denote the forgetful functors from **DL** to **JSL** and **MSL** by $U_\vee$ and $U_\wedge$, respectively. Sometimes we omit the forgetful functors $U_\wedge$ and $U_\vee$ for notational brevity and just write the respective tensor products of the underlying semilattices as $D \otimes D'$ and $D \boxtimes D'$.

Remark 5.2. The monad induced by the dual of $F_\wedge \dashv U_\wedge$ sends a Priestley space X to its hyperspace $\mathbb{V}_\uparrow X$ of closed *upsets* [1]. The comonads of the adjunctions $F_\wedge \dashv U_\wedge$ and $F_\vee \dashv U_\vee$ are not isomorphic but *conjugate*: $F_\wedge U_\wedge \cong (F_\vee U_\vee(-)^\circ)^\circ$. Their restrictions to the category of boolean algebras are isomorphic since their dual monads satisfy $\mathbb{V}_\downarrow = \mathbb{V} = \mathbb{V}_\uparrow$ – trivially so, as the order on their dual Priestley space is discrete. On the category of finite Priestley spaces, which are simply posets, $\mathbb{V}_\downarrow$ restricts to the downset monad, which further restricts to the finite powerset monad on the category of finite sets (i.e., discrete finite posets).

Remark 5.3 (Adjunctions on Lattices). By the adjoint functor theorem [25, Thm. V.6.1] a monotone function $f \colon D \to D'$ between complete lattices preserves all joins iff it has a right adjoint $f_* \colon D' \to D$, which is then given by $f_*(d') = \bigvee_{f(d) \leq d'} d$; dually, it preserves all meets iff it has a left adjoint $f^* \colon D' \to D$, given by $f^*(d') = \bigwedge_{d' \leq f(d)} d$. Finite lattices are complete, so every lattice homomorphism f between finite lattices has a left and a right adjoint. The join-irreducibles $\mathcal{J}D$ of a finite lattice D are precisely those elements $p \in D$ whose characteristic function $\chi_p \colon D \to 2$ (mapping $x \in D$ to 1 iff $p \leq x$) is a lattice morphism. The left adjoint of χ_p, also denoted $p \colon 2 \to D$, maps $1 \mapsto p$.

Lemma 5.4. (1) *The join- and meet-semilattice tensor products of distributive lattices D, E are isomorphic, that is, there is an isomorphism $\omega \colon D \otimes E \cong D \boxtimes E$.*
(2) *Adjunctions on lattices "compose horizontally": Given adjunctions $f \colon D \dashv E \colon g$ and $f' \colon D' \dashv E' \colon g'$ on lattices, the following composites are adjoints:*

$$
\begin{array}{cccc}
E \boxtimes E' \xrightarrow{\overset{g \boxtimes g'}{\underleftarrow{\ \ }}} D \boxtimes D' &
E \boxtimes E' \xrightarrow{g \boxtimes g'} D \boxtimes D' &
E \boxtimes E' \xrightarrow{g \boxtimes g'} D \boxtimes D' &
E \boxtimes E' \xrightarrow{g \boxtimes g'} D \boxtimes D' \\
\omega \uparrow \quad\quad \downarrow \omega^{-1} &
\omega \uparrow \quad\quad \downarrow \omega^{-1} &
\omega \uparrow \quad\quad \downarrow \omega^{-1} &
\omega \uparrow \quad\quad \downarrow \omega^{-1} \\
E \otimes E' \xleftarrow{f \otimes f'} D \otimes D' &
E \otimes E' \xleftarrow{f \otimes f'} D \otimes D' &
E \otimes E' \xleftarrow{f \otimes f'} D \otimes D' &
E \otimes E' \xrightarrow[f \otimes f']{\overrightarrow{\ \ }} D \otimes D'
\end{array}
$$

Construction 5.5. For every finite lattice D the map $x \otimes (-) \colon D \to D \otimes D$ preserves all joins, so it admits a right adjoint $x \multimap (-) \colon U_\wedge(D \otimes D) \to U_\wedge D$ which we call *tensor implication*. By Remark 5.3, it is given by $x \multimap T = \bigvee_{x \otimes y \leq T} y$. Analogously, we let $(-) \circ\!\!- x$ denote the right adjoint of $(-) \otimes x$.

Definition 5.6. A *(boolean) residuation algebra* consists of a (boolean) lattice $R \in \mathbf{DL}$ equipped with $\mathbf{MSL}$-morphisms $\backslash \colon R^\circ \boxtimes R \to R$ and $/ \colon R \boxtimes R^\circ \to R$, the *left* and *right residual*, satisfying the *residuation property*: $b \leq a \backslash c \iff a \leq c / b$. We call R *associative* if it satisfies $x \backslash (z / y) = (x \backslash z) / y$ for all $x, y, z \in R$. A join-irreducible element $e \in \mathcal{J}R$ is a *unit* if it satisfies $e \backslash z = z = z / e$.

Residuals may be thought of as algebraic generalizations of language derivatives, but as the following examples indicate they are not limited to this interpretation.

Examples 5.7. (1) Every distributive Heyting algebra is an associative residuation algebra with residuals $a \backslash c = a \to c$ and $c / b = b \to c$.

(2) Every boolean algebra B is a non-associative residuation algebra with $x \backslash 1 = 1$ and $x \backslash z = \neg x$ for $z \neq 1$. If $|B| > 1$ it does not have a unit.

(3) The dual boolean algebra $\hat{X}$ of a continuous algebra $\cdot \colon X \times X \to X$ on a Stone space X forms a residuation algebra: given clopens $A, C \subseteq X$, put

$$A \backslash C = \{x \in X \mid \forall(a \in A) \colon a \cdot x \in C\},$$
$$C / A = \{x \in X \mid \forall(b \in B) \colon x \cdot b \in C\}.$$

(4) The regular languages $\mathrm{Reg}\,\Sigma$ over a finite alphabet Σ form an associative boolean residuation algebra with residuals given by (extended) left and right *derivatives*: $K \backslash L = \{v \in \Sigma^* \mid Kv \subseteq L\}$ and $L / K = \{v \in \Sigma^* \mid vK \subseteq L\}$. The singleton empty word $\{\varepsilon\}$ is a unit. This example is a special case of item (3) obtained by taking as Stone algebra the *free profinite monoid* $\widehat{\Sigma^*}$.

We now introduce the notion of a *residuation morphism* between residuation algebras and also its *relational* generalization.

Definition 5.8. (1) A lattice morphism $f \colon R \to S$ between unital residuation algebras is a *(pure) residuation morphism* if it satisfies the conditions

$$f(x \backslash z) \leq f(x) \backslash f(z) \qquad\qquad\qquad\qquad\qquad\qquad \text{(Forth)}$$
$$\forall(y, z) \in S \times R \colon \exists(x_{y,z} \in R) \colon y \leq f(x_{y,z}) \quad\wedge\quad y \backslash f(z) = f(x_{y,z} \backslash z) \quad \text{(Back)}$$
$$\forall x \colon e \leq x \Leftrightarrow e' \leq f(x) \qquad\qquad\qquad\qquad\qquad\qquad\quad \text{(Unit)}$$

The morphism f is *open* if, additionally, it has a left adjoint. The category of unital residuation algebras with residuation morphisms is denoted $\mathbf{Res}$.

(2) A *(lax) relational residuation morphism* from a unital residuation algebra R to a unital residuation algebra S is a morphism $\rho \in \mathbf{JSL}_1(R, S)$ satisfying

$$\rho(x \backslash z) \leq \rho(x) \backslash \rho(z) \quad \text{and} \quad e' \leq \rho(e).$$

Unital residuation algebras with relational residuation morphisms form a category **RelRes**.

We use the convention that for a subcategory **C** of **Res** or **RelRes** we denote the full subcategory of **C** with boolean carriers by **BC**.

Remark 5.9. Let us provide some intuition behind the choices made in Definition 5.8. Recall that a *relational monoid morphism* from a finite monoid M to N is a total relation $\rho \colon M \to \mathcal{P}N$ such that $\rho(x)\rho(y) \subseteq \rho(xy)$ and $1_N \in \rho(1_M)$.

(1) The notion of residuation morphism is derived from a result by Gehrke [14, Theorem 3.19], where it is shown to capture precisely the conditions satisfied by the duals of morphisms of binary Stone algebras.

(2) We speak about *relational* morphisms of residuation algebras since for finite algebras these will dualize precisely to *relational morphisms* of finite monoids, which model inverses of surjective monoids homomorphisms [29, p. 38]: on finite monoids the inverse relation $e^{-1} \colon N \to \mathcal{P}M$ of a surjective homomorphism $e \colon M \twoheadrightarrow N$ is the *right adjoint* $e \dashv e^{-1}$ in the order-enriched category **Rel** with sets as objects and relations as morphisms, i.e. as relations they satisfy $\mathrm{id} \leq e^{-1} \cdot e$ and $e \cdot e^{-1} \leq 1$. Under duality the composition is reversed, so e^{-1} dualizes to a *left adjoint* $\widehat{e^{-1}} \dashv \hat{e}$. As left adjoints between finite lattices are precisely the join-preserving functions this suggests the choice that relational morphisms of residuation algebras preserve finite joins (and not necessarily meets). Surjectivity of e is equivalent to totality of e^{-1}, which by Corollary 4.5 is equivalent to $\widehat{e^{-1}}$ preserving the top element.

(3) This is also the reasoning behind the naming for *open* residuation morphisms: if $e \colon M \twoheadrightarrow N$ is a continuous surjection between *profinite* monoids (that is, topological monoids in **Stone**), then $e^{-1} \colon N \to \mathbb{V}M$ is continuous precisely iff e is an open map.

For open residuation morphisms the conditions (Back) and (Forth) can be combined into a much simpler condition. Over finite residuation algebras this is particularly convenient since every residuation morphism is open.

Lemma 5.10. *Let R, S be residuation algebras. A lattice morphism $f \colon R \to S$ is an open residuation morphism iff $f^*(e') = e$ and it satisfies the condition*

$$y \setminus f(z) = f(f^*(y) \setminus z). \qquad \text{(Open)}$$

Example 5.11. Let Σ, Δ be finite alphabets. Every substitution $f_0 \colon \Sigma \to \Delta^*$ can be extended to a monoid homomorphism $f \colon \Sigma^* \to \Delta^*$, and for regular languages $L \in \mathrm{Reg}\,\Sigma$ and $K \in \mathrm{Reg}\,\Delta$ both $f[L]$ and $f^{-1}[K]$ are also regular. Then $f^{-1} \colon \mathrm{Reg}\,\Delta \to \mathrm{Reg}\,\Sigma$ is an open residuation morphism. Indeed, its left adjoint is $f[-]$, and we have $f[\{\varepsilon\}] = \{f(\varepsilon)\} = \{\varepsilon\}$ and

$$K \setminus f^{-1}[L] = \{w \mid Kw \subseteq f^{-1}[L]\} = \{w \mid f[K]f(w) \subseteq L\} = f^{-1}(f[K] \setminus L).$$

5.2   Finite Residuation Algebras

We will start by characterizing *finite* residuation algebras, and then generalize the results to *locally finite* residuation algebras This approach allows us to first introduce the key concepts and constructions of the duality on a finite level, and then extend them to more general structures by forming appropriate free completions. All results of this section apply more generally to structures with a complete and completely distributive lattice as carrier.

Construction 5.12. In a finite residuation algebra R the partially applied residuals $(x \setminus -), (- / y)$ have respective left adjoints $\mu(x, -) \dashv (x \setminus -)$ and $\mu(-, y) \dashv (- / y)$ that can be combined, by the universal property of $\otimes$, into a $U_\vee$-algebra $\mu \colon U_\vee R \otimes U_\vee R \to U_\vee R$ called *multiplication*. Every algebra $U_\vee D \otimes U_\vee D \to U_\vee D$ on a finite lattice D has a right adjoint $\gamma \colon U_\wedge D \to U_\wedge(D \otimes D)$ that can, by using the isomorphism ω from Lemma 5.4, be extended to a $U_\wedge$-coalgebra

$$\hat{\gamma} = U_\wedge \omega \cdot \gamma \colon U_\wedge D \to U_\wedge(D \otimes D) \cong U_\wedge(D \boxtimes D) = U_\wedge D \boxtimes U_\wedge D.$$

Since γ and $\hat{\gamma}$ are essentially the same function (differing only by the isomorphism ω) we refer to both as *comultiplication* or *coalgebra structure*. Conversely, we obtain a $U_\vee$-algebra from a comultiplication $\gamma \colon U_\wedge D \to U_\wedge(D \otimes D)$ by taking its left adjoint. In summary, each of $/, \setminus, \mu, \gamma$ determine each other uniquely:

$$x \leq z / y \iff y \leq x \setminus z \iff \mu(x \otimes y) \leq z \iff x \otimes y \leq \gamma(z),$$

Lemma 5.13. *In a finite residuation algebra R the residuals can be expressed via comultiplication γ and tensor implication as $x \setminus z = x \multimap \gamma(z)$ and $z / y = \gamma(z) \circ\!\!- y$. Conversely, the comultiplication can be expressed via residuals as*

$$\gamma(z) = \bigvee_{x \in R} x \otimes (x \setminus z) = \bigvee_{p \in \mathcal{J}R} p \otimes (p \setminus z).$$

First we investigate when the comultiplication is a *pure*, i.e. lifts to a lattice morphism $R \to R + R$.

Lemma 5.14. *For a finite residuation algebra R, the following are equivalent:*
(1) *The comultiplication is pure, i.e., $\gamma(0) = 0$ and $\gamma(x \vee y) = \gamma(x) \vee \gamma(y)$.*
(2) *For all $p \in \mathcal{J}R$ we have $p \setminus 0 = 0 = 0 / p$, and the following equations hold:*

$$p \setminus (x \vee y) = p \setminus x \vee p \setminus y \qquad and \qquad (x \vee y) / p = x / p \vee y / p.$$

(3) *For all $x, y \in R$: $\mu(x \otimes y) = 0 \iff x = 0 \vee y = 0$, and $\mu[\mathcal{J}(R + R)] \subseteq \mathcal{J}R$.*

Next we inspect how structural identities like (co-)associativity or unitality translate to the other operations. Note that while the statements are to be expected, the proof is non-trivial due to the complication introduced by the seemingly innocent isomorphism $\omega \colon R \otimes R \cong R \boxtimes R$. Recall that a coalgebra $c \colon U_\wedge R \to U_\wedge R \boxtimes U_\wedge R$ is *coassociative* if $(c \boxtimes \mathrm{id}) \cdot c = (\mathrm{id} \boxtimes c) \cdot c$ and *counital* if it is equipped with a *counit* $\varepsilon \in \mathbf{DL}(R, 2)$ such that $(\varepsilon \boxtimes \mathrm{id}) \cdot c = \mathrm{id} = (\mathrm{id} \boxtimes \varepsilon) \cdot c$.

Lemma 5.15. *The following are equivalent for a finite residuation algebra R:*

(1) *The comultiplication on R is coassociative and has a counit.*

(2) *The residuals are associative and R has a unit.*

(3) *The multiplication μ is associative and has a unit, i.e. a join-irreducible $e \in \mathcal{J}R$ satisfying $\mu(e \otimes -) = \mathrm{id} = \mu(- \otimes e)$.*

These lemmas suggest the following definitions.

Definition 5.16. (1) A finite residuation algebra R is *pure* if it satisfies (one of) the equivalent conditions of Lemma 5.14.

(2) A finite residuation algebra R is a *finite derivation algebra* if it is pure, associative and has a unit. The respective full subcategories of $\mathbf{Res_f}$ and $\mathbf{RelRes_f}$ are denoted by $\mathbf{Der_f}$ and $\mathbf{RelDer_f}$.

(3) A (not necessarily finite) $U_\wedge$-coalgebra $\hat{\gamma}\colon U_\wedge C \to U_\wedge C \boxtimes U_\wedge C$ is a $U_\wedge$-*comonoid* if its coassociative and counital, and a *(lattice) comonoid* if $\hat{\gamma}$ is pure.

In order to extend the correspondence of (finite) residuation algebras and $U_\wedge$-coalgebras to a categorical equivalence we introduce appropriate morphisms.

Definition 5.17. (1) A *pure morphism* from a counital $U_\wedge$-coalgebra $(C, \hat{\gamma}, \epsilon)$ to $(C', \hat{\gamma}', \epsilon')$ is a lattice morphism $f\colon C \to D$ satisfying $(f \boxtimes f) \cdot \hat{\gamma} = \hat{\gamma}' \cdot f$ and $\epsilon = \epsilon' \cdot f$.

$$
\begin{array}{ccc}
U_\wedge C & \xrightarrow{\;U_\wedge f\;} & U_\wedge C' \\
\downarrow{\scriptstyle \hat{\gamma}} & & \downarrow{\scriptstyle \hat{\gamma}'} \\
U_\wedge C \boxtimes U_\wedge C & \xrightarrow{\;U_\wedge f \boxtimes U_\wedge f\;} & U_\wedge C' \boxtimes U_\wedge C'
\end{array}
\qquad\qquad
\begin{array}{ccc}
U_\wedge C & \xrightarrow{\;U_\wedge f\;} & U_\wedge C' \\
& {\scriptstyle U_\wedge \epsilon}\searrow & \downarrow{\scriptstyle U_\wedge \epsilon'} \\
& & U_\wedge 2
\end{array}
$$

The category of counital $U_\wedge$-coalgebras with pure morphisms is denoted by $\mathbf{Coalg}(U_\wedge)$ and its full subcatgegory of $U_\wedge$-comonoids by $\mathbf{Comon}(U_\wedge)$, again with the full subcategory $\mathbf{Comon}$ of comonoids.

(2) Let C and C' be comonoids. A *(lax) relational morphism* from C to C' is a morphism $\rho \in \mathbf{JSL}_1(C, C')$ satisfying $(\rho \otimes \rho) \cdot \gamma \le \gamma' \cdot \rho$ and $\epsilon \le \epsilon' \cdot \rho$, i.e. the following diagrams in $\mathbf{JSL}$ commute laxly:

$$
\begin{array}{ccc}
U_\vee C & \xrightarrow{\;\rho\;} & U_\vee C' \\
\downarrow{\scriptstyle U_\vee \gamma} & {\scriptstyle \le} & \downarrow{\scriptstyle U_\vee \gamma'} \\
U_\vee C \otimes U_\vee C & \xrightarrow{\;\rho \otimes \rho\;} & U_\vee C' \otimes U_\vee C'
\end{array}
\qquad\qquad
\begin{array}{ccc}
U_\vee C & \xrightarrow{\;\rho\;} & U_\vee C' \\
& {\scriptstyle U_\vee \epsilon}\searrow\ {\scriptstyle \le} & \downarrow{\scriptstyle U_\vee \epsilon'} \\
& & U_\vee 2
\end{array}
$$

Comonoids with relational morphisms form a category $\mathbf{RelComon}$.

Theorem 5.18. *The following categories are isomorphic:*

$$\mathbf{Coalg_f}(U_\wedge) \cong \mathbf{Res_f}, \qquad \mathbf{Comon_f} \cong \mathbf{Der_f} \quad and \quad \mathbf{RelComon_f} \cong \mathbf{RelDer_f}.$$

Proof (Sketch). On objects the isomorphism swaps between residuals and comultiplication; the residual unit is left adjoint of the counit. The first isomorphism restricts to the second by Lemmas 5.14 and 5.15. On morphisms one proves that a lattice morphism $f: C \to C'$ is a pure coalgebra morphism iff it is an (open) residuation morphism, and if C and C' are comonoids, then $\rho \in \mathbf{JSL}_1(C, C')$ is a relational comonoid morphism iff it is a relational residuation morphism.

From Theorem 5.18 we obtain the following dual characterization of finite ordered monoids; it restricts to the order-discrete setting of ordinary finite monoids and finite boolean derivation algebras.

Theorem 5.19. (1) *The category of finite ordered monoids is dually equivalent to the category of finite derivation algebras (or finite lattice comonoids):*

$$\mathbf{OrdMon}_f \simeq^{\mathrm{op}} \mathbf{Comon}_f \cong \mathbf{Der}_f.$$

(2) *The category of finite ordered monoids with relational morphisms is dually equivalent to the category of finite derivation algebras (or finite lattice comonoids) with relational morphisms.*

$$\mathbf{RelOrdMon}_f \simeq^{\mathrm{op}} \mathbf{RelComon}_f \cong \mathbf{RelDer}_f.$$

Proof. The first statement is a trivial extension of Theorem 5.18 by (finite) Priestley duality since finite ordered monoids dualize to finite lattice comonoids. For item (2) note that a relational ordered monoid morphism $(M, \cdot_M, 1_M) \to (N, \cdot_N, 1_N)$ is a total relation $\rho: M \to \mathcal{D}N$ (where $\mathcal{D}$ is the downset monad) making the following diagrams commute laxly:

$$
\begin{array}{ccc}
M \times M \xrightarrow{\quad\quad \cdot_M \quad\quad} M & \qquad & 1 \xrightarrow{\ 1_M\ } M \\
{\scriptstyle \rho \times \rho}\big\downarrow \quad {\scriptstyle \nearrow} \quad \big\downarrow {\scriptstyle \rho} & & {\scriptstyle 1_N}\big\downarrow \ {\scriptstyle \nearrow} \ \big\downarrow {\scriptstyle \rho} \\
\mathcal{D}N \times \mathcal{D}N \xrightarrow{\ \delta\ } \mathcal{D}(N \times N) \xrightarrow{\mathcal{D}(\cdot_N)} \mathcal{D}N & & N \xrightarrow{\ \eta\ } \mathcal{D}N
\end{array}
$$

If we view N as a finite Priestley space, then $\mathcal{D}N = \mathbb{V}_\downarrow N$, so the dual of ρ under (order-enriched) extended duality is a relational morphism $\hat{\rho}^- \in \mathbf{JSL}_1(\hat{N}, \hat{M})$ of finite lattice comonoids, or equivalently, a relational residuation morphism.

5.3   Locally Finite Residuation Algebras

The main complication in the generalization from finite to infinite structures comes from the reliance on adjoints, as these may not exist anymore on infinite lattices. The prime example of a residuation algebra in automata theory suggests a *local* translation between residuals and comultiplication:

Example 5.20. It is well-known that the boolean algebra $\mathrm{Reg}\,\Sigma$ of regular languages dualizes under Stone duality to the *free profinite monoid* $\widehat{\Sigma^*}$ (see Pippenger [27]). The multiplication $\mu: \widehat{\Sigma^*} \times \widehat{\Sigma^*} \to \widehat{\Sigma^*}$ of profinite words dualizes under Stone duality to a comultiplication $\mu^{-1}\colon \operatorname{Reg}\varSigma \to \operatorname{Reg}\varSigma + \operatorname{Reg}\varSigma$ on regular languages defined on $L \in \operatorname{Reg}\varSigma$ by

$$\mu^{-1}(L) = \bigvee\nolimits_{[v]\in \operatorname{Syn}_L} [v] \otimes [v] \setminus L. \tag{5.1}$$

Here Syn_L is the *syntactic monoid* of L, whose elements are the equivalence classes of the equivalence relation on $\varSigma^*$ defined by $v \equiv_L w$ iff v, w belong to the same residuals $K \setminus L / M$. Gehrke [13, Thm. 15] has shown that, under Stone duality, Syn_L dualizes to the *residuation ideal* generated by $L \in \operatorname{Reg}\varSigma$.

Definition 5.21. A *residuation ideal* of a residuation algebra R is a sublattice $I \hookrightarrow R$ such that for all $z \in I$ and $x \in R$ one has $x \setminus z, z / x \in R$. We denote the residuation ideal generated by a subset $X \subseteq R$ by $\setminus X /$.

Residuation ideals were used by Gehrke [14] to characterize quotients of Priestley topological algebras. Note that in the formula (5.1) for the comultiplication on regular languages it is crucial that the residuation ideal $\setminus \{L\} /$ generated by a single regular language L is *finite*, as otherwise the join might not exist. This leads to the following restriction.

Definition 5.22. (1) A residuation algebra R is *locally finite* if every finite subset of R is contained in a finite residuation ideal of R.

(2) A $U_\wedge$-coalgebra C is *locally finite* if every finite subset of C is contained in a finite subcoalgebra of C. The category of locally finite comonoids is denoted **Comon**$_{\mathsf{lf}}$.

Note that not every residuation algebra is locally finite, consider for example an infinite boolean algebra in Example 5.7(2).

Proposition 5.23. (1) *Every locally finite residuation algebra R yields a locally finite $U_\wedge$-coalgebra $\gamma\colon U_\wedge R \to U_\wedge(R \otimes R)$ with comultiplication given by*

$$\gamma(z) = (\iota_A \otimes \iota_A)(\gamma_A(z)) = \bigvee\nolimits_{x\in A} \iota_A(x) \otimes \iota_A(x \setminus z) = \bigvee\nolimits_{p\in \mathcal{J}A} \iota_A(p) \otimes \iota_A(p \setminus z)$$

for any finite residuation ideal $\iota_A\colon A \hookrightarrow R$ containing z (here γ_A is the comultiplication on A as in Construction 5.12).

(2) *Every locally finite $U_\wedge$-coalgebra (C, γ) yields any locally finite residuation algebra with the left residual given by $x \setminus_\gamma z = \iota_A(x \setminus_A z) = \iota_A(x \multimap \gamma(z))$ for any finite subcoalgebra $\iota_A\colon A \hookrightarrow C$ containing x, z (here $\setminus_A$ is the residual on A as given by Construction 5.12). The residual has a canonical presentation as $x \setminus_\gamma z = \iota_z(\iota_z^*(x) \setminus z)$, where $\iota_z\colon \langle z \rangle \to C$ is the smallest (finite) subcoalgebra containing z. The right residual is defined analogously.*

(3) *These translations are mutually inverse.*

Proposition 5.23 shows that every locally finite residuation algebra carries a unique $U_\wedge$-coalgebra structure and vice versa. We may thus translate at will between the residuals and comultiplication as in the finite case and omit the subscripts. We extend Lemmas 5.14 and 5.15 to locally finite structures:

Lemma 5.24. *Let R be a locally finite residuation algebra.*

(1) *Finite residuation ideals correspond to finite subcoalgebras.*

(2) *The residuals are associative iff the comultiplication is coassociative.*

(3) *The residuals have a unit iff the comultiplication is counital.*

(4) *The comultiplication is pure iff every finite residuation ideal is pure (see Definition 5.16).*

Remark 5.25. Lemma 5.24(4) characterizes locally finite residuation algebras with a pure comultiplication. By extended duality, their dual Priestley relation is functional. We note that Gehrke [14, Proposition 3.15] presented a necessary and sufficient condition for a general residuation algebra R to have a functional dual relation, namely *join-preservation at primes*:

$$\forall F \in \mathbf{DL}(R,2)\colon \forall (a \in F), \forall (b,c \in R)\colon \exists a' \in F\colon a \setminus (b \vee c) \leq (a' \setminus b) \vee (a' \setminus c).$$

One can show that every locally finite residuation algebra satisfying Lemma 5.24(4) is join-preserving at primes.

Definition 5.26. A residuation algebra R is a *derivation algebra* if it is locally finite, associative, unital and every finite residuation ideal I is pure. The ensuing full subcategories of **Res** and **RelRes** are denoted **Der** and **RelDer**.

Theorem 5.27. (1) *The category of locally finite residuation algebras and residuation morphisms is isomorphic to the category of locally finite unital $U_\wedge$-coalgebras and pure coalgebra morphisms.*

(2) *The isomorphism restricts to the full subcategories of derivation algebras and locally finite comonoids.*

(3) *The categories of derivation algebras and relational residuation morphisms and locally finite comonoids with relational morphisms are isomorphic.*

Combining this characterization with our approach to extended Priestley duality we establish a duality between profinite ordered monoids and derivation algebras, and extend it to relational morphisms. Conceptually, this general duality is an extension of the finite duality $\mathbf{OrdMon_f} \simeq^{\mathrm{op}} \mathbf{Comon_f} \cong \mathbf{Der_f}$ by forming suitable completions: Profinite ordered monoids are the *Pro-completion* (the free completion under cofiltered limits) of the category of finite ordered monoids; dually a routine verification establishes that lattice comonoids (and therefore also derivation algebras by Theorem 5.27(2)) form *Ind-completions* (free completions under directed colimits) of their respective subcategories of finite objects.

Proposition 5.28. *The category of locally finite comonoids forms the Ind-completion of the category of finite comonoids:*

$$\mathbf{Comon_{lf}} \simeq \mathrm{Ind}(\mathbf{Comon_f}).$$

We define a *Priestley relational morphism* between profinite ordered monoids X, Y to be a Priestley relation $\rho\colon X \to \mathbb{V}_\downarrow Y$ such that $\rho(x)\rho(x') \subseteq \rho(xx')$ and $1_N \in \rho(1_M)$.

Theorem 5.29. (1) *The category of derivation algebras is dually equivalent to the category of profinite ordered monoids:*

$$\mathbf{Der} \cong \mathbf{Comon}_{lf} \simeq^{op} \mathbf{ProfOrdMon}.$$

(2) *The category of derivation algebras and relational residuation morphisms is dually equivalent to the category of profinite ordered monoids and Priestley relational morphisms:*

$$\mathbf{RelDer} \cong \mathbf{RelComon}_{lf} \simeq^{op} \mathbf{RelProfOrdMon}.$$

Remark 5.30. (1) Theorem 5.29 clearly restricts to profinite monoids with Stone relational morphisms and boolean derivation algebras. It is well-known that every Stone monoid is profinite (see e.g. [20]). So dually, every boolean comonoid is locally finite.

(2) All results of Section 5 hold analogously for the extension of the "discrete" duality between posets (or sets) and algebraic completely distributive lattices (or completely atomic boolean algebras) along the free-forgetful adjunction between completely distributive lattices and *complete join-semilattices*. This yields a duality between the category of all (ordered) monoids and (completely distributive lattices) completely atomic boolean residuation algebras with open residuation morphisms. Moreover, this duality also can be extended to relational morphisms.

6   Conclusion and Future Work

We have presented an abstract approach to extending Stone-type dualities based on adjunctions between monoidal categories and instantiated it to recover and generalize extended Priestley duality. Guided by these foundations we have investigated residuation and derivation algebras and proved a duality between the latter and (ordered) profinite monoids, Moreover, we have extended this duality to relational morphisms.

Relational morphisms are an important tool in algebraic language theory, notably for charaterizing language operations algebraically. For instance, aperiodic relational morphisms are tightly connected to the concatenation product and the star operation on regular languages. In future work we intend to apply the new duality-theoretic results on relational morphisms to illuminate such connections, much in the spirit of the duality-theoretic persepective of Eilenberg's Variety Theorem by Gehrke et. al. [15].

Another goal is to apply our abstract duality framework beyond classical Stone and Priestley dualities. Specifically, we aim to develop an extended duality theory for the recently developed *nominal* Stone duality [4], which would allow to generalize our present results on residuation algebras to the nominal setting and uncover new results about data languages.

A conceptually rather different dual characterization of the category of profinite monoids and continuous monoid morphisms in terms of semi-Galois categories has been provided by Uramoto [35]. Extending this result to relational morphisms, similar to our Theorem 5.29, is another interesting point for future work.

References

1. Bezhanishvili, G., Harding, J., Morandi, P.: Remarks on hyperspaces for priestley spaces. Theoretical Computer Science **943**, 187–202 (2023). https://doi.org/https://doi.org/10.1016/j.tcs.2022.12.001
2. Bezhanishvili, N., Fontaine, G., Venema, Y.: Vietoris bisimulations. J. Log. Comput. **20**(5), 1017–1040 (2010). https://doi.org/10.1093/logcom/exn091
3. Birkhoff, G.: Rings of sets. Duke Mathematical Journal **3**, 443–454 (1937)
4. Birkmann, F., Milius, S., Urbat, H.: Nominal topology for data languages. In: 50th International Colloquium on Automata, Languages, and Programming, ICALP 2023, July 10-14, 2023, Paderborn, Germany. LIPIcs, vol. 261, pp. 114:1–114:21. Schloss Dagstuhl - Leibniz-Zentrum für Informatik (2023). https://doi.org/10.4230/LIPIcs.ICALP.2023.114
5. Birkmann, F., Urbat, H., Milius, S.: Monoidal extended stone duality (2024), https://arxiv.org/abs/2401.08219
6. Bonsangue, M., Kurz, A., Rewitzky, I.: Coalgebraic representations of distributive lattices with operators. Topology and its Applications **154**(4), 778–791 (2007). https://doi.org/https://doi.org/10.1016/j.topol.2005.10.010
7. Cignoli, R., Lafalce, S., Petrovich, A.: Remarks on priestley duality for distributive lattices. Order **8**(3), 299–315 (1991). https://doi.org/10.1007/BF00383451
8. Eilenberg, S.: Automata, Languages, and Machines, vol. 2. Academic Press, New York (1976)
9. Fraser, G.A.: The semilattice tensor product of distributive lattices. Transactions of the American Mathematical Society **217**, 183–194 (1976). https://doi.org/https://doi.org/10.1007/BF02485362
10. Fussner, W., Palmigiano, A.: Residuation algebras with functional duals. Algebra universalis **80**(4), 40 (2019). https://doi.org/10.1007/s00012-019-0613-5
11. Gehrke, M., Priestley, H.: Canonical extensions of double quasioperator algebras: An algebraic perspective on duality for certain algebras with binary operations. Journal of Pure and Applied Algebra **209**(1), 269–290 (2007). https://doi.org/https://doi.org/10.1016/j.jpaa.2006.06.001
12. Gehrke, M.: Stone duality and the recognisable languages over an algebra. In: Algebra and Coalgebra in Computer Science. pp. 236–250. Springer Berlin Heidelberg (2009). https://doi.org/https://doi.org/10.1007/978-3-642-03741-2_17
13. Gehrke, M.: Duality in computer science. In: Proceedings of the 31st Annual ACM/IEEE Symposium on Logic in Computer Science, LICS '16, New York, NY, USA, July 5-8, 2016. pp. 12–26. ACM (2016). https://doi.org/10.1145/2933575.2934575
14. Gehrke, M.: Stone duality, topological algebra, and recognition. Journal of Pure and Applied Algebra **220**(7), 2711–2747 (2016). https://doi.org/https://doi.org/10.1016/j.jpaa.2015.12.007
15. Gehrke, M., Grigorieff, S., Pin, J.É.: Duality and equational theory of regular languages. In: Automata, Languages and Programming. pp. 246–257. Springer Berlin Heidelberg, Berlin, Heidelberg (2008). https://doi.org/https://doi.org/10.1007/978-3-540-70583-3_21
16. Goldblatt, R.: Varieties of complex algebras. Annals of Pure and Applied Logic **44**(3), 173–242 (1989). https://doi.org/10.1016/0168-0072(89)90032-8
17. Halmos, P.R.: Algebraic logic, i. monadic boolean algebras. Journal of Symbolic Logic **23**(2), 219–222 (1958). https://doi.org/10.2307/2964417

18. Hofmann, D., Nora, P.: Dualities for modal algebras from the point of view of triples. Algebra universalis **73**(3), 297–320 (2015). https://doi.org/10.1007/s00012-015-0324-5

19. Hofmann, K., Mislove, M., Stralka, A.: The Pontryagin Duality of Compact 0-dimensional Semilattices and Its Applications. Lecture notes in mathematics, Springer-Verlag (1974)

20. Johnstone, P.: Stone Spaces. Cambridge Studies in Advanced Mathematics, Cambridge University Press (1982)

21. Jónsson, B., Tarski, A.: Boolean algebras with operators. part i. American Journal of Mathematics **73**(4), 891–939 (1951)

22. Jónsson, B., Tarski, A.: Boolean algebras with operators. American Journal of Mathematics **74**(1), 127–162 (1952)

23. Jónsson, B., Tsinakis, C.: Relation algebras as residuated boolean algebras. Algebra Universalis **30**(4), 469–478 (1993). https://doi.org/10.1007/BF01195378

24. Kupke, C., Kurz, A., Venema, Y.: Stone coalgebras. Theoretical Computer Science **327**(1), 109–134 (2004). https://doi.org/https://doi.org/10.1016/j.tcs.2004.07.023

25. MacLane, S.: Categories for the Working Mathematician. Springer-Verlag (1971)

26. Pin, J.: Relational morphisms, transductions and operations on languages. In: Formal Properties of Finite Automata and Applications, LITP Spring School on Theoretical Computer Science, Ramatuelle, France, May 23-27, 1988, Proceedings. Lecture Notes in Computer Science, vol. 386, pp. 34–55. Springer (1988). https://doi.org/10.1007/BFb0013110

27. Pippenger, N.: Regular languages and Stone duality. Theory Comput. Syst. **30**(2), 121–134 (1997). https://doi.org/https://doi.org/10.1007/BF02679444

28. Priestley, H.A.: Representation of distributive lattices by means of ordered stone spaces. Bulletin of The London Mathematical Society **2**, 186–190 (1970). https://doi.org/https://doi.org/10.1112/blms/2.2.186

29. Rhodes, J., Steinberg, B.: The q-theory of Finite Semigroups. Springer Monographs in Mathematics, Springer US (2009). https://doi.org/https://doi.org/10.1007/b104443

30. Salamanca, J.: Unveiling eilenberg-type correspondences: Birkhoff's theorem for (finite) algebras + duality. CoRR (2017)

31. Sambin, G., Vaccaro, V.: Topology and duality in modal logic. Annals of Pure and Applied Logic **37**(3), 249–296 (1988). https://doi.org/10.1016/0168-0072(88)90021-8

32. Schwede, S., Shipley, B.: Equivalences of monoidal model categories. Algebraic and Geometric Topology **3** (10 2002). https://doi.org/10.2140/agt.2003.3.287

33. Seal, G.J.: Tensors, monads and actions. Theory and Applications of Categories **28**(15), 70–71 (1953)

34. Stone, M.H.: The theory of representation for boolean algebras. Transactions of the American Mathematical Society **40**(1), 37–111 (1936)

35. Uramoto, T.: Semi-galois categories i: The classical eilenberg variety theory. In: Proceedings of the 31st Annual ACM/IEEE Symposium on Logic in Computer Science. p. 545–554. LICS '16, Association for Computing Machinery (2016). https://doi.org/10.1145/2933575.2934528

36. Urbat, H., Adámek, J., Chen, L., Milius, S.: Eilenberg theorems for free. In: MFCS 2017. LIPIcs, vol. 83, pp. 43:1–43:15. Schloss Dagstuhl - Leibniz-Zentrum für Informatik (2017)

Towards a Compositional Framework for Convex Analysis (with Applications to Probability Theory)

Dario Stein[1(✉)] and Richard Samuelson[2]

[1] Radboud University Nijmegen, Nijmegen, The Netherlands
dario.stein@ru.nl
[2] University of Florida, Gainesville, USA
rsamuelson@ufl.edu

Abstract. We introduce a compositional framework for convex analysis based on the notion of *convex bifunction* of Rockafellar. This framework is well-suited to graphical reasoning, and exhibits rich dualities such as the Legendre-Fenchel transform, while generalizing formalisms like graphical linear algebra, convex relations and convex programming. We connect our framework to probability theory by interpreting the Laplace approximation in its context: The exactness of this approximation on normal distributions means that logdensity is a functor from Gaussian probability (densities and integration) to concave bifunctions and maximization.

Keywords: convex analysis · category theory · categorical probability

1 Introduction

Convex analysis is a classical area of mathematics with innumerous applications in engineering, economics, physics, statistics and information theory. The central notion is that of a convex function $f : \mathbb{R}^n \to \mathbb{R}$, satisfying the inequality $f(tx + (1-t)y) \leq tf(x) + (1-t)f(y)$ for all $t \in [0,1]$. Convexity is a useful property for optimization problems: Every local minimum of f is automatically a global minimum. Convex functions furthermore admit a beautiful duality theory; the ubiquitous Legendre-Fenchel transform (or convex conjugation) defined as

$$f^*(x^*) = \sup_x \left\{ \langle x^*, x \rangle - f(x) \right\}$$

encodes f in terms of all affine functions $\langle x^*, x \rangle - c$ majorized by f (here $\langle -, - \rangle$ denotes the standard inner product on $\mathbb{R}^n$). The function f^* is convex regardless of f, and under a closedness assumption we recover $f^{**} = f$.

While convex analysis is a rich field, its compositional structure is not readily apparent; the central notion, convex functions, is not closed under composition. The notion which *does* compose is less well known: a *convex bifunction*, due to [27], is a jointly convex function $F : \mathbb{R}^m \times \mathbb{R}^n \to \mathbb{R}$ of two variables. Such bifunctions compose via infimization

$$(F \circ G)(x, z) = \inf_y \left\{ F(y, z) + G(x, y) \right\}$$

© The Author(s) 2024
N. Kobayashi and J. Worrell (Eds.): FoSSaCS 2024, LNCS 14574, pp. 166–187, 2024.
https://doi.org/10.1007/978-3-031-57228-9_9

Categorical Methods In this work, we will study bifunctions and their associated dualities in the framework of category theory. Graphical methods are ubiquitous in engineering and statistics, and can used to derive efficient algorithms by making use of the factorized structure of a problem. The language of props and string diagrams unifies these methods, as a large body of work on graphical linear algebra and applied category theory shows [2, 1, 19, 7]. We extend these methods to problems of convex analysis and optimization. Our category of bifunctions subsumes an array of mathematical structures, such as linear maps and relations, convex relations, and (surprisingly) multivariate Gaussian probability.

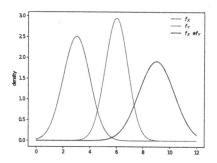 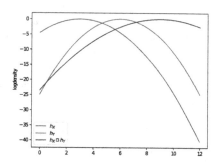

Fig. 1. Addition of independent normal variables X, Y. Left: pdf and convolution, right: logpdf and sup-convolution

Applications to Probability Theory Convex analysis offers a rich perspective on Gaussian (multivariate normal) probability distributions: The log-density $h(x) = \log f(x)$ of a Gaussian random variable is a concave function of the form[3]

$$h(x) = -\frac{(x-\mu)^2}{2\sigma^2}$$

It turns out that anything we can do with Gaussian densities and integration can instead be done with logdensities and maximization. For example, to compute the density of a sum of independent variables, we may take a convolution of densities, or instead compute a sup-convolution of logdensities (see Fig. 1), as

$$\log \int f_X(x) f_Y(z-x) \mathrm{d}x = \sup_x \{ h_X(x) + h_Y(z-x) \}$$

This is highly particular to Gaussians. We can elegantly formalize this statement in categorical terms, as our main theorem states: Logprobability defines a functor from Gaussian probability to concave bifunctions (Theorem 5)

In this sense, the essence of Gaussians is captured by concave quadratic functions. By extending our viewpoint to partial concave quadratic functions, we

---
[3] we intentionally disregard a scalar $+C$

obtain a generalized notion of *Gaussian relation* which includes improper priors. Such entities are subtle to describe measure-theoretically, but straightforward in the convex analytic view. The duality theory of bifunctions generalizes the duality of precision and covariance, and more generally connects to the notion of cumulant-generating function in probability theory.

We elegantly formalize the connections between convex analysis and probability theory using the language of Markov categories [17], which are a categorical formalism for probability theory, and have close connections to the semantics of probabilistic programs [30].

Contribution and Outline This paper is intended to serve as a high-level roadmap to a categorical treatment of convex analysis. Our aim is to spell out the underlying structures, and present a diverse range of connections, especially with diagrammatic methods and categorical probability. For the sake of presentation, we choose to stick to general statements and keep some technical notions (such as regularity conditions) informal. Spelling out the details in a concrete setting is a starting point for future developments. We elaborate one such particular setting in detail, namely Gaussian probability.

We begin §2 by recalling the relevant notions of convex analysis, and proceed to define and study the categorical structure of bifunctions in §3. This includes two structures as a hypergraph category and the duality theory of §3.1.

In §4, we elaborate different examples of categories which embed in bifunctions, such as linear and affine algebra, linear algebra, convex relations and convex optimization problems. In each case, the embedding preserves the relevant categorical structures and dualities. In particular, we show that the theory of bifunctions is a conservative extension of graphical linear algebra [25].

In §5 we begin making connections to probability theory. We recall Gaussian probability from a categorical point of view, and construct the embedding functor to bifunctions. We discuss how partial quadratic functions can be seen as an extension of Gaussian probability beyond measure theory.

We conclude with §6-7 discussing the wider context of this work, elaborating connections of probability and convex analysis such as the Laplace approximation and cumulant generating functions, and the idea of idempotent analysis as a 'tropical limit' of ordinary analysis.

2   Overview of Convex Analysis

The following section is a brief overview of standard material in convex analysis; all propositions and conventions are taken from [27].

Caveat: An important feature of convex analysis is that it deals with formal infinities $+\infty, -\infty$ in a consistent fashion. This is crucial because optimization problems may be unbounded. Traditionally, one considers the extended real numbers $\bar{\mathbb{R}} = [-\infty, +\infty]$ and extends the usual laws of arithmetic to them. The case $(+\infty) + (-\infty)$ is left undefined and carefully avoided like $0/0$ in real analysis.

A more systematic approach [37, 18] is based on enriched category theory, and endows $\overline{\mathbb{R}}$ with the structure of a commutative quantale, which gives it totally defined operations with a particular arithmetic.

A more serious caveat is that many results in convex analysis require specific regularity assumptions to hold. As these assumptions are not the focus of the present paper, so we will state some big picture theorems in §3 under reservation of these conditions. We then elaborate an array of concrete examples §4-5 where we make sure that all regularity conditions are indeed satisfied. We discuss this drawback in §7.

A subset $A \subseteq \mathbb{R}^n$ is *convex* if for all $x, y \in A$ and $t \in [0, 1]$, we have $tx + (1 - t)y \in A$. The *epigraph* of a function $f : \mathbb{R}^n \to \overline{\mathbb{R}}$ is the set epi$(f) = \{(x, y) \in \mathbb{R}^{n+1} : y \geq f(x)\}$. We say that f is convex if epi(f) is a convex subset of $\mathbb{R}^{n+1}$. This is equivalent to the well-known definition from the introduction, while accounting for infinities. We say that f is *concave* if $(-f)$ is convex.

Example 1. The following functions are convex: linear functions, $|x|$, x^2, $\exp(x)$, $-\ln(x)$. For a convex subset $A \subseteq \mathbb{R}^n$, the *convex indicator function* $\delta_A : \mathbb{R}^n \to \overline{\mathbb{R}}$ is defined by

$$\delta_A(x) = \begin{cases} 0, & x \in A \\ +\infty, & x \notin A \end{cases}$$

We also write indicator functions using modified Iverson brackets as $\{|x \in A|\} = \delta_A(x)$. The *concave indicator function* of A is $-\delta_A(x)$.

The *infimal convolution* of convex functions $f, g : \mathbb{R}^n \to \overline{\mathbb{R}}$ is defined by $(f \,\square\, g)(x) = \inf_y \{f(x - y) + g(y)\}$. The convex function f is called *closed* if epi(f) is a closed subset of $\mathbb{R}^{n+1}$; this is equivalent to f being lower semicontinuous.

2.1   Conjugacy – the Legendre-Fenchel transform

Definition 1. *For a convex function* $f : \mathbb{R}^n \to \overline{\mathbb{R}}$, *its convex conjugate* $f^* : \mathbb{R}^n \to \overline{\mathbb{R}}$ *is the convex function*

$$f^*(x^*) = \sup_x \{\langle x^*, x \rangle - f(x)\}$$

For a concave function $g : \mathbb{R}^n \to \overline{\mathbb{R}}$, *its concave conjugate* $g^* : \mathbb{R}^n \to \overline{\mathbb{R}}$ *is the concave function*

$$g^*(x^*) = \inf_x \{\langle x^*, x \rangle - g(x)\}$$

Note that if $g = -f$ *then* $g^*(x^*) = -f^*(-x^*)$

Geometrically, f^* encodes information about which affine functions $\langle x^*, - \rangle - c$ are majorized by f. It is thus natural to view f^* as a function on covectors $x^* \in (\mathbb{R}^n)^*$. This is for example done in [37], but in order to keep notation consistent with [27], we make the traditional identification $(\mathbb{R}^n)^* \cong \mathbb{R}^n$ via the inner product, and the notation x^* is purely decoration. The Legendre-Fenchel transform has applications in many areas of mathematics and physics [34], such as the Hamiltonian formalism in mechanics, statistical mechanics or large deviation theory (e.g. §6.2).

A closed convex function f is the pointwise supremum of all affine functions $h \leq f$ [27, 12.1]. This allows them to be recovered by their Legendre transform

Proposition 1 ([27, Theorem 12.2]). *For any convex function $f : \mathbb{R}^n \to \overline{\mathbb{R}}$, f^* is a closed convex function. We have $f^{**} = f$ if and only if f is closed.*

For arbitrary functions f, the operation $f \mapsto f^{**}$ is a closure operator which we denote by $\mathrm{cl}(f)$. This is the largest closed convex function majorized by f.

Example 2. The absolute value function $f(x) = |x|$ is convex and closed. The supremum $\sup_x \{cx - |x|\}$ equals 0 if $|c| \leq 1$, and ∞ otherwise. Hence $f^*(c) = \{|c| \leq 1\}$, and $f^{**} = f$.

Example 3. Let $f(x) = ax^2$ for $a > 0$. Then $x \mapsto c \cdot x - ax^2$ is differentiable and has a maximum at $x = c/2a$. We obtain $f^*(c) = \frac{1}{4a}c^2$. In particular, we see that the function $f(x) = \frac{1}{2}x^2$ is a fixed point of the Legendre transform.

Proposition 2 ([27, Theorem 16.4]). *If f, g are closed convex functions, then under certain regularity conditions $(f \,\square\, g)^* = f^* + g^*$ and $(f + g)^* = f^* \,\square\, g^*$.*

3   Categories of Convex Bifunctions

We now come to the central definition of this article, namely that of convex (or concave) *bifunctions*. This concept is due to [27] and scattered throughout his book.

A *bifunction* F from $\mathbb{R}^m$ to $\mathbb{R}^n$ is the convex analysis terminology for a curried function $\mathbb{R}^m \to (\mathbb{R}^n \to \overline{\mathbb{R}})$. The uncurried function $\underline{F} : \mathbb{R}^{m+n} \to \overline{\mathbb{R}}$ is referred to as the *graph function* of F. We will suppress the partial application and write $F(x)(y)$ and $F(x, y)$ interchangeably.

Definition 2. *A bifunction F from $\mathbb{R}^m$ to $\mathbb{R}^n$ is called* convex *(or* concave, closed*) if its graph function $\underline{F} : \mathbb{R}^{m+n} \to \overline{\mathbb{R}}$ has that property. The closure operation $\mathrm{cl}(F)$ is applied on the level of graph functions. We denote a convex bifunction by $F : \mathbb{R}^m \rightharpoonup \mathbb{R}^n$ and a concave bifunction by $F : \mathbb{R}^m \rightharpoondown \mathbb{R}^n$.*

Bifunction composition is known as *product* in [27, § 38].

Definition 3 (Categories of bifunctions). *We define a category* CxBiFn *of convex bifunctions as follows*

    – *objects are the spaces $\mathbb{R}^n$*
    – *morphisms are convex bifunctions $\mathbb{R}^m \rightharpoonup \mathbb{R}^n$*
    – *the identity $\mathbb{R}^n \rightharpoonup \mathbb{R}^n$ is given by the indicator function*

$$\mathrm{id}_n(x, y) = \{x = y\}$$

- *composition is infimization over the middle variable*

$$(F \overset{\rightarrow}{\circ} G)(x, z) = \inf_y \{G(x, y) + F(y, z)\}$$

Analogously, the category CvBiFn *of concave bifunctions is defined as*

- *objects are the spaces* $\mathbb{R}^n$
- *morphisms are concave bifunctions* $\mathbb{R}^m \rightarrow \mathbb{R}^n$
- *the identity* $\mathbb{R}^n \rightarrow \mathbb{R}^n$ *is given by the concave indicator function*

$$-\mathrm{id}_n(x, y) = -\{\!| x = y |\!\}$$

- *composition is supremization over the middle variable*

$$(F \overset{\rightarrow}{\circ} G)(x, z) = \sup_y \{G(x, y) + F(y, z)\}$$

Proof (of well-definedness). This construction is a subcategory of the the category of weighted relations $\mathtt{Rel}(Q)$ taking values in a commutative quantale Q [3, 12, 23], where $Q = \overline{\mathbb{R}}$ are the extended reals. It suffices to verify that convex bifunctions are closed under composition, tensor (addition) and contain the identities ([27, p. 408]).

We will write bifunction composition as $F \circ G$ when it is clear from context whether we use the convex or concave variety. We will write I for the unit space $\mathbb{R}^0$, and $\mathbf{0}$ for its unique element.

Example 4. The *states* (morphisms $I \rightarrow \mathbb{R}^n$ out of the unit) are in bijection with convex functions $f : \mathbb{R}^n \rightarrow \overline{\mathbb{R}}$, as are the *effects* $\mathbb{R}^n \rightarrow I$. States and effects in CvBiFn are in bijection with concave functions $f : \mathbb{R}^n \rightarrow \overline{\mathbb{R}}$.

3.1   Duality for Bifunctions

Unless otherwise stated, theorems phrased for convex bifunctions will hold for concave bifunctions by selecting the appropriate versions of the operations.

The duality theory of convex functions extends to bifunctions as follows.

Definition 4 ([27, §30]). *The* adjoint *of a convex bifunction* $F : \mathbb{R}^m \rightarrow \mathbb{R}^n$ *is the* concave bifunction $F^* : \mathbb{R}^n \rightarrow \mathbb{R}^m$ *defined by*

$$F^*(y^*, x^*) = \inf_{x,y} \{F(x, y) + \langle x^*, x \rangle - \langle y^*, y \rangle\}$$

The adjoint of a concave bifunction is convex and uses sup *instead of* inf. *The adjoint of the convex bifunction F is related to the conjugate of its graph function $\underline{F}$ using the formula $F^*(y^*, x^*) = -\underline{F}^*(-x^*, y^*)$. (Note the slight asymmetry that one input is negated)*

The analogue of Proposition 1 for bifunctions is as follows

Proposition 3 ([27, Theorem 30.1]). *For any convex bifunction F, the adjoint F^* is a closed concave bifunction, and we have $F^{**} = \mathrm{cl}(F)$. In particular, if F is a closed convex bifunction, then $F^{**} = F$.*

Theorem 1 ([27, Theorem 38.5]). *Under regularity assumptions, the adjoint operation respects composition. That is, for $F : \mathbb{R}^m \rightharpoonup \mathbb{R}^n$ and $G : \mathbb{R}^n \rightharpoonup \mathbb{R}^k$, we have*

$$(G \overset{\rightarrow}{\circ} F)^* = F^* \overset{\leftarrow}{\circ} G^*$$

That is, the adjoint operation defines a pair of mutually inverse functors

$$\texttt{CxBiFn}^{\mathrm{op}} \xrightarrow{\;\;(-)^*\;\;} \texttt{CvBiFn}$$
$$\texttt{CxBiFn}^{\mathrm{op}} \xleftarrow{\;\;(-)^*\;\;} \texttt{CvBiFn}$$

We indicate with dashed arrows that the functoriality depends on regularity assumptions.

For the interested reader, the regularity assumptions in Theorem 1 include closedness, as well as properness and certain (relative interiors of) domains of the involved bifunctions intersecting [27, § 38]. These assumptions are not necessary conditions.

As a corollary of functoriality, we can derive the following well-known fact:

Corollary 1 (Fenchel duality). *Let $f : \mathbb{R}^n \to \overline{\mathbb{R}}$ be convex, $g : \mathbb{R}^n \to \overline{\mathbb{R}}$ concave, and let f^*, g^* be their convex and concave conjugates respectively. Then under sufficient regularity assumptions, we have*

$$\inf_x \{f(x) - g(x)\} = \sup_{x^*} \{g^*(x^*) - f^*(x^*)\}$$

Proof. Consider the convex function $h = -g$ and form the state $s_f : I \rightharpoonup \mathbb{R}^n$, $s_f(0, x) = f(x)$ and effect $e_h : \mathbb{R}^n \rightharpoonup I$, $e_h(x, 0) = h(x)$. The proof proceeds by using functoriality to compute the scalar $(e_h \overset{\rightarrow}{\circ} s_f)^* = (s_f^* \overset{\rightarrow}{\circ} e_h^*)$ in two ways: On the one hand, we have

$$(e_h \overset{\rightarrow}{\circ} s_f)(0, 0) = \inf_x \{s_f(0, x) + e_h(x, 0)\} = \inf_x \{f(x) - g(x)\}$$

On the other hand, we express the adjoints in terms of the conjugates f^*, g^*

$$s_f^*(x^*, 0) = \inf_x \{s_f(0, x) - \langle x^*, x \rangle\} = -f^*(x^*)$$
$$e_h^*(0, x^*) = \inf_x \{e_h(x, 0) + \langle x^*, x \rangle\} = g^*(x^*)$$

The adjoint acts as the identity on scalars, so we obtain

$$\inf_x \{f(x) - g(x)\} = (e_h \overset{\rightarrow}{\circ} s_f)^*(0, 0) = (s_f^* \overset{\rightarrow}{\circ} e_h^*)(0, 0) = \sup_x \{g^*(x^*) - f^*(x^*)\}$$

3.2   Hypergraph Structure and Symmetries

Bifunctions can not only be composed in sequence, but also in parallel. The relevant structure is that of a symmetric monoidal category $(\mathbb{C}, \otimes, I)$. In this work, we are dealing with a particular simple form of such categories called a *prop*. A prop $\mathbb{C}$ is a strict monoidal category which is generated by a single object R so that every object is of the form $R^{\otimes n}$ for some $n \in \mathbb{N}$. The monoid of objects $(\mathrm{ob}(\mathbb{C}), \otimes, I)$ is thus isomorphic to $(\mathbb{N}, +, 0)$.

Proposition 4. *Convex bifunctions have the structure of a prop, generated by the object $\mathbb{R}$*

1. *The tensor is $\mathbb{R}^m \otimes \mathbb{R}^n = \mathbb{R}^{m+n}$*
2. *The unit is $I = \mathbb{R}^0$.*
3. *The tensor of bifunctions is given by addition: If $F : \mathbb{R}^{m_1} \rightharpoonup \mathbb{R}^{n_1}$, $G : \mathbb{R}^{m_2} \rightharpoonup \mathbb{R}^{n_2}$ then $F \otimes G : \mathbb{R}^{m_1+m_2} \rightharpoonup \mathbb{R}^{n_1+n_2}$ is defined as*

$$(F \otimes G)((x_1, x_2), (y_1, y_2)) = F(x_1, y_1) + G(x_2, y_2)$$

Proof (of well-definedness). General fact about categories of weighted relations $\mathrm{Rel}(Q)$ ([23]). $\qquad\blacksquare$

Symmetric monoidal categories are widely studied and admit a convenient graphical language using string diagrams [28]. It is useful to consider further pieces of structure on such a category

1. in a *copy-delete category* [11], every object carries the structure of a commutative comonoid $\mathrm{copy}_X : X \to X \otimes X$ and $\mathrm{discard}_X : X \to I$. This lets information be used in a non-linear way (in the sense of linear logic).
2. in a *hypergraph category* [14], every object carries the structure of a special commutative Frobenius algebra

Every hypergraph category is in particular a copy-delete category. The pieces of structure of a hypergraph category are often rendered as cups and caps in string diagrams

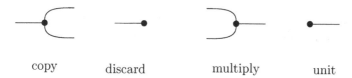

copy	discard	multiply	unit

subject to equations such as the Frobenius law

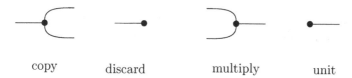

This gives rise to a rich graphical calculus, which has been explored for a number of engineering applications like signal-flow diagrams or electrical circuits [25, 8, 7, 9, 2, 1]

Proposition 5. CxBiFn *has the structure of a hypergraph category in two different ways, which we call the additive and co-additive structure. That is, every object carries two different structures as a special commutative Frobenius algebra*

1. *The additive structure is given by*

$$\begin{aligned}
\text{unit} &: I \rightharpoonup \mathbb{R}^n, & \text{unit}(\mathbf{0}, x) &= 0 \\
\text{discard} &: \mathbb{R}^n \rightharpoonup I, & \text{discard}(x, \mathbf{0}) &= 0 \\
\text{copy} &: \mathbb{R}^n \rightharpoonup \mathbb{R}^n \otimes \mathbb{R}^n & \text{copy}(x, y, z) &= \{|x = y = z|\} \\
\text{comp} &: \mathbb{R}^n \otimes \mathbb{R}^n \rightharpoonup \mathbb{R}^n, & \text{comp}(x, y, z) &= \{|x = y = z|\}
\end{aligned}$$

2. *The co-additive structure is given by*

$$\begin{aligned}
\text{zero} &: I \rightharpoonup \mathbb{R}^n, & \text{zero}(\mathbf{0}, x) &= \{|x = 0|\} \\
\text{cozero} &: \mathbb{R}^n \rightharpoonup I, & \text{cozero}(x, \mathbf{0}) &= \{|x = 0|\} \\
\text{add} &: \mathbb{R}^n \otimes \mathbb{R}^n \rightharpoonup \mathbb{R}^n, & \text{add}(x, y, z) &= \{|x + y = z|\} \\
\text{coadd} &: \mathbb{R}^n \rightharpoonup \mathbb{R}^n \otimes \mathbb{R}^n, & \text{coadd}(z, x, y) &= \{|x + y = z|\}
\end{aligned}$$

The analogous structures on CvBiFn *use concave indicator functions instead.*

We can motivate the names of the hypergraph structures by observing how multiplications acts on states. This duality is clarified in what follows.

Example 5. Let $f, g : I \rightharpoonup \mathbb{R}^n$ be two states. Then

$$(\text{copy} \circ (f \otimes g))(z) = \inf_{x,y} \{f(x) + g(y) + \{|x = y = z|\}\} = f(z) + g(z)$$

$$(\text{add} \circ (f \otimes g))(z) = \inf_{x,y} \{f(x) + g(y) + \{|x + y = z|\}\} = f(z) \,\square\, g(z)$$

Definition 5. *The* dagger *of a bifunction* $F : \mathbb{R}^m \rightharpoonup \mathbb{R}^n$ *is given by reversing its arguments*

$$F^\dagger : \mathbb{R}^n \rightharpoonup \mathbb{R}^m, F^\dagger(y, x) = F(x, y)$$

The inverse *of a bifunction* $F : \mathbb{R}^m \rightharpoonup \mathbb{R}^n$ *is the concave bifunction [27, p. 384]*

$$F_*(x, y) = -F(y, x)$$

Both these operations define involutive[4] functors

$$(-)^\dagger : \text{CxBiFn}^{\text{op}} \to \text{CxBiFn}, \qquad (-)_* : \text{CxBiFn}^{\text{op}} \to \text{CvBiFn}$$

The functor $(-)^\dagger$ *is a dagger functor in the sense of [29].*

Proposition 6 ([27, p. 384]). *The operations of inverse and adjoint commute, i.e. for* $F : \mathbb{R}^m \rightharpoonup \mathbb{R}^n$ *we have* $(F^*)_* = (F_*)^*$.

---

[4] i.e. applying the appropriate version of these operations twice is the identity

The composite operation F_*^* defines another covariant functor $\mathsf{CxBiFn} \to \mathsf{CxBiFn}$, which we now interpret: As is customary in graphical linear algebra, we render the two hypergraph structures as follows

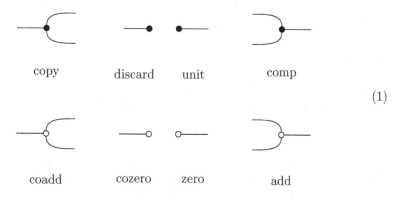

$$\text{(1)}$$

We refer to the additive structure as 'black' ($\bullet$) and the co-additive one as 'white' ($\circ$). This presentation reveals an array of symmetries (mirror-image and color-swap[5]), which we are relating now:

Theorem 2. *The adjoint operation interchanges the additive and co-additive structure. That is we have functors of hypergraph categories*

$$(-)^* : (\mathsf{CxBiFn}^{\mathrm{op}}, \bullet) \to (\mathsf{CvBiFn}, \circ)$$
$$(-)^* : (\mathsf{CxBiFn}^{\mathrm{op}}, \circ) \to (\mathsf{CvBiFn}, \bullet)$$

Note that the opposite of a hypergraph category is again a hypergraph category where cups and caps are interchanged.

Proof. Follows from the results in §4.1, as the hypergraph structures are induced by linear maps.

In terms of the generators (1), the mirror image is given by the $(-)^\dagger$ functor. Both hypergraph structures consist of †-Frobenius algebras, meaning that $(-)^\dagger$ is a functor of hypergraph categories $\mathsf{CxBiFn}^{\mathrm{op}} \to \mathsf{CxBiFn}$.

The color-swap operation is given by the inverse adjoint F_*^*, which gives a hypergraph equivalence $(\mathsf{CxBiFn}, \bullet) \to (\mathsf{CxBiFn}, \circ)$. This equivalence does however not commute with †, i.e. is not an equivalence of dagger hypergraph categories.

4 Example Categories of Bifunctions

We now elaborate example subcategories of bifunctions on which functoriality and duality work as desired (that is, all regularity conditions apply).

---

[5] we will discuss these symmetries in more detail in Section 4.1

4.1 Linear Algebra

The identities and dualities of convex bifunctions generalize those of linear algebra. Let $A : \mathbb{R}^m \to \mathbb{R}^n$ be a linear map. The convex indicator bifunction of A is defined as

$$F_A(x, y) = \{y = Ax\}$$

The following facts hold [27, p 310]:

1. For composable linear maps, A, B we have $F_{AB} = F_A \circ F_B$
2. The adjoint F_A^* is the concave indicator bifunction of the transpose A^T

$$F_A^*(y^*, x^*) = -\{x^* = A^T y^*\}$$

3. if A is invertible, then the inverse $(F_A)_*$ is the concave indicator bifunction associated to the inverse A^{-1}. In that case, Proposition 6 generalizes the identity $(A^{-1})^T = (A^T)^{-1}$.

In more categorical terms, let Vect denote the prop of the vector spaces $\mathbb{R}^n$ and linear maps. This is a copy-delete category equipped with the linear maps $\Delta : \mathbb{R}^n \to \mathbb{R}^n \oplus \mathbb{R}^n$ and $! : \mathbb{R}^n \to \mathbb{R}^0$. For a linear map $A : \mathbb{R}^m \to \mathbb{R}^n$, define

$$F_A : \mathbb{R}^m \rightharpoonup \mathbb{R}^n, F_A(x, y) = \{y = Ax\}$$
$$G_A : \mathbb{R}^n \rightharpoonup \mathbb{R}^m, G_A(y, x) = -\{x = A^T y\}$$

Theorem 3. *We have a commutative diagram of functors between copy-delete categories*

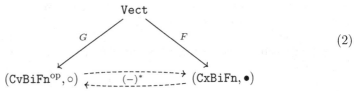

$$(2)$$

Proof. Functoriality and commutativity follow from the above facts. For the copy-delete structures, notice that copy, delete, add, zero are the indicator bifunctions of the linear maps Δ and $!$. The transpose of Δ is summation $(x, y) \mapsto x + y$.

We call a diagram like (2) a *duality situation*. The dashed arrows indicate that, while $(-)^*$ is neither a functor nor idempotent on all bifunctions without further conditions, everything works out on the image of F, G respectively. We could thus obtain a genuine commutative diagram of functors by characterizing these images exactly (which we refrain from doing here for the sake of simplicity).

Linear Relations Graphical Linear Algebra [25] is the diagrammatic study of the prop LinRel of linear relations, which are relations $R \subseteq \mathbb{R}^m \times \mathbb{R}^n$ that are also vector subspaces. This category is a hypergraph category using the two structures shown in (1), and the operations mirror-image and color-swap are defined for linear relations via relational converse and a twisted orthogonal complement

$$R^\dagger = \{(y,x) : (x,y) \in R\}$$
$$R^c = \{(x^*, y^*) : \forall (x,y) \in R, \langle x^*, x \rangle - \langle y^*, y \rangle = 0\}$$

The operations $(-)^\dagger$ and $(-)^c$ commute and define a composite contravariant involution $(-)^* : \mathtt{LinRel}^{\mathrm{op}} \to \mathtt{LinRel}$. The following theorem shows that bifunctions are a conservative extension of graphical linear algebra.

Theorem 4. *If we embed a linear relation $R \subseteq \mathbb{R}^m \times \mathbb{R}^n$ via its indicator function as a bifunction $I_R : \mathbb{R}^m \to \mathbb{R}^n$, then we have a commutative diagram*

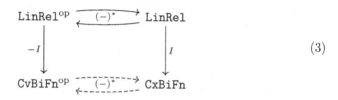

$$(3)$$

In addition, the functor I preserves both hypergraph structures.

Affine Relations Graphical linear algebra has been extended to affine relations [6]; those are affine subspaces $R \subseteq \mathbb{R}^m \times \mathbb{R}^n$. This still forms a hypergraph category with both structures $\bullet, \circ$, however the color-swap symmetry of linear relations is broken. That is because the affine generator $\underline{1} : 0 \to 1$ representing the affine relation $\{(0,1)\}$ does not have an obvious color-swapped dual; affine subspaces are not recovered by their orthogonal complements.

The embedding into bifunctions suggests an avenue to recover such a symmetry: Taking the embedding (3) as a starting point, the indicator bifunction of $\underline{1}$ is $f : I \to \mathbb{R}, f(0,x) = \{|x = 1|\}$. Its adjoint is $f^*(x^*, 0) = -x^*$, which is a perfectly well-defined bifunction but not the indicator bifunction of any affine relation. This suggests that an extension of affine relations with color-swap symmetry can be obtained using a category of partial affine function (e.g. [27, p. 107]) but details are to left for future work. We will discuss the case of partial quadratic functions in §5.2.

4.2   Convex Relations

Generalizing the previous example even further, a *convex relation* $R \subseteq \mathbb{R}^m \times \mathbb{R}^n$ is a relation which is also a convex subset of $\mathbb{R}^{m+n}$. Convex relations are closed under the usual relation composition and thus form a prop $\mathtt{CxRel}$ [3, 12, 23].

Every linear relation is in particular convex, and like linear relations, convex relations embed into convex bifunctions via the indicator function.

We sketch a certain converse to this embedding: The space $(\mathbb{R}, +, 0)$ is a monoid object in $\mathtt{CxRel}$. We consider the 'writer' monad $T : \mathtt{CxRel} \to \mathtt{CxRel}$

associated to that monoid, i.e. $T(\mathbb{R}^m) = \mathbb{R}^{m+1}$. If $S \subseteq \mathbb{R}^m \times \mathbb{R}^{n+1}$ and $R \subseteq \mathbb{R}^n \times \mathbb{R}^{k+1}$ are Kleisli arrows, then Kleisli composition takes the following form

$$R \bullet S = \{(x, z, t_1 + t_2) : (x, y, t_1) \in S, (y, z, t_2) \in R\}$$

Given a convex bifunction $F : \mathbb{R}^m \rightharpoonup \mathbb{R}^n$, the epigraph of its graph function $\mathrm{epi}(\underline{F}) \subseteq \mathbb{R}^m \times \mathbb{R}^{n+1}$ is thus a Kleisli arrow for T. Under sufficient regularity assumptions, this is functorial, and we have an embedding epi : CxBiFn $\to$ CxRel$_T$.

4.3   Ordinary Convex Programs

We briefly discuss the historical origins of bifunctions in convex optimization [27, § 29-30]: For simplicity, we say that a *ordinary convex program* P is a minimization problem of the form

$$\inf\{f(x) : x \in \mathbb{R}^n, g_1(x) \leq 0, \ldots, g_k(x) \leq 0\}$$

where the objective function f and the constraints $g_1, \ldots, g_k : \mathbb{R}^n \to \overline{\mathbb{R}}$ are finite convex functions. The *bifunction associated to* P is defined as

$$F_P : \mathbb{R}^k \rightharpoonup \mathbb{R}^n, F_P(v, x) = f(x) + \sum_{i=1}^{k} \{\!| f_i(x) \leq v_i |\!\}$$

The inputs of $v \in \mathbb{R}^k$ can be thought of as perturbations of the constraints. The so-called *perturbation function* of P is the parameterized minimization problem $(\inf F_P)(v) = \inf_x \{F_P(v, x)\}$. The convex function $F_P(0, -)$ represents the unperturbed problem and $(\inf F_P)(0)$ is the desired solution. Note that in categorical language, the perturbation function is straightforwardly obtained as the bifunction composite $(\mathrm{discard} \circ F_P) : \mathbb{R}^k \rightharpoonup I$, or graphically

$$\mathbb{R}^k \quad \boxed{F_P} \!\!-\!\!\bullet$$

The associated bifunction F_P contains all information about the problem P, and allows one to find the dual problem P^* by taking its adjoint. This way one can think of *any* bifunction $\mathbb{R}^k \rightharpoonup \mathbb{R}^n$ as a *generalized convex program* ([27, p. 294]).

Example 6 ([27, p. 312]). Consider a linear minimization problem P of the form

$$\inf\{\langle c, x \rangle : b - Ax \leq 0\}$$

The associated bifunction and its adjoint are

$$F(v, x) = \langle c, x \rangle + \{\!| x \geq 0, b - Ax \leq v |\!\}$$
$$F^*(x^*, v^*) = \langle b, v^* \rangle - \{\!| v^* \geq 0, c - A^T v^* \geq x^* |\!\}$$

which is the concave bifunction associated to the dual maximization problem

$$\sup\{\langle b, y \rangle : y \geq 0, c - A^T y \geq 0\}$$

5 Gaussian Probability and Convexity

We now study the probabilistic applications of our categorical framework: Recently, a sizeable body of work in categorical probability theory has been developed in terms of copy-delete and Markov categories. A Markov category [17] is a copy-delete category $(\mathbb{C}, \otimes, I)$ where every morphism $f : X \to Y$ is *discardable* in the sense that $\text{discard}_Y \circ f = \text{discard}_X$. Classic examples of Markov categories are the category $\mathtt{FinStoch}$ of finite sets and stochastic matrices, and the category $\mathtt{Stoch}$ of measurable spaces and Markov kernels. Discardability expresses that probability measures are normalized (integrate to 1). Markov categories provide a natural semantic domain for probabilistic programs [30].

In this section, we will focus on *Gaussian probability*, by which we mean the study of multivariate normal (Gaussian) distributions and affine-linear maps. This is a small but expressive fragment of probability, which suffices for a range of interesting application from linear regression and Gaussian processes to Kalman filters. The univariate normal distribution $\mathcal{N}(\mu, \sigma^2)$ is defined on $\mathbb{R}$ via the density function

$$f(x) = \frac{1}{\sqrt{2\pi\sigma^2}} \exp\left(-\frac{(x-\mu)^2}{2\sigma^2}\right)$$

Multivariate Gaussian distributions are easiest described as the laws of random vectors $A \cdot X + \mu$ where $A \in \mathbb{R}^{n \times k}$ and $X_1, \ldots, X_k \sim \mathcal{N}(0, 1)$ are independent variables. The law is fully characterized by the mean μ and the covariance matrix $\Sigma = AA^T$. Conversely, for every vector $\mu \in \mathbb{R}^n$ and positive semidefinite matrix $\Sigma \in \mathbb{R}^{n \times n}$, there exists a unique Gaussian law $\mathcal{N}(\mu, \Sigma)$. If $X \sim \mathcal{N}(\mu, \Sigma)$ and $Y \sim \mathcal{N}(\mu', \Sigma')$ are independent then $X + Y \sim \mathcal{N}(\mu + \mu', \Sigma + \Sigma')$ and $AX \sim \mathcal{N}(A\mu, A\Sigma A^T)$. Gaussians are self-conjugate: If (X, Y) are jointly Gaussian, then so is the conditional distribution $X|Y = y$ for any constant $y \in \mathbb{R}^k$.

If the covariance matrix Σ is positive definite, then the Gaussian has a density with respect to the Lebesgue measure on $\mathbb{R}^n$ given by

$$f(x) = \frac{1}{\sqrt{(2\pi)^n \det(\Sigma)}} \exp\left(-\frac{1}{2}(x-\mu)^T \Omega (x-\mu)\right) \tag{4}$$

where $\Omega = \Sigma^{-1}$ is known as the *precision matrix*. This suggests two equivalent representations of Gaussians with different advantages (e.g. [20, 31]):

- In *covariance representation* Σ, pushforwards (addition, marginalization) are easy to compute. Conditioning requires solving an optimization problem
- In *precision representation* Ω, conditioning is straightfoward. Pushforwards require solving an optimization problem.

If Σ is singular, the Gaussian distribution is only supported on the affine subspace $\mu + S$ where $S = \text{im}(\Sigma)$. In that case, the distribution has a density only with respect to the Lebesgue measure on S. This variability of base measure makes it complicated to work with densities, and by extension the precision representation.

The situation becomes clearer if we represent Gaussians by the quadratic functions induced by their covariance and precision matrices. These functions are convex (concave), and turn out to be adjoints of each other. This explains the duality of the two representations, and paves the way for generalizations of Gaussian probability like improper priors [31] which correspond to partial quadratic functions (§5.2).

5.1   Embedding Gaussians in Bifunctions

We now give a categorical account of Gaussian probability (in covariance representation). A *Gaussian morphism* $\mathbb{R}^m \to \mathbb{R}^n$ is a stochastic map of the form $x \mapsto Ax + \mathcal{N}(\mu, \Sigma)$, that is a linear map with Gaussian noise.

Definition 6 ([17, §6]). *The Markov prop* Gauss *is given as follows*

1. *objects are the spaces* $\mathbb{R}^n$, *and* $\mathbb{R}^m \otimes \mathbb{R}^n = \mathbb{R}^{m+n}$
2. *morphisms* $\mathbb{R}^m \to \mathbb{R}^n$ *are tuples* (A, μ, Σ) *with* $A \in \mathbb{R}^{n \times m}, \mu \in \mathbb{R}^n$ *and* $\Sigma \in \mathbb{R}^{n \times n}$ *positive semidefinite*
3. *composition and tensor are given by the formulas*

$$(A, \mu, \Sigma) \circ (B, \mu', \Sigma') = (AB, \mu + A\mu', \Sigma + A\Sigma' A^T)$$
$$(A, \mu, \Sigma) \otimes (B, \mu', \Sigma') = (A \oplus B, \mu \oplus \mu', \Sigma \oplus \Sigma'))$$

   where $\oplus$ *is block-diagonal composition.*
4. *the copy-delete structure is given by the linear maps* $\Delta, !$

We have a Markov functor Gauss $\to$ Stoch which sends $\mathbb{R}^n$ to the measurable space $(\mathbb{R}^n, \mathrm{Borel}(\mathbb{R}^n))$ and assigns (A, μ, Σ) to the probability kernel given by $x \mapsto \mathcal{N}(Ax + \mu, \Sigma)$. Functoriality expresses that the formulas of Definition 6 agree with composition of Markov kernels given by integration of measures. Our main theorem shows that, surprisingly, the representation of Gaussians by quadratic functions is also functorial, i.e. we have an embedding Gauss $\to$ CxBiFn.

Theorem 5. *We have functors of copy-delete categories in a duality situation*

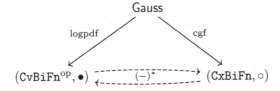

The functors are defined as follows: Let $f = (A, \mu, \Sigma) \in$ Gauss$(\mathbb{R}^m, \mathbb{R}^n)$, *and define bifunctions*

$$\mathrm{logpdf}_f : \mathbb{R}^n \dashrightarrow \mathbb{R}^m, \quad \mathrm{logpdf}_f(y, x) = -\frac{1}{2}\langle z, \Sigma^- z \rangle - \{|z \in S|\}$$

$$\mathrm{cgf}_f : \mathbb{R}^m \dashrightarrow \mathbb{R}^n, \quad \mathrm{cgf}_f(x, y) = \frac{1}{2}\langle y, \Sigma y \rangle + \langle \mu, y \rangle + \{|x = A^T y|\}$$

where $z = y - (Ax + \mu), S = \mathrm{im}(\Sigma)$ *and* Σ^- *denotes any generalized inverse of* Σ.

Proof (Sketch). Functoriality of cgf follows from a straightforward computation, and one can check that $\text{cgf}_f^* = \text{logpdf}_f$ using the formula of [27, p. 109]. Functoriality for logpdf then follows from Theorem 1. The full proof is elaborated in the extended version of this paper [32].

The value $\text{logpdf}_f(y, x)$ is indeed the conditional log-probability (4) minus a scalar. The name cgf is short for cumulant-generating function, which we elaborate in §6.2. For now, we can see cgf as a generalized covariance representation.

5.2   Outlook: Gaussian Relations

Measure-theoretically, there is no uniform probability distribution over the real line. Such a distribution, if it existed, would be useful to model complete absence of information about a point X – in Bayesian inference, this is called an uninformative prior. Intuitively, such a distribution should have density 1, but this would not integrate to 1. On the other hand, a formal logdensity of 0 makes sense – this is simply the indicator function of the full subset $\mathbb{R} \subseteq \mathbb{R}$.

An *extended Gaussian distribution*, as described in [31], is a formal sum $\mathcal{N}(\mu, \Sigma) + D$ of a Gaussian distribution and a vector subspace $D \subseteq \mathbb{R}^n$, called a fibre, thereby blending relational and probabilistic nondeterminism. Such entities were considered by Willems in the control theory literature, under the name of linear open stochastic systems [35, 36]; he identifies them with Gaussian distributions on the quotient space $\mathbb{R}^n/D$. A categorical account based on decorated cospans is developed in [31].

It is straightforward to embed extended Gaussian distributions into convex bifunctions, by taking the sum of the interpretations from Theorems 4 and 5. The distribution $\psi = \mathcal{N}(\mu, \Sigma) + D$ has a convex interpretation given by

$$\text{cgf}_\psi(x) = \frac{1}{2}\langle x, \Sigma x \rangle + \langle \mu, x \rangle + \{\!| x \in D^\perp |\!\}$$

Functions of this form are *partial convex quadratic functions*, which are known to form a well-behaved class of convex functions (see appendix of the extended version [32]). The theory of such functions can be understood as an extension of Gaussian probability with relational nondeterminism and conditioning, which we term *Gaussian relations*. In Gaussian relations, we achieve full symmetry between covariance and density representation (that is, there exists a color-swap symmetry).

Partiality is necessary to be able to interpret all generators of (1); on the upside, the presence of partiality makes conditioning a first-class operation: For example, if $f : \mathbb{R}^2 \rightharpoonup I$ is the joint logdensity of Gaussian variables (X, Y), then conditioning on $Y = 0$ is the same as computing the bifunction composite with the zero map, which is a simple restriction of logdensity $f_{X|Y=0}(x) = f(x, 0)$. On the other hand, conditioning in the covariance representation f^* requires solving the infimization problem $\inf_{y^*} \{f^*(x^*, y^*)\}$. Graphically, we have

6 A Wider Perspective

The example of Gaussian probability was particular situation in which we could map probabilistic concepts to concepts of convex analysis in a functorial way. In this section, we will take an even wider perspective and view convex bifunctions as a categorical model of probability on its own. We will then point out known connections between probability theory and convex analysis, such as the Laplace approximation and cumulant generating functions.

6.1 The Laplace Approximation

For every copy-delete category $\mathbb{C}$, the subcategory of discardable morphisms is a Markov category, and can therefore be seen as a generalized model of probability theory. We investigate this notion for categories of bifunctions.

Proposition 7. *Let $F : \mathbb{R}^m \rightharpoonup \mathbb{R}^n, G : \mathbb{R}^n \rightharpoonup \mathbb{R}^m$ be bifunctions. Then*

1. *F is discardable in $(\texttt{CxBiFn}, \bullet)$ if $\forall x, \inf_y F(x, y) = 0$*
2. *G is discardable in $(\texttt{CvBiFn}^{\mathrm{op}}, \circ)$ if $\forall x, G(0, x) = \{|x = 0|\}$*

and the adjoint $(-)^*$ defines a bijection between the two.

Proof. Direct calculation.

The embedding of Theorem 5 takes values in discardable bifunctions and hence preserve Markov structure. Functoriality means that the composition of Gaussians (integration) and the composition of bifunctions (optimization) coincide. For general probability distributions, this will no longer be the case. We can however understand bifunction composition as an approximation of ordinary probability theory under the so-called *Laplace approximation*. In its simplest instance, Laplace's method (or method of steepest ascent) is a method to approximate certain integrals by finding the maxima of its integrand (e.g. [34])

$$\int e^{n(\langle c,x\rangle - f(x))} \mathrm{d}x \approx \exp\left(n \sup_x \{\langle c, x\rangle - f(x)\} \right) \text{ for } n \to \infty$$

A wide class of commonly used probability distributions is *log-concave*, including Gaussian, Laplace, Dirichlet, exponential and uniform distributions. Laplace's approximation (e.g. [22, §27]) is a way of approximating such distributions around their mode x_0 by a normal distribution, as the Taylor expansion of their logdensity resembles a Gaussian one

$$h(x) \approx h(x_0) + \frac{1}{2} h''(x_0)(x - x_0)^2$$

We can attempt to reduce questions about such distributions to mode-finding (maximization). The Laplace approximation is fundamental in many applications such as neuroscience [15, 16] and has been generalized to a large body of literature on so-called saddle-point methods [10, 24]. The existence of the functor from Gaussians to bifunctions expresses that, as desired, the Laplace approximation is exact on Gaussians. We give an example of the approximation *not* being exact (ironically) on Laplacian distributions.

Example 7. The standard Laplacian distribution has the density function $f(x) = \frac{1}{2}\exp(|x|)$ on the real line. The logpdf $h(x) = |x|$ is a convex function whose convex conjugate is given by $h^*(x^*) = \{||x^*| \le 1\}$ (see Example 2). The latter function is idempotent under addition, and conversely $h \,\square\, h = h$, so h is idempotent under infimal convolution. In contrast, the density $f(x)$ is not idempotent under integral convolution: The sum of independent standard Laplacians is *not* itself Laplacian.

6.2   Convex Analysis in Probability Theory

For a random variable X on $\mathbb{R}^n$, the moment generating function M_X is defined by the following expectation (provided that it exists) $M_X(x^*) = \mathbb{E}[e^{\langle x^*, X \rangle}]$. The *cumulant-generating function* is defined as its logarithm $c_X(x^*) = \log M_X(x^*)$. The function c_X is always convex. The cumulant-generating function of a multivariate Gaussian $X \sim \mathcal{N}(\mu, \Sigma)$ is precisely

$$c_X(x^*) = \frac{1}{2}\langle x^*, \Sigma x^* \rangle + \langle x^*, \mu \rangle \tag{5}$$

which explains our choice of the convex bifunction cgf associated to a Gaussian morphism in Theorem 5. The notion of cumulant-generating function has a central place in the study of exponential families.

It is a particular fact about Gaussians that the cumulant-generating function is the convex conjugate of the logdensity. In the general case, the convex conjugate $c_X^*(x)$ does have a probabilistic interpretations as a so called-rate function in large deviations theory (Cramér's theorem, [13]). It has also been used to formulate a variational principle [38].

6.3   Idempotent Mathematics

We zoom out to an even wider perspective: This subsection briefly outlines some further background of the connections between convex and probabilistic world: The logarithm of base $t < 1$ defines an isomorphism of semirings $([0, \infty), \times, +) \to (\mathbb{R} \cup \{+\infty\}, +, \oplus_t)$ where $\oplus_t$ is $x \oplus_t y = \log_t(t^x + t^y)$. In the 'tropical limit' $t \searrow 0$, we have $x \oplus_t y \approx \min(x, y)$, so we can consider working in the semiring $(\mathbb{R}, +, \min)$ as a limit or deformation of the usual operations on the reals. The semiring $\overline{\mathbb{R}}$ is idempotent, meaning $x \oplus x = \min(x, x) = x$, hence this field of study is also known as *idempotent mathematics* [26], and the limiting procedure has been called *Maslov dequantization* [21]. Our definition of convex bifunctions in terms of the idempotent semiring $\overline{\mathbb{R}}$ thus carries a strong flavor of idempotent mathematics.

Idempotent analogues of measure theory are discussed in [26, 21], and many theorems in classical probability theory are mirrored by theorems of idempotent probability theory. For example, the idempotent analogue of integration is infimization; under this view, the tropical analogue of the Laplace transform (cf. moment-generating function) is the Legendre transform [21, §7]

$$\int e^{\langle x^*, x \rangle} f(x)\,dx \quad \leftrightarrow \quad \inf_x \{\langle x^*, x \rangle + f(x)\}$$

which explains the appearance of the cumulant-generating function in our work. Theorem 5 means that for Gaussians, it makes no difference whether we work in the real-analytic or idempotent world. Idempotent Gaussians have been defined in [26, 1.11.10] using the same formula (5).

7   Related and Future Work

We have described categories of bifunctions as a compositional setting for convex analysis which subsumes a variety of formalisms like linear functions and relations, as well as convex optimization problems, and has a rich duality theory and an elegant graphical language. We have then explored connections between convex analysis and probability theory, and showed that Gaussian probability can be equivalently described in a measure-theoretic and a convex-analytic language. The equivalence of these two perspectives is elegantly formalized as a structure-preserving functor between copy-delete categories. It will be interesting to see how this approach can be generalized to larger classes of distributions such as exponential families.

Concurrently to our work, the categorical structure of convex bifunctions has been exploited by [19] to compositionally build up objective functions for MPC in control theory. That work does not explore Legendre duality and the connections with categorical models of probability theory. The language of props has a history of applications in engineering [2, 1, 7], and our work was directly inspired by the semantics of probabilistic programming [33, 30].

A starting point for future work is to flesh out the outlook given in §5.2, that is to define a hypergraph category of partial quadratic convex functions, which generalizes Gaussian and extended Gaussian probability. It is also interesting to give a presentation for this prop in the style of [25]: We believe that this is achieved by the addition of a single generator $\nu : I \to \mathbb{R}$ to graphical affine algebra [6] which represents the quadratic function $f(x) = \frac{1}{2}x^2$, and that its equational theory is essentially given by invariance under the orthogonal groups $O(n)$. A similar equational theory has been attempted in [33] though no completeness has been proven. Diagrammatic presentations of concepts from geometry and optimization such as polyhedral algebra and Farkas lemma have been given in [4, 5].

We realize that the dependence on regularity assumptions (the caveat of §2) makes general theorems about categories of bifunctions like Theorem 1 somewhat awkward to state. We still believe that using a general categorical language is a useful way of structuring the field and making connections, but see the following avenues of improving the technical situation

1. Identifying specific, well-behaved subcategories of bifunctions (such as convex relations, (partial) linear and (partial) quadratic functions) on which everything behaves as desired. This was pursued in §4 and §5.
2. The Legendre-Fenchel transform has been phrased in terms of enriched adjunctions in [37]. It stands to hope that developing this enriched-categorical approach may take care of some regularity conditions in a systematic way.

Acknowledgements We thank the anonymous reviewers for their careful reviews and suggestions for this work.

References

1. Baez, J.C., Coya, B., Rebro, F.: Props in network theory (2018)
2. Baez, J.C., Erbele, J.: Categories in control. Theory Appl. Categ. **30**, 836–881 (2015)
3. Bolt, J., Coecke, B., Genovese, F., Lewis, M., Marsden, D., Piedeleu, R.: Interacting conceptual spaces i: Grammatical composition of concepts. Conceptual spaces: Elaborations and applications pp. 151–181 (2019)
4. Bonchi, F., Di Giorgio, A., Sobociński, P.: Diagrammatic Polyhedral Algebra. In: Bojańczyk, M., Chekuri, C. (eds.) 41st IARCS Annual Conference on Foundations of Software Technology and Theoretical Computer Science (FSTTCS 2021). Leibniz International Proceedings in Informatics (LIPIcs), vol. 213, pp. 40:1–40:18. Schloss Dagstuhl – Leibniz-Zentrum für Informatik, Dagstuhl, Germany (2021). https://doi.org/10.4230/LIPIcs.FSTTCS.2021.40, https://drops.dagstuhl.de/entities/document/10.4230/LIPIcs.FSTTCS.2021.40
5. Bonchi, F., Di Giorgio, A., Zanasi, F.: From Farkas' Lemma to Linear Programming: an Exercise in Diagrammatic Algebra. In: Gadducci, F., Silva, A. (eds.) 9th Conference on Algebra and Coalgebra in Computer Science (CALCO 2021). Leibniz International Proceedings in Informatics (LIPIcs), vol. 211, pp. 9:1–9:19. Schloss Dagstuhl – Leibniz-Zentrum für Informatik, Dagstuhl, Germany (2021). https://doi.org/10.4230/LIPIcs.CALCO.2021.9, https://drops.dagstuhl.de/entities/document/10.4230/LIPIcs.CALCO.2021.9
6. Bonchi, F., Piedeleu, R., Sobocinski, P., Zanasi, F.: Graphical affine algebra. In: Proc. LICS 2019 (2019)
7. Bonchi, F., Sobociński, P., Zanasi, F.: A categorical semantics of signal flow graphs. In: CONCUR 2014–Concurrency Theory: 25th International Conference, CONCUR 2014, Rome, Italy, September 2-5, 2014. Proceedings 25. pp. 435–450. Springer (2014)
8. Bonchi, F., Sobocinski, P., Zanasi, F.: The calculus of signal flow diagrams I: linear relations on streams. Inform. Comput. **252** (2017)
9. Bonchi, F., Sobociński, P., Zanasi, F.: Interacting Hopf algebras. Journal of Pure and Applied Algebra **221**(1), 144–184 (2017)
10. Butler, R.W.: Saddlepoint approximations with applications, vol. 22. Cambridge University Press (2007)
11. Cho, K., Jacobs, B.: Disintegration and Bayesian inversion via string diagrams. Mathematical Structures in Computer Science **29**, 938 – 971 (2019)
12. Coecke, B., Genovese, F., Lewis, M., Marsden, D., Toumi, A.: Generalized relations in linguistics & cognition. Theoretical Computer Science **752**, 104–115 (2018). https://doi.org/https://doi.org/10.1016/j.tcs.2018.03.008, https://www.sciencedirect.com/science/article/pii/S0304397518301476, quantum structures in computer science: language, semantics, retrieval
13. Cramér, H.: Sur un nouveau theoreme-limite de la theorie des probabilités. Scientifiques et Industrielles **736**, 5–23 (1938)
14. Fong, B., Spivak, D.I.: Hypergraph categories. Journal of Pure and Applied Algebra **223**(11), 4746–4777 (2019)

15. Friston, K., Kiebel, S.: Predictive coding under the free-energy principle. Philosophical transactions of the Royal Society B: Biological sciences **364**(1521), 1211–1221 (2009)
16. Friston, K., Mattout, J., Trujillo-Barreto, N., Ashburner, J., Penny, W.: Variational free energy and the laplace approximation. Neuroimage **34**(1), 220–234 (2007)
17. Fritz, T.: A synthetic approach to markov kernels, conditional independence and theorems on sufficient statistics. Advances in Mathematics **370**, 107239 (2020)
18. Fujii, S.: A categorical approach to l-convexity. arXiv preprint arXiv:1904.08413 (2019)
19. Hanks, T., She, B., Hale, M., Patterson, E., Klawonn, M., Fairbanks, J.: A compositional framework for convex model predictive control. arXiv preprint arXiv:2305.03820 (2023)
20. JAMES, A.: The variance information manifold and the functions on it. In: Multivariate Analysis–III, pp. 157–169. Academic Press (1973). https://doi.org/https://doi.org/10.1016/B978-0-12-426653-7.50016-8, https://www.sciencedirect.com/science/article/pii/B9780124266537500168
21. Litvinov, G.L.: Maslov dequantization, idempotent and tropical mathematics: A brief introduction. Journal of Mathematical Sciences **140**, 426–444 (2007)
22. MacKay, D.J.: Information theory, inference and learning algorithms. Cambridge university press (2003)
23. Marsden, D., Genovese, F.: Custom hypergraph categories via generalized relations. arXiv preprint arXiv:1703.01204 (2017)
24. McCullagh, P.: Tensor methods in statistics. Courier Dover Publications (2018)
25. Paixão, J., Rufino, L., Sobociński, P.: High-level axioms for graphical linear algebra. Science of Computer Programming **218**, 102791 (2022). https://doi.org/https://doi.org/10.1016/j.scico.2022.102791, https://www.sciencedirect.com/science/article/pii/S0167642322000247
26. Puhalskii, A.: Large deviations and idempotent probability. CRC Press (2001)
27. Rockafellar, R.T.: Convex Analysis, vol. 11. Princeton University Press (1997)
28. Selinger, P.: A Survey of Graphical Languages for Monoidal Categories, pp. 289–355. Springer Berlin Heidelberg, Berlin, Heidelberg (2011). https://doi.org/10.1007/978-3-642-12821-9˙4
29. Selinger, P.: Dagger compact closed categories and completely positive maps. Electronic Notes in Theoretical computer science **170**, 139–163 (2007)
30. Stein, D.: Structural foundations for probabilistic programming languages. University of Oxford (2021)
31. Stein, D., Samuelson, R.: A category for unifying gaussian probability and nondeterminism. In: 10th Conference on Algebra and Coalgebra in Computer Science (CALCO 2023). Schloss Dagstuhl-Leibniz-Zentrum für Informatik (2023)
32. Stein, D., Samuelson, R.: Towards a compositional framework for convex analysis (with applications to probability theory) (2023)
33. Stein, D., Staton, S.: Compositional semantics for probabilistic programs with exact conditioning. In: 2021 36th Annual ACM/IEEE Symposium on Logic in Computer Science (LICS). pp. 1–13. IEEE (2021)
34. Touchette, H.: Legendre-fenchel transforms in a nutshell. Unpublished Report (Queen Mary University of London) (2005)
35. Willems, J.C.: Constrained probability. In: 2012 IEEE International Symposium on Information Theory Proceedings. pp. 1049–1053 (2012). https://doi.org/10.1109/ISIT.2012.6283011
36. Willems, J.C.: Open stochastic systems. IEEE Transactions on Automatic Control **58**(2), 406–421 (2013). https://doi.org/10.1109/TAC.2012.2210836

37. Willerton, S.: The Legendre-Fenchel transform from a category theoretic perspective. arXiv preprint arXiv:1501.03791 (2015)
38. Zajkowski, K.: A variational formula on the cramér function of series of independent random variables. Positivity **21**(1), 273–282 (2017)

Automata and Synthesis

Determinization of Integral Discounted-Sum Automata is Decidable***

Shaull Almagor [ID] and Neta Dafni [✉]

Technion, Haifa, Israel
shaull@technion.ac.il
netad@campus.technion.ac.il

Abstract. Nondeterministic Discounted-Sum Automata (NDAs) are non-deterministic finite automata equipped with a discounting factor $\lambda > 1$, and whose transitions are labelled by weights. The value of a run of an NDA is the discounted sum of the edge weights, where the i-th weight is divided by λ^i. NDAs are a useful tool for modelling systems where the values of future events are less influential than immediate ones.

While several problems are undecidable or open for NDA, their deterministic fragment (DDA) admits more tractable algorithms. Therefore, determinization of NDAs (i.e., deciding if an NDA has a functionally-equivalent DDA) is desirable.

Previous works establish that when $\lambda \in \mathbb{N}$, then every *complete* NDA, namely an NDA whose states are all accepting and its transition function is complete, is determinizable. This, however, no longer holds when the completeness assumption is dropped.

We show that the problem of whether an NDA has an equivalent DDA is decidable when $\lambda \in \mathbb{N}$ (in particular, it is in EXPSPACE and is PSPACE−hard).

Keywords: Discounted Sum Automata · Determinization · Quantitative Automata

1 Introduction

Traditional methods of modelling systems rely on Boolean automata, where every word is assigned a Boolean value (i.e., accepted or rejected). This setting is often generalized into a richer, quantitative one, where every word is assigned a numerical value, and thus the Boolean concept of a language, i.e., a set of words, is lifted to a more general function, namely a function from words to values.

A particular instance of quantitative automata is that of *discounted-sum automata*. There, the weight function sums the weights along the run, but discounts the future. Discounting as a general notion is a well studied concept in game theory and various social choice models [9]. Computational models with discounting, such as discounted-payoff games [21, 3, 1], discounted-sum Markov

---

* This research was supported by the ISRAEL SCIENCE FOUNDATION (grant No. 989/22)
** The full version can be found on https://arxiv.org/abs/2310.09115

N. Kobayashi and J. Worrell (Eds.): FoSSaCS 2024, LNCS 14574, pp. 191–211, 2024.
https://doi.org/10.1007/978-3-031-57228-9_10

Decision Processes [17, 19, 14] and discounted-sum automata [15, 12, 13, 11], are therefore useful to model settings where the far future has less influence than the immediate future.

In this work we focus on non-deterministic discounted-sum automata (NDAs). An NDA is a quantitative automaton equipped with a *discounting factor* $\lambda > 1$. The value of a run is the discounted sum of the transitions along the run, where the value of transition i is divided by λ^i. The value of a word is then the value of the minimal accepting run on it. We also allow *final weights* that are added to the run at its end (with appropriate discounting).

Unlike Boolean automata, NDAs are strictly more expressive than their deterministic counterpart (DDAs) [11]. In particular, certain decision problems for NDAs are undecidable, but become decidable for DDAs [5]. There is, however, a subclass of NDAs that always admit an equivalent DDA: the *complete integral NDAs* [6]. An automaton is *complete* if its transition function is total and all its states are accepting with final weight 0. This means that runs never "die", and that all runs are accepting. An NDA is *integral* if its discounting factor λ is an integer. It is further shown in [6] that if the completeness requirement is removed then for every discounting factor there is an integral NDA that is not determinizable.

The existence of NDAs that are not determinizable implies that the determinization problem is not trivial. However, its decidability and complexity have not been studied. In this work, we show that determinization of integral NDAs is decidable. Specifically, we show that determinization is in EXPSPACE and is PSPACE – hard.

Example 1. We demonstrate the determinization problem, as well as some intricacies involved in its analysis. Consider the NDA in Figure 1a. Intuitively, the NDA either reads only a's, or reads a word of the form a^*b. However, it guesses in q_0 whether it is going to read many a's, in which case it may be worthwhile incurring weight 3 to q_2 in order to read the remaining a's at cost 0.

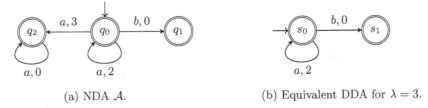

(a) NDA $\mathcal{A}$.   (b) Equivalent DDA for $\lambda = 3$.

Fig. 1: The NDA $\mathcal{A}$ on the left is determinizable with $\lambda = 3$, with an equivalent DDA depicted on the right. However, $\mathcal{A}$ is not determinizable with $\lambda = 2$.

We now ask if this NDA has a deterministic equivalent. As it turns out, this is dependent on the discounting factor. Indeed, consider the discounting factor $\lambda = 3$, then when reading the word a^k, the run that remains in q_0 has weight $\sum_{i=0}^{k} 2 \cdot 3^{-i} = 3 - 3^{-k} < 3$, whereas a run that moves to q_2 at step $j \geq 0$ has weight $\sum_{i=0}^{j-1} 2 \cdot 3^{-i} + 3 \cdot 3^{-j} = 3 - 3^{-j} + 3^{-j} = 3$. Thus, it is always beneficial to remain in q_0. In this case, we do have a deterministic equivalent, depicted in Figure 1b. We remark that the fact that this deterministic equivalent is obtained by removing transitions is not a standard behaviour, and typically determinization involves a blowup.

Next, consider the discounting factor $\lambda = 2$. Similar analysis shows that for the word a^k, the weight of the run that stays in q_0 is $\sum_{i=0}^{k} 2 \cdot 2^{-i} = 4 - 2 \cdot 2^{-k}$, whereas leaving to state q_2 at step 0 yields cost 3, so the latter is preferable for large k. Intuitively, this means that nondeterminism is necessary in this setting, since the NDA does not "know" whether b will be seen. Indeed, for $\lambda = 2$ this NDA is not determinizable.

Observe that in the case of $\lambda = 2$, the two "extreme" runs on a^k, namely the one that stays in q_0 and the one that immediately leaves to q_2, create a "gap" between their values that tends toward 1 as k increases. Keeping in mind that for large k the transition value is multiplied by 2^{-k}, intuitively this gap becomes huge. As we show in this work, this concept of gaps exactly characterizes whether an NDA can be determinized.   □

We remark that for non-integral NDAs, many problems, including the determinization problem, are open due to number-theoretic difficulties [8]. Therefore, it is unlikely that progress is made there, pending breakthroughs in number theory.

Related Work Discounted-sum automata have been studied in various contexts. Specifically, certain algorithmic problems for them are still open, and are closely related to longstanding open problems [8]. In addition, they are not closed under standard Boolean operations [7] (which is often the case in quantitative models, due to the "minimum" semantics which conflicts with notions of conjunction).

Recently, discounted sum automata were also studied in the context of two-player games [10]. Of particular interest are "regret-minimizing strategies", where the concept of regret minimization is closely related to determinization of automata [16].

An extension of discounted-sum automata to multiple discounting factors (NMDAs) was studied in [4, 5], where NMDAs are NDAs where every transition is allowed a different discounting factor. NMDAs are generally non-determinizable, but imposing certain restrictions on the choice of discounting factors can ensure determinizability [4]. We remark that the study of NMDAs is still only with respect to complete automata.

Determinization of other quantitative models has also received some attention in recent years. A major open problem is the decidability of determinization for weighted automata over the tropical semiring (for some subclasses it is known to be decidable [20, 18]). Interestingly, a tropical weighted automaton can be seen as the "limit" of NDAs where $\lambda \to 1$. This, however, does not seem to help in resolving the decidability of the former.

In [2], the determinization problem for one-counter nets (OCNs) is studied. OCNs are automata equipped with a *counter* that cannot decrease below zero. They can be thought of as pushdown automata with a singleton stack alphabet.

Most notions of determinizability introduced in [2] are undecidable, with one case being open (and seemingly related to the setting of weighted automata).

Due to space constraints, some proofs appear in the full version.

2 Preliminaries

A *nondeterministic integral discounted-sum automaton* (NDA) is a tuple $\mathcal{A} = (\Sigma, Q, Q_0, \alpha, \delta, \mathtt{val}, \mathtt{fval}, \lambda)$, where Σ is a finite alphabet, Q is a finite set of states, $Q_0 \subseteq Q$ is a set of initial states, $\alpha \subseteq Q$ is a set of accepting states, $\delta \subseteq Q \times \Sigma \times Q$ is a transition relation, $\mathtt{val} : \delta \to \mathbb{Z}$ is a *weight function* that assigns to each transition $(p, \sigma, q) \in \delta$ a *weight* $\mathtt{val}((p, \sigma, q)) \in \mathbb{Z}$, $\mathtt{fval} : \alpha \to \mathbb{Z}$ is a *final weight function* that assigns a *final weight*[1] to every accepting state, and $1 < \lambda \in \mathbb{N}$ is an integer *discounting factor*.

The existence of a transition $(p, \sigma, q) \in \delta$ means that when $\mathcal{A}$ is in state p and reads the letter σ it can move to state q. If there exists q such that $(p, \sigma, q) \in \delta$, we say that p has a σ-*transition*. If p does not have a σ-transition, that means that when in state p and reading the letter σ, $\mathcal{A}$'s run cannot continue.

Consider a word $w = w_1 \cdots w_n \in \Sigma^*$. A *run* of $\mathcal{A}$ on w is a sequence of states $\rho = \rho_0, \rho_1, \ldots, \rho_n$ such that $\rho_0 \in Q_0$ and $(\rho_{i-1}, w_i, \rho_i) \in \delta$ for every $1 \le i \le n$. The run is *accepting* if $\rho_n \in \alpha$. The *weight* of ρ is the discounted sum $\mathtt{val}(\rho) = \Sigma_{i=0}^{n-1} \lambda^{-i} \mathtt{val}(\rho_i, w_{i+1}, \rho_{i+1})$.

The *value* of w by $\mathcal{A}$, denoted $\mathcal{A}^*(w)$, is $\min\{\mathtt{val}(\rho) + \lambda^{-n}\mathtt{fval}(\rho_n) \mid \rho = \rho_0, \ldots, \rho_n$ is an accepting run on $w\}$, that is, the minimal weight of a run on w including final weights, or ∞ if no such run exists. Two NDAs $\mathcal{A}, \mathcal{B}$ are *equivalent* if $\mathcal{A}^*(w) = \mathcal{B}^*(w)$ for every $w \in \Sigma^*$.

We say that $\mathcal{A}$ is *deterministic* (DDA, for short) if $|Q_0| = 1$ and $\{q \in Q | (p, \sigma, q) \in \delta\} \le 1$ for every $p \in Q, \sigma \in \Sigma$. Note that if $\mathcal{A}$ is deterministic then for every word there is at most one run starting in each state. For a DDA we define the partial function $\delta^* : Q \times \Sigma^* \hookrightarrow Q$ such that $\delta^*(q, w)$ is the final state in the run on w starting in q, if such a run exists. We say that an NDA $\mathcal{A}$ is *determinizable* if it has an equivalent DDA.

It will also be useful to consider non-accepting runs and runs that start and end in specific states of $\mathcal{A}$. For sets of states $P, P' \subseteq Q$ we define $\mathcal{A}_{[P \to P']}(w)$ to be the weight of a minimal run of $\mathcal{A}$ on w from some state in P to some state in P'. Similarly, $\mathcal{A}_{[P \to_f P']}(w)$ is the minimal weight of an accepting run including final weights. When P or P' is a singleton $\{p\}$ we omit the parenthesis. When $P = Q_0$ and $P' = Q$ (or α, for the setting of including final weights) we omit the sets and write e.g., $\mathcal{A}(w)$ instead of $\mathcal{A}_{[Q_0 \to Q]}(w)$, and $\mathcal{A}^*(w)$ instead of $\mathcal{A}_{[Q_0 \to_f \alpha]}(w)$. Under these notations, if a run does not exist, the assigned value is ∞. For the remainder of the paper, fix an integral NDA $\mathcal{A}$.

---

[1] In some works, the weights are assumed to be rational. For determinizability we can assume all weights are integers, since we can always multiply every weight by a common denominator.

3 Gaps and Separation of Runs

In this section we lay down the basic definitions we use throughout the paper, concerning the ways several runs on the same word accumulate different weights.

Denote by $m_{\mathcal{A}}$ the maximal absolute value of a weight of a transition or a final weight in $\mathcal{A}$. Recall that the geometric sum (for $\lambda > 1$) satisfies $\sum_{i=0}^{\infty} \lambda^{-i} = \frac{\lambda}{1-\lambda}$. Therefore, $\frac{\lambda}{\lambda-1} m_{\mathcal{A}}$ is an upper bound on $|\text{val}(\rho)|$ for any run ρ. Indeed, we have

$$|\text{val}(\rho_0, \ldots, \rho_n)| = |\Sigma_{i=0}^{n-1} \lambda^{-i} \text{val}(\rho_i, w_{i+1}, \rho_{i+1})| \leq \Sigma_{i=0}^{n-1} \lambda^{-i} m_{\mathcal{A}} < \frac{\lambda}{\lambda - 1} m_{\mathcal{A}}$$

Clearly, the same bound holds when including final weights.

Let $\mathcal{M} = 2 \frac{\lambda}{\lambda-1} m_{\mathcal{A}}$, then for every two runs ρ^1, ρ^2 we have $|\text{val}(\rho^1) - \text{val}(\rho^2)| < \frac{\lambda}{\lambda-1} m_{\mathcal{A}} - (-\frac{\lambda}{\lambda-1} m_{\mathcal{A}}) = \mathcal{M}$. The constant $\mathcal{M}$ is central in our study of gaps between runs.

Consider a word $w \in \Sigma^*$. The run attaining the minimal value $\mathcal{A}^*(w)$ might not be minimal while reading prefixes of w. The *gap* between the value of an eventually-minimal run and minimal runs on prefixes of w is central to characterizing determinizability of NDAs [6]. This gap is captured by the following definition.

Definition 1 (Recoverable gap). *Consider words $w, z \in \Sigma^*$ and states $q_u, q_l \in Q$. the tuple (w, q_u, q_l) is called a recoverable gap with respect to z, or simply a recoverable gap, if the following hold:*

1. $\mathcal{A}_{[Q_0 \to q_l]}(w) \leq \mathcal{A}_{[Q_0 \to q_u]}(w)$, and
2. $\mathcal{A}_{[Q_0 \to q_u]}(w) + \lambda^{-|w|} \mathcal{A}_{[q_u \to f_Q]}(z) = \mathcal{A}^*(wz) < \infty$.

Intuitively, in a recoverable gap (w, q_u, q_l) there are runs ρ_1 and ρ_2 of $\mathcal{A}$ on w that end in q_u and q_l, respectively, where ρ_1 attains a higher value than ρ_2, but there is a suffix z that "recovers" this gap: when reading z from q_u starting with weight $\text{val}(\rho_1)$, the resulting minimal run including final weight attains the minimal value of a run of $\mathcal{A}$ on $w \cdot z$. This is depicted in Figure 2.

For a recoverable gap (w, q_u, q_l) we define $\text{gap}(w, q_u, q_l) = \lambda^{|w|}(\mathcal{A}_{[Q_0 \to q_u]}(w) - \mathcal{A}_{[Q_0 \to q_l]}(w))$. The normalizing factor $\lambda^{|w|}$ eliminates the effect of the length of w on the gap, allowing us to study gaps independently of the length of their corresponding words.

We say that $\mathcal{A}$ has *finitely/infinitely many recoverable gaps* if the set $\{\text{gap}(w, q_u, q_l) \mid w \in \Sigma^*, q_u, q_l \in Q\}$ is finite/infinite, respectively. Note that since $\mathcal{A}$ is integral, $\lambda^{|w|}(\mathcal{A}_{[Q_0 \to q_u]}(w) - \mathcal{A}_{[Q_0 \to q_l]}(w))$ is always an integer and so the existence of infinitely many recoverable gaps is equivalent to the existence of unboundedly large recoverable gaps.

While gaps refer to two distinct runs, we sometimes need a more global view of gaps. To this end, we lift the definition to all the reachable states, as follows.

Definition 2 (n-separation). *For a word w and $n \in \mathbb{N}$, we say that w has the n-separation property if there exists a partition of Q into two non-empty sets of states U, L such that the following holds:*

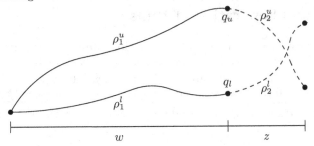

Fig. 2: The run ρ_1^l, ending in state q_l, is the minimal run of $\mathcal{A}$ on w. The higher run ρ_1^u is the minimal run on w that ends in state q_u, thus creating a gap between q_u and q_l. However, the concatenation $\rho_1^u \cdot \rho_2^u$ is the minimal run on the concatenated word wz, while the concatenation $\rho_1^l \cdot \rho_2^l$, where ρ_2^l is the minimal run on z starting in q_l, is not smaller. Therefore, the gap is recoverable. Note that here the final weights are zero.

1. For every $q_u \in U$ and $q_l \in L$, $\lambda^{|w|}(\mathcal{A}_{[Q_0 \to q_u]}(w) - \mathcal{A}_{[Q_0 \to q_l]}(w)) > n$.
2. There exist $q_u \in U$ and $z \in \Sigma^*$ such that for every $q_l \in L$, (w, q_u, q_l) is a recoverable gap with respect to z.

We sometimes explicitly specify that w has the n-separation property with respect to (U, L, q_u), or with respect to (U, L, q_u, z). If there exists w with the n-separation property, we say that $\mathcal{A}$ has the n-separation property.

See Figure 3 for a depiction of n-separation.

4 Determinizability of Integral NDAs is Decidable – Proof Overview

Recall that our goal is to show the decidability of the determinization problem.

As showed in [6], determinizability is closely related to recoverable gaps. More precisely, a DDA $\mathcal{D}$ that "attempts" to be equivalent to $\mathcal{A}$ must keep track of all the relevant runs of $\mathcal{A}$. If two runs end in the same state, it is clearly enough to track only the minimal one. However, this may still require keeping track of runs that attain unboundedly high values (when normalized). Therefore, in order for $\mathcal{D}$ to be finite, it must discard information on runs that get too high. The main issue is whether we can give a bound above which runs are no longer relevant.

For *complete* integral NDAs, there are always finitely many recoverable gaps, and this is used to show that complete NDAs are always determinizable [6]. For a general integral NDA $\mathcal{A}$, we similarly show in Section 5 that if there are only finitely many recoverable gaps, then $\mathcal{A}$ is determinizable.

There are now two main challenges: First, to show that if $\mathcal{A}$ has infinitely many recoverable gaps, then it is not determinizable, and second, that it is decidable whether $\mathcal{A}$ has finitely many recoverable gaps.

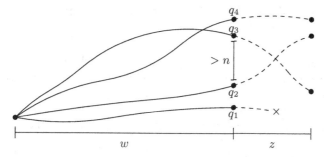

Fig. 3: Depicted are minimal runs of an NDA on a word w that end in each of four states, q_1, q_2, q_3, q_4, and minimal runs on z starting in each of them. The run from q_1 (lowest) "gets stuck", i.e., such a run from q_1 on z does not exist. The states are partitioned into two sets $L = \{q_1, q_2\}$ and $U = \{q_3, q_4\}$, with a gap larger than n between them after reading w; additionally, one of the upper runs then becomes minimal after reading z, since each of the lower runs either ends higher or "gets stuck". This means that the word w has the n-separation property with respect to (U, L, q_3, z).

We start by showing in Section 6.1 that we can compute a bound $\mathcal{N}$ such that $\mathcal{A}$ has infinitely many recoverable gaps if and only if it has a recoverable gap larger than $\mathcal{N}$. Next, in Section 6.2, we show that the existence of a gap larger than $\mathcal{N}$ is also equivalent to some word having the $\mathcal{N}$-separation property.

We then turn to exhibit a small-model property on witnesses for $\mathcal{N}$-separation. Specifically, we show in Section 7 that if there exist w, z such that w has the $\mathcal{N}$-separation property with respect to (U, L, q_u, z), then we can bound the length of the shortest w, z.

Using the above, we obtain the decidability of whether $\mathcal{A}$ has infinitely many recoverable gaps. In addition, we use these results to prove (in Lemma 11) that if $\mathcal{A}$ has infinitely many recoverable gaps, then it is not determinizable. This allows us to conclude the decidability of determinization in Theorem 1.

Conceptually, our approach can be viewed as a "standard" one when treating determinization of quantitative models, in the sense that considering gaps between runs generally characterizes when a deterministic equivalent exists [16, 2]. The crux is showing that this condition is decidable. To this end, our work greatly differs from other works on weighted automata in that we establish the decidability of the condition. Technically, this involves careful analysis of the behaviours of runs under discounting.

5   Finitely Many Recoverable Gaps Imply Determinizability

The main result of this section is an adaptation of the determinization techniques in [6] from complete to general automata. While the construction itself is similar, the correctness proof requires finer analysis. We remark that in the case that $\mathcal{A}$ is a complete NDA and all final weights are zero, the construction obtains a complete DDA with all final weights zero, thus generalizing the result in [6].

Lemma 1. *If an NDA $\mathcal{A}$ has finitely many recoverable gaps, then it is determinizable.*

Proof. Let $\mathcal{A} = (\Sigma, Q, Q_0, \alpha, \delta, \text{val}, \text{fval}, \lambda)$ be an NDA with finitely many recoverable gaps. We construct a DDA $\mathcal{D} = (\Sigma, Q_D, \{v_0\}, \alpha_D, \delta_D, \text{val}_D, \text{fval}_D, \lambda)$ that is equivalent to $\mathcal{A}$.

Since $\mathcal{A}$ has finitely many recoverable gaps, there exists a bound $B \in \mathbb{N}$ on the size of those gaps. The states of $\mathcal{D}$ are then $Q_D = \{0, \dots, B, \infty\}^Q$. Intuitively, a run of $\mathcal{D}$ tracks, for each $q \in Q$, the gap between the minimal run of $\mathcal{A}$ on w ending in q and the minimal run on w overall. When this gap becomes too large to be recoverable, the states corresponding to the higher run are assigned ∞. For $v \in Q_D$ and $q \in Q$, we denote by (v_q) the entry in v corresponding to q.

The initial state is therefore $(v_0)_q = \begin{cases} 0 & q \in Q_0 \\ \infty & q \notin Q_0 \end{cases}$, assigning for each $q \in Q$ the weight of the minimal run of $\mathcal{A}$ on the empty word ending in q.

We now turn to define δ_D. Intuitively, when taking a transition, $\mathcal{D}$ D first updates the vector entry of every state with the value of the minimal run on the new word ending in it, using the values specified in the last vector. Then, if the minimal entry is not 0, the entries are shifted so that it becomes 0, and the value subtracted from every entry is assigned to the transition weight. Finally, the entries are all multiplied by λ to account for the word length. Thus, the actual value of the minimal run is exactly the value attained by $\mathcal{D}$, and the vector entries correctly represent the normalized gaps. The construction is demonstrated in Figure 4. Formally:

- For every $v \in Q_D$, and for every $\sigma \in \Sigma$ such that there exists $q \in Q$ with $v_q < \infty$ and q has a σ-transition, define $u \in \{0, \dots, B, \infty\}^Q$ as follows.
  - Define the intermediate vector u': For every $q \in Q$, $u'_q = \min_{q' \in Q}(v_{q'} + \text{val}(q', \sigma, q))$, where $\text{val}(q', \sigma, q)$ is regarded as ∞ if $(q', \sigma, q) \notin \delta$.
  - Define $r = \min_{q \in Q} u'_q$, the offset of the vector from 0. Note that r is finite due to the requirement that there exists $q \in Q$ with $v_q < \infty$ and q has a σ-transition.
  - For every $q \in Q$, $u_q = \begin{cases} \lambda(u'_q - r) & \lambda(u'_q - r) \leq B \\ \infty & \text{otherwise} \end{cases}$

Where ∞ is handled using the standard semantics. Note that $u \in \{0, \dots, B, \infty\}^Q$ as $\mathcal{A}$ is integral. The manipulations done on the intermediate vector u'_q when defining u_q should be viewed as normalization – first subtracting r

so that the gap represented by u_q is with respect to the minimal run overall over w; then multiplying by λ to account for the length of w. Note that the subtraction of r also implies that $\min_{q \in Q} u_q = 0$, as is expected since $\min_{q \in Q} \mathcal{A}_{[Q_0 \to q]}(w) = \mathcal{A}(w)$.

- We now introduce the transition $(v, \sigma, u) \in \delta_D$.
- We set $\text{val}_D(v, \sigma, u) = r$. This can be viewed, together with the subtraction of r from every entry of u', as transferring the weight from each entry of u' to the transition.

(a) an NDA $\mathcal{A}$.   (b) An equivalent DDA $\mathcal{D}$.

Fig. 4: An example of an NDA $\mathcal{A}$ (on the left) and the resulting DDA $\mathcal{D}$ (on the right), with $\lambda = 2$. The name of each state of $\mathcal{D}$ corresponds to a vector whose first entry tracks q_0 and the second q_1. We demonstrate the construction using the a-transition from $(2,0)$ to itself. First we construct the intermediate vector u': $(u')_{q_0} = \min(2 + \text{val}(q_0, a, q_0), 0 + \text{val}(q_1, a, q_0)) = \min(2 + 1, 0 + \infty) = 3$ and $(u')_{q_1} = \min(2 + \text{val}(q_0, a, q_1), 0 + \text{val}(q_1, a, q_1)) = \min(2 + 0, 0 + \infty) = 2$, and so $u' = (3, 2)$. We then have $r = 2$, which is assigned to the weight of the transition, and $u = 2(3 - 2, 2 - 2) = (2, 0)$.

We next define α_D and fval_D. We set α_D to include every vector v such that $v_q < \infty$ for some $q \in \alpha$. We note that the construction can be viewed as a generalization of the standard subset construction, where for a vector v, the states q that satisfy $v_q < \infty$ represent the states that can be reached by $\mathcal{A}$ when reading w, ignoring those states whose gap is unrecoverable. For $v \in \alpha_D$, we set $\text{fval}_D(v) = \min_{q \in \alpha}(v_q + \text{fval}(q))$. Figure 4 depicts an example for an NDA and the DDA constructed from it (with no final weights). Note that we do not yet actually provide an algorithm for constructing $\mathcal{D}$ from $\mathcal{A}$, since that requires computing B.

The correctness of this construction is proved in the full version.   □

6   Recoverable Gaps and n-separation

6.1   A Large Gap is Equivalent to Infinitely Many Gaps

In this section we show that the existence of infinitely many recoverable gaps is characterized by the existence of a (computable) large-enough recoverable gap.

Consider a run $\rho = \rho_0 \ldots \rho_n$, and recall that $\texttt{val}(\rho)$ is the weight of ρ and that $\mathcal{M} = 2\frac{\lambda}{\lambda - 1} m_{\mathcal{A}}$, where $m_{\mathcal{A}}$ is the maximal absolute value of a weight of a transition or a final weight in $\mathcal{A}$. We denote by $\Gamma(\rho) = \lambda^n \texttt{val}(\rho)$ the normalized "un-discounted" value of ρ. For two runs ρ^1, ρ^2 on the same word w, we are interested in the value $\Gamma(\rho^1) - \Gamma(\rho^2)$, as it captures how far the runs are from each other, in the sense of how difficult it is to recover their gap. We claim that if two runs get too far from each other, the gap between them from that point on can only increase. Intuitively, this is because at each step the value is multiplied by λ, and so beyond a certain gap size, this multiplication separates the runs further even if their added values pull them closer before multiplying by λ.

Lemma 2. *Let $\rho^1 = \rho_0^1, \ldots, \rho_{n+1}^1$ and $\rho^2 = \rho_0^2, \ldots, \rho_{n+1}^2$ be two runs of $\mathcal{A}$, such that $\Gamma(\rho_0^1, \ldots, \rho_n^1) - \Gamma(\rho_0^2, \ldots, \rho_n^2) > \mathcal{M}$. Then $\Gamma(\rho_0^1, \ldots, \rho_{n+1}^1) - \Gamma(\rho_0^2, \ldots, \rho_{n+1}^2) > \Gamma(\rho_0^1, \ldots, \rho_n^1) - \Gamma(\rho_0^2, \ldots, \rho_n^2)$.*

In particular, once the gap between ρ^1, ρ^2 is larger than $\mathcal{M}$, concatenating any runs to ρ^1, ρ^2 can only increase the gap and therefore cannot result in ρ^1 "bypassing" ρ^2.

Corollary 1. *Let $\rho^1 = \rho_0^1, \ldots, \rho_n^1$ and $\rho^2 = \rho_0^2, \ldots, \rho_n^2$ be two runs such that $\texttt{val}(\rho^1) \leq \texttt{val}(\rho^2)$. Then for every $0 \leq i \leq n$, it holds that $\Gamma(\rho_0^1, \ldots, \rho_i^1) - \Gamma(\rho_0^2, \ldots, \rho_i^2) \leq \mathcal{M}$.*

On the other hand, the gap between two runs cannot increase too much within a small number of steps. We capture the contra-positive of this, by showing that if two runs reach a large enough gap, then the runs have been far from each other for a long suffix.

Lemma 3. *Consider $n_{steps}, n_{gap} \in \mathbb{N}$, there exists an effectively computable number N such that for any two runs $\rho^1 = \rho_0^1, \ldots, \rho_n^1$ and $\rho^2 = \rho_0^2, \ldots, \rho_n^2$, if $\Gamma(\rho^1) - \Gamma(\rho^2) > N$ then $n > n_{steps}$ and $\Gamma(\rho_0^1, \ldots, \rho_{n-n_{steps}}^1) - \Gamma(\rho_0^2, \ldots, \rho_{n-n_{steps}}^2) > n_{gap}$.*

We also need a version of Corollary 1 where the inequality between the weights of the runs includes final weights. We claim that concatenating any runs to runs that are far from each other cannot result in the lower run "bypassing" the upper run, including final weights:

Lemma 4. *Let ρ^u, ρ^l be two runs of $\mathcal{A}$ on w, ending in states q_u, q_l respectively, such that $\Gamma(\rho^u) - \Gamma(\rho^l) > \mathcal{M}$. Let ρ^{u_f}, ρ^{l_f} be accepting runs on z starting in q_u, q_l respectively and ending in q_{u_f}, q_{u_l} respectively. Then $\texttt{val}(\rho^u \rho^{u_f}) + \lambda^{-|wz|} \texttt{fval}(q_{u_f}) > \texttt{val}(\rho^l \rho^{l_f}) + \lambda^{-|wz|} \texttt{fval}(q_{l_f})$.*

In particular, once a gap becomes too large, the only way to recover from it is if the lower run cannot continue at all.

Lemma 5. *Consider a recoverable gap (w, q_u, q_l) with respect to z such that $\texttt{gap}(w, q_u, q_l) > \mathcal{M}$, then $\mathcal{A}_{[q_u \to_f \alpha]}(z) < \infty$ and $\mathcal{A}_{[q_l \to_f \alpha]}(z) = \infty$.*

Proof. From the second condition in the definition of recoverable gap (Definition 1), we have $\mathcal{A}_{[q_u \to_f \alpha]}(z) < \infty$. Let ρ^u, ρ^l be minimal runs on $|w|$ ending in q_u, q_l respectively. Assume by way of contradiction that $\mathcal{A}_{[q_l \to_f \alpha]}(z) < \infty$, that is, there exists an accepting run $\rho^{l'}$ on z starting in q_l. Let $\rho^{u'}$ be a minimal accepting run on z starting in q_u. Since $\lambda^{|w|}(\text{val}(\rho^u) - \text{val}(\rho^l)) = \text{gap}(w, q_u, q_l) > \mathcal{M}$, Lemma 4 contradicts the fact that $\rho^1 \rho^{l'}$ is a minimal accepting run on wz by the definition of recoverable gap. □

We can now prove the main result of this section.

Lemma 6. *There exists an effectively computable number N (depending on $\mathcal{A}$) such that $\mathcal{A}$ has infinitely many recoverable gaps if and only if there exists a recoverable gap (w, q_u, q_l) such that $\text{gap}(w, q_u, q_l) > N$.*

Proof Overview We start with an overview of the more complex direction – the existence of a large recoverable gap implies the existence of infinitely many recoverable gaps. Assume that (w, q_u, q_l) is a large recoverable gap with respect to z. We consider two minimal runs ρ^{q_u}, ρ^{q_l} on w ending in q_u and q_l, respectively. These two runs end "far" from each other, so we can use Lemma 3 to claim that for a large enough N, they have already been far from each other for a while. Specifically, for the last n_{steps} steps the gap between the runs was at least n_{gap} for some large $n_{\text{steps}}, n_{\text{gap}}$ that we choose to fit our needs.

We now look for two indices $i < j$ among the last n_{steps} indices of w such that pumping the infix of w between i and j generates words that induce unboundedly large recoverable gaps. To do so, we choose n_{steps} such that Q can be partitioned into two sets of states – an upper set U and a lower set L, that are far from each other and "separate" the runs ρ^{q_u}, ρ^{q_l} at step i. In particular, pumping the infix does not interleave the runs, and maintains the growing gap. The above is depicted in Figure 5. We require the following properties:

1. Every two runs on the prefix $w_1 \cdots w_i$ of w ending in U and in L, respectively, that are minimal runs ending in their respective states, are far enough from each other to satisfy the condition of Lemma 2;
2. Every run on w that is minimal among the runs ending in q_u has to visit U at the i'th step;
3. Every run on w that is minimal among the runs ending in q_l has to visit L at the i'th step;

As we show, finding such a partition is possible by choosing $n_{\text{gap}} = (|Q| - 1)\mathcal{M}$.

Next, we show that in fact, U and L induce a certain separation of the run *trees* emanating from them on the pumped words. Specifically, we show that:

(i) There exist runs of $\mathcal{A}$ on the pumped words (denoted $w^{(*)}$) ending in q_u, q_l.
(ii) Every run on $w^{(*)}$ (ending in any state) that is a prefix of a minimal run on $w^{(*)}z$ has to visit U at the i'th step. That is, a variant of Condition (2), where instead of q_u we consider any state p reached after reading $w^{(*)}$ along a minimal run on $w^{(*)}z$.

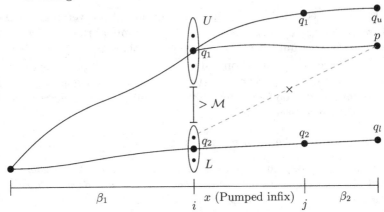

Fig. 5: At step i of $\mathcal{A}$'s run on w, the states are partitioned into an upper set U and a lower set L that are separated by a large gap. The runs ρ^{q_u}, ρ^{q_l} visit U, L respectively, meaning the gap between them can only grow after step i. The indices i, j are chosen such that both runs ρ^{q_u}, ρ^{q_l} repeat states and the sets of ancestors $\mathrm{Anc}_q(i), \mathrm{Anc}_q(j)$ are identical for each $q \in Q$. The state p, which is visited after reading a pumped word $w^{(*)}$ by a minimal run on $w^{(*)}z$, is not reachable from L on any of the pumped suffixes.

(iii) Condition (3) above holds not only for w but for the pumped words $w^{(*)}$ as well.

Note that (ii) and (iii) imply that runs on $w^{(*)}$ also induce a recoverable gap.

From this, it follows from Condition 1 and Lemma 2 that the pumped words induce unboundedly large recoverable gaps.

In order to ensure (i), we require $n_{\text{steps}} \geq |Q|^2$ (which is the length of the "large gaps" suffix) such that both runs ρ^{q_u}, ρ^{q_l} must repeat their pair of respective states at some indices i, j. Consequently, the runs ρ^{q_u}, ρ^{q_l} can be pumped to achieve the desired runs.

To ensure (ii), it follows from Corollary 1 and the fact that ρ^{q_u} is a prefix of a minimal run on wz that any state p reached along a run on $w^{(*)}z$ after reading $w^{(*)}$ is not reachable from L when reading $w_{i+1} \cdots w_{|w|}$, and we want to ensure that p is not reachable from L when reading the pumped suffix as well. For that, for each state q and for each index of w we consider the set of states $\mathrm{Anc}_q(i)$ from which q is reachable when reading the respective suffix (from index i), called the *ancestors* of q at index i, and it is enough to require that for each state q this set is identical for indices i and j. This, in turn, requires to increase n_{steps} by a factor of $2^{|Q|^2}$. Combined with the previous requirement on i, j, we choose $n_{\text{steps}} = |Q|^2 2^{|Q|^2}$.

Finally, for condition (3) in (iii), we use the fact that there exists a run ending in q_l that visits L at the i'th step (namely the pumped run) and apply

Corollary 1. Indeed, any run that does not visit L at the i'th step must visit U instead, and by Corollary 1 and the gap between U and L, it must be larger than the run we have that visits L and therefore not minimal. $\qquad \square$

Proof (of Lemma 6). Consider runs $\rho_0^1, \ldots, \rho_n^1$ and $\rho_0^2, \ldots, \rho_n^2$. From Lemma 3, we can effectively compute N such that if $\Gamma(\rho_0^1, \ldots, \rho_n^1) - \Gamma(\rho_0^2, \ldots, \rho_n^2) > N$, then $n > |Q|^2 2^{|Q|^2}$ and $\Gamma(\rho_0^1, \ldots, \rho_{n-|Q|^2 2^{|Q|^2}}^1) - \Gamma(\rho_0^2, \ldots, \rho_{n-|Q|^2 2^{|Q|^2}}^2) > (|Q|-1)\mathcal{M}$.

Assume that (w, q_u, q_l) is a recoverable gap with respect to z and $\mathsf{gap}(w, q_u, q_l) > N$. Let $\rho^{q_u} = \rho_0^{q_u} \ldots \rho_{|w|}^{q_u}$ be a run on w ending in q_u that is minimal among the runs on w ending in q_u, and similarly $\rho^{q_l} = \rho_0^{q_l} \ldots \rho_{|w|}^{q_l}$ for q_l. Since these runs are minimal runs ending in their respective states, it holds that $\Gamma(\rho^{q_u}) - \Gamma(\rho^{q_l}) = \mathsf{gap}(w, q_u, q_l) > N$, and so we have $|w| > |Q|^2 2^{|Q|^2}$ and $\Gamma(\rho_0^1, \ldots, \rho_{|w|-|Q|^2 2^{|Q|^2}}^1) - \Gamma(\rho_0^2, \ldots, \rho_{|w|-|Q|^2 2^{|Q|^2}}^2) > (|Q| - 1)\mathcal{M}$.

For every $q \in Q$ and $1 \leq i \leq |w|$, let $\mathrm{Anc}_q(i) = \{q' \in Q \mid \mathcal{A}_{[q' \to q]}(w_{i+1} \cdots w_{|w|}) < \infty\}$ be the set of *ancestors* of q at step i, i.e., states from which q is reachable when reading the input $w_{i+1}, \cdots, w_{|w|}$. By the pigeonhole principle there exist $|w| - |Q|^2 2^{|Q|^2} \leq i < j \leq |w|$ such that

- For every $q \in Q$, $\mathrm{Anc}_q(i) = \mathrm{Anc}_q(j)$,
- $\rho_i^{q_u} = \rho_j^{q_u}$ and $\rho_i^{q_l} = \rho_j^{q_l}$.

Write $w = \beta_1 x \beta_2$ where $\beta_1 = w_1 \cdots w_i$, $x = w_{i+1} \cdots w_j$ and $\beta_2 = w_{j+1} \cdots w_{|w|}$. We now turn to show that by pumping x, we can obtain unboundedly large recoverable gaps.

Let $k \in \mathbb{N}$. We can easily show that $\beta_1 x^k \beta_2$ induces unboundedly large gaps between q_u and q_l, but that would not be sufficient: We also need those gaps to be recoverable with respect to z, that is, the minimal run on $\beta_1 x^k \beta_2 z$ has to visit q_u after reading $\beta_1 x^k \beta_2$. However, this is not necessarily true: It can visit a different state, and we need to show that that state is also far enough from q_l. The runs $\rho_0^{q_u} \cdots \rho_i^{q_u} (\rho_{i+1}^{q_u} \cdots \rho_j^{q_u})^k \rho_{j+1}^{q_u} \cdots \rho_{|w|}^{q_u}$ and $\rho_0^{q_l} \cdots \rho_i^{q_l} (\rho_{i+1}^{q_l} \cdots \rho_j^{q_l})^k \rho_{j+1}^{q_l} \cdots \rho_{|w|}^{q_l}$ are runs on $\beta_1 x^k \beta_2$ ending in q_u, q_l respectively. In particular, since there exists a run on z starting in q_u, we have that $\mathcal{A}$ has a run on $\beta_1 x^k \beta_2 z$. Let ρ be minimal among those runs, and let $q_{\min_k}$ be the state ρ visits after reading $\beta_1 x^k \beta_2$. Let $\rho^{q_{\min_k}, k}, \rho^{q_l, k}$ be runs on $\beta_1 x^k \beta_2$ that are minimal among the runs ending in $q_{\min_k}, q_l$ respectively. Note that $\rho^{q_{\min_k}, k}$ can be obtained as a prefix of ρ, since ρ is minimal. Then we have $\mathsf{gap}(\beta_1 x^k \beta_2, q_{\min_k}, q_l) = \Gamma(\rho^{q_{\min_k}, k}) - \Gamma(\rho^{q_l, k})$, and it remains to show that $\Gamma(\rho^{q_{\min_k}, k}) - \Gamma(\rho^{q_l, k})$ can get unboundedly large for a large enough k.

We already know that the runs ρ^{q_u}, ρ^{q_l} are far enough from each other at their i'th step to satisfy the condition of Lemma 2, and we want to show that the same is true for $\rho^{q_{\min_k}, k}, \rho^{q_l, k}$.

To do so, we intuitively show that after reading β_1, the runs ρ^{q_u}, ρ^{q_l} have become so far apart that they now stem from disjoint sets of states with a large gap between them. Formally, consider the sets

$$U' = \{q \in Q \mid \text{there exists a run } \rho \text{ on } w \text{ with } \text{val}(\rho) = \mathcal{A}_{[Q_0 \to q_u]}(w) \text{ and } \rho_i = q\}$$
$$L' = \{q \in Q \mid \text{there exists a run } \rho \text{ on } w \text{ with } \text{val}(\rho) = \mathcal{A}_{[Q_0 \to q_l]}(w) \text{ and } \rho_i = q\}$$

That is, U' (resp. L') is the set of states that appear at step i in a minimal run to q_u (resp. q_l). For every $q \in Q$, let $v(q) = \lambda^i \mathcal{A}_{[Q_0 \to q]}(\beta_1)$ be the "undiscounted" value of a minimal run of $\mathcal{A}$ on β_1 ending in q. Then, from Lemma 3 and the constants we chose, for every $q'_u \in U', q'_l \in L'$ we have $v(q'_u) - v(q'_l) > (|Q|-1)\mathcal{M}$.

In particular, there is a partition of Q into two sets U, L such that $U' \subseteq U, L' \subseteq L$ and $v(p) - v(q) > \mathcal{M}$ for every $p \in U$ and $q \in L$. Indeed, otherwise the maximal gap between two states is less than $(|Q|-1)\mathcal{M}$. We next show that (i) $\rho_i^{q_{\min_k}, k} \in U$, and (ii) $\rho_i^{q_l, k} \in L$.

For (i), we note that $\mathcal{A}_{[L \to q_{\min_k}]}(x\beta_2) = \infty$: Otherwise, since $\mathcal{A}_{[q_{\min_k} \to f\alpha]}(z) < \infty$, there exists an accepting run on wz that visits L after reading β_1 and $q_{\min_k}$ after reading w. By Lemma 4, such a run must be of lower weight than any run that visits U after reading β_1, in contradiction to the fact that ρ^{q_u} is a prefix of a minimal run on wz by the definition of recoverable gap. Since $\text{Anc}_{q_{\min_k}}(i) = \text{Anc}_{q_{\min_k}}(j)$, we also have that $\mathcal{A}_{[L \to q_{\min_k}]}(x^k\beta_2) = \infty$. In particular, $\rho_i^{q_{\min_k}, k} \in U$.

For (ii), the run $\rho_0^{q_l} \cdots \rho_i^{q_l} (\rho_{i+1}^{q_l} \cdots \rho_j^{q_l})^k \rho_{j+1}^{q_l} \cdots \rho_{|w|}^{q_l}$ satisfies $\rho_i^{q_l} \in L$ (since it is in L'), and in particular $\mathcal{A}_{[L \to q_l]}(x^k\beta_2) < \infty$. By Corollary 1, any run whose i'th state is in U results in a higher weight than any run whose i'th state is in L, and so since $\rho^{q_l, k}$ is minimal we have $\rho_i^{q_l, k} \in L$.

It remains to show that the runs $\rho^{q_{\min_k}, k}, \rho^{q_l, k}$, being far from each other at the i'th step, get unboundedly far from each other as k increases. Let $f_u, f_l : \{i, i+1, \ldots\} \to \mathbb{N}$ be defined as follows:

- $f_u(i) = \min_{q \in U} v(q)$
- For $m \geq i$, $f_u(m+1) = \lambda(f_u(m) - m_\mathcal{A})$
- $f_l(i) = \max_{q \in L} v(q)$
- For $m \geq i$, $f_l(m+1) = \lambda(f_l(m) + m_\mathcal{A})$

Intuitively, f_u (resp. f_l) represents a lower (resp. upper) bound on the "undiscounted" weight of runs visiting U (resp. L) in their i'th step. That is, for every $m \geq i$ we have $\Gamma(\rho_0^{q_{\min_k}, k} \cdots \rho_m^{q_{\min_k}, k}) \geq f_u(m)$, and $\Gamma(\rho_0^{q_l, k} \cdots \rho_m^{q_l, k}) \leq f_l(m)$. Additionally, $f_u(i) - f_l(i) > \mathcal{M}$ and so the function $f_u(m) - f_l(m)$ increases with m. Thus, for every $M \in \mathbb{N}$, taking a large enough k, we can obtain $\Gamma(\rho^{q_{\min_k}, k}) - \Gamma(\rho^{q_l, k}) \geq f_u(|\beta_1 x^k \beta_2|) - f_l(|\beta_1 x^k \beta_2|) > M$. This concludes the proof that if $\mathcal{A}$ has a large recoverable gap, then it has infinitely many recoverable gaps.

For the converse direction, assume $\mathcal{A}$ has infinitely many recoverable gaps. Since $\mathcal{A}$ is integral, the term $\lambda^{|w|}(\mathcal{A}_{[Q_0 \to p]}(w) - \mathcal{A}_{[Q_0 \to q]}(w))$ is always an integer, therefore infinitely many recoverable gaps imply the existence of unboundedly large recoverable gaps, and in particular one larger than N.   □

Remark 1. Following the arguments in the proofs of Lemmas 3 and 6, the number N provided by Lemma 6 equals $\lambda^{|Q|^2 2^{|Q|^2}}((|Q|-1)\mathcal{M} - \mathcal{M}) + \mathcal{M}$. We denote this value by $\mathcal{N}$.

6.2 A Large Gap is Equivalent to Separation

Recall that a gap refers to minimal runs that end in two specific states, but ignores the remaining states (to an extent). A more "holistic" view of gaps is via separations (Definition 2). In this section we show that the two views are equivalent, and that both characterize when $\mathcal{A}$ has infinitely many gaps.

Lemma 7. *$\mathcal{A}$ has a recoverable gap larger than $\mathcal{N}$ if and only if $\mathcal{A}$ has the $\mathcal{N}$-separation property.*

Proof. Assume that $\mathcal{A}$ has a recoverable gap larger than $\mathcal{N}$. By Lemma 6, there exist unboundedly large recoverable gaps, and in particular there exists a recoverable gap (w, q_u, q_l) with $\mathsf{gap}(w, q_u, q_l) > (|Q|-1)\mathcal{N}$.

Intuitively, when ordering the states by the weight of the minimal run that reaches each state, such a gap implies a gap of at least $\mathcal{N}$ between two successive states, leading to the desired partition. We then claim that the sets are separated by the same suffix z that separates the states from the original gap.

Write $Q = \{q_1, \ldots, q_{|Q|}\}$ such that $\mathcal{A}_{[Q_0 \to q_1]}(w) \leq \cdots \leq \mathcal{A}_{[Q_0 \to q_{|Q|}]}(w)$ (recall that if there is no run on w ending in q, then $\mathcal{A}_{[Q_0 \to q]}(w) = \infty$), and let $i_l < i_u$ be indices such that $q_l = q_{i_l}, q_u = q_{i_u}$. Then there exists $j \in \{i_l, \ldots, i_{u-1}\}$ such that $\mathcal{A}_{[Q_0 \to q_{j+1}]}(w) - \mathcal{A}_{[Q_0 \to q_j]}(w) > \mathcal{N}$. Let $U = \{q_{j+1}, \ldots, q_{|Q|}\}$ and $L = \{q_1, \ldots, q_j\}$. Then for every $q'_u \in U, q'_l \in L$ we have $\lambda^{|w|}(\mathcal{A}_{[Q_0 \to q'_u]}(w) - \mathcal{A}_{[Q_0 \to q'_l]}(w)) > \mathcal{N}$.

Consider $z \in \Sigma^*$ such that the gap (w, q_u, q_l) is recoverable with respect to z. Note that $\mathcal{N} > \mathcal{M}$, and so it follows from Lemma 4 that $\mathcal{A}_{[q'_l \to_f \alpha]}(z) = \infty$ for every $q'_l \in L$. Indeed, if $\mathcal{A}$ had a run on z starting in q'_l, concatenating it to a minimal run on w ending in q'_l would result in a run of lower weight than any run on wz that visits q_u after reading w, contradicting the fact that (w, q_u, q_l) is a recoverable gap. Additionally, it follows from (w, q_u, q_l) being a recoverable gap that there exists a minimal run on wz that visits q_u after reading w. Then (w, q_u, q'_l) is a recoverable gap with respect to z, and so w has the $\mathcal{N}$-separation property with respect to (U, L, q_u, z).

For the converse direction, assume that w has the $\mathcal{N}$-separation property with respect to some (U, L, q_u, z). In particular, $\mathcal{A}_{[Q_0 \to q_u]}(w) + \lambda^{-|w|}\mathcal{A}_{[q_u \to_f \alpha]}(z) < \infty$. Let $q'_u \in U$ be such that $\mathcal{A}_{[Q_0 \to q'_u]}(w) + \lambda^{-|w|}\mathcal{A}_{[q'_u \to_f \alpha]}(z)$ is minimal. Let some $q'_l \in L$. Then (w, q'_u, q'_l) is a recoverable gap with respect to z, and it is larger than $\mathcal{N}$, as needed. □

7 Bounding the Witnesses for Separation

In Section 6 we show that $\mathcal{A}$ has infinitely many recoverable gaps if and only if there exists a word w with the $\mathcal{N}$-separation property. Expanding Definition 2, this happens if and only if there exist a partition of Q into two sets U, L and there exist words w, z that "separate" U from L. In this section we can bound the length of such minimal w, z. We start with w (see the full version for the detailed proof).

Lemma 8. *Let $C = \frac{\lambda}{\lambda-1}(\mathcal{N}|Q| + 2m_A)$. Assume that w has the $\mathcal{N}$-separation property for some $w \in \Sigma^*$. Then there exists w' such that w' has the $\mathcal{N}$-separation property and $|w'| \leq (C+2)^{|Q|}$.*

Proof (Sketch). Assume that w has the $\mathcal{N}$-separation property with respect to (U, L, q_u, z).

We start by using an identical construction to that of Lemma 1, with bound C, in order to define a sequence of vectors $v_0, \ldots, v_{|w|}$ with $v_i \in \{0, \ldots, C, \infty\}^Q$ for every $0 \leq i \leq |w|$ that, intuitively, keep track of the runs of $\mathcal{A}$ on w, as follows.

- For every $q \in Q$ set $(v_0)_q = \begin{cases} 0 & q \in Q_0 \\ \infty & \text{otherwise} \end{cases}$

- For every $i > 0, q \in Q$ let $v'_{i,q} = \min_{q' \in Q}((v_{i-1})_{q'} + \texttt{val}(q', w_i, q))$, where $\texttt{val}(q', \sigma, q)$ is regarded as ∞ if $(q', \sigma, q) \notin \delta$ (the $v'_{i,q}$ are "intermediate" values).

- For every $i > 0$ let $r_i = \min_{q \in Q} v'_{i,q}$ (the r_i are the offset of the vector from 0).

- For every $i > 0, q \in Q$ set $(v_i)_q = \begin{cases} \lambda(v'_{i,q} - r_i) & \lambda(v'_{i,j} - r_i) \leq C \\ \infty & \text{otherwise} \end{cases}$

Recall that intuitively, (v_i) tracks, for each $q \in Q$, the gap between the minimal run on $w_1 \cdots w_i$ ending in q and the minimal run on this prefix overall. When this gap becomes large enough that recovering from it implies the existence of $\mathcal{N}$-separation, it is denoted ∞.

Denote the normalized difference $\lambda^i(\mathcal{A}_{[Q_0 \to q]}(w_1 \cdots w_i) - \mathcal{A}(w_1 \cdots w_i))$ by $\Delta_{q,i}(w)$. It is easy to show that v_i keeps the correct weight of runs whose gap from the minimal one remains always under C. However, if a gap of a run goes over C but then comes back down, then v_i no longer tracks it correctly. To account for this, we claim that since w has the $\mathcal{N}$-separation property, for every q, i at least one of the following must hold:

- $(v_i)_q = \begin{cases} \Delta_{q,i}(w) & \Delta_{q,i}(w) \leq C \\ \infty & \text{otherwise} \end{cases}$.

- There exists $i' < i$ such that $w_1 \cdots w_{i'}$ has the $\mathcal{N}$-separation property.

That is, either v_i tracks the runs correctly, or there is some shorter prefix that already has the $\mathcal{N}$-separation property.

The proof is by induction on i, with the only problematic case arising when $(v_{i-1})_{q'} = \infty$, and so the information about the exact value of the gap represented by $(v_{i-1})_{q'}$ is gone. We consider the normalization value r_i (i.e., the offset of the minimal run from 0): if r_i is small, then the gap represented by $(v_i)_q$ is still very large, and we show that marking it as ∞ is sound. Otherwise, if r_i is large, then the above gap might indeed be wrongly marked as ∞. However, we show that in this case, r_i is so large that we can actually obtain an $\mathcal{N}$-separation property "below" r_i, using a shorter witness. More precisely:

- If $(v_{i-1})_{q'} = \infty$ and $r_i \leq C\frac{\lambda-1}{\lambda} - m_{\mathcal{A}}$, then since $(v_{i-1})_{q'} = \infty$, we have $(v_i)_q = \infty$. It remains to show that $\lambda^i(\mathcal{A}_{[Q_0 \to q]}(w_1 \cdots w_i) - \mathcal{A}(w_1 \cdots w_i)) > C$. Indeed,

$$\lambda^i(\mathcal{A}_{[Q_0 \to q]}(w_1 \cdots w_i) - \mathcal{A}(w_1 \cdots w_i))$$
$$\geq \lambda^i(\mathcal{A}_{[Q_0 \to q']}(w_1 \cdots w_{i-1}) - \mathcal{A}(w_1 \cdots w_{i-1}) - (m_{\mathcal{A}} + r_i) \cdot \lambda^{-(i-1)})$$
$$= \lambda(\lambda^{i-1}(\mathcal{A}_{[Q_0 \to q']}(w_1 \cdots w_{i-1}) - \mathcal{A}(w_1 \cdots w_{i-1})) - r_i - m_{\mathcal{A}})$$
$$> \lambda(C - (C\frac{\lambda-1}{\lambda} - m_{\mathcal{A}}) - m_{\mathcal{A}}) = \lambda(\lambda^{-1}C + m_{\mathcal{A}} - m_{\mathcal{A}}) > C$$

where the first transition follows by observing that when reading w_i, in the worst case, the weight of a specific run can decrease by $\lambda^{-(i-1)}m_{\mathcal{A}}$, and the overall weight of the word can increase by $\lambda^{-(i-1)}r_i$.

- $r_i > C\frac{\lambda-1}{\lambda} - m_{\mathcal{A}}$. This is only possible if for every q_l such that $(v_{i-1})_{q_l} < C\frac{\lambda-1}{\lambda} - 2m_{\mathcal{A}} = \mathcal{N}|Q|$, q_l has no w_i-transition. Let $L'' = \{q_l \in Q \mid (v_{i-1})_{q_l} < \mathcal{N}|Q|\}$. Write $Q = q_1, \ldots, q_{|Q|}$ such that $(v_{i-1})_{q_1} \leq \ldots \leq (v_{i-1})_{q_{|Q|}}$, and so $L'' = \{q_1, \ldots, q_{|L''|}\}$. Since w has the $\mathcal{N}$-separation property, in particular $\mathcal{A}$ has a run on w and so $L'' \subsetneq Q$. Then, there exists $1 \leq r \leq |L''|$ such that $(v_{i-1})_{q_{r+1}} - (v_{i-1})_{q_r} > \mathcal{N}$. Let $U' = \{q_{r+1}, \ldots, q_{|Q|}\}, L' = \{q_1, \ldots, q_r\}$, and note that for every $q'_l \in L'$, q'_l has no w_i-transition. For every $q'_l \in L', q'_u \in U'$, we have $\lambda^{i-1}(\mathcal{A}_{[Q_0 \to q'_u]}(w_1 \cdots w_{i-1}) - \mathcal{A}_{[Q_0 \to q'_l]}(w_1 \cdots w_{i-1}) = (v_{i-1})_{q'_u} - (v_{i-1})_{q'_l} > \mathcal{N}$. Let $q'_u \in U'$ be such that $\mathcal{A}_{[Q_0 \to q'_u]}(w_1 \cdots w_{i-1}) + \lambda^{-(i-1)}\mathcal{A}_{[q'_u \to f\alpha]}(w_i)$ is minimal. Then for every $q'_l \in L'$, $(w_1 \cdots w_{i-1}, q'_u, q'_l)$ is a recoverable gap with respect to w_i, and so $w_1 \cdots w_{i-1}$ has the $\mathcal{N}$-separation property with respect to (U', L', q'_u, w_i), and we are done.

Now, it remains to show that if $|w| > (C + 2)^{|Q|}$, there exists w' such that $|w'| < |w|$ and w' has the $\mathcal{N}$-separation property.

To this end, we use the induction hypothesis and the pigeonhole principle to remove an infix of w, and argue that the resulting word w' also has the $\mathcal{N}$-separation property with respect to some (U', L', q'_u): Either all of the minimal runs ending in the states of L have values far enough (below) of C, in which case U', L' can be chosen to be U, L respectively; or some state of L attains a high value, in which case there must be a large gap between two consecutive states of L, and the resulting lower set can be chosen as L'. As for q'_u, it is simply enough to consider the state in U' that the minimal run on $w'z$ visits after reading w' (see the full version for the details). □

Next, we give a bound on the length of the minimal separating suffix z from Definition 2. Recall that by Lemma 5, a large gap can only be recoverable if the smaller runs cannot continue at all. Following that, we can now limit the search to suffixes that separate runs in a Boolean sense (i.e., making one accept and another reject). This yields a bound from standard arguments about Boolean automata, as follows.

Lemma 9. *Consider a word w that has the $\mathcal{N}$-separation property with respect to (U, L, q_u, z). Then there exists z' such that w has the $\mathcal{N}$-separation property with respect to (U, L, q_u, z') and $|z'| \leq 2^{2|Q|}$.*

8   Determinizability of Integral NDAs is Decidable

In this section we establish the decidability of determinization. To this end, we start by completing the characterization of determinizable NDAs by means of gaps, and then use the results from previous sections to conclude the decidability of this characterization.

Recall that in Lemma 1 we show that finitely many recoverable gaps imply determinizability. In this section we show the converse, thus completing the characterization of determinizable integral NDAs as exactly those that have finitely many recoverable gaps.

We first need the following lemma which is proved in [6, Lemma 5].

Lemma 10. *Consider an NDA $\mathcal{A}$ for which there is an equivalent DDA $\mathcal{D}$. If there is a state q of $\mathcal{A}$ and words w, w', z such that:*

 – *$\mathcal{A}$ has runs on w and w' ending in q;*
 – *$\mathsf{gap}(w, q, p) \neq \mathsf{gap}(w', q, p')$, where p, p' are the last states of some minimal runs of $\mathcal{A}$ on w, w' respectively;*
 – *both gaps (w, q, p) and (w', q, p') are recoverable with respect to z;*

then the runs of $\mathcal{D}$ on w and w' end in different states.

We now show the converse of Lemma 1.

Lemma 11. *If an NDA $\mathcal{A}$ has infinitely many recoverable gaps, it is not determinizable.*

Proof. Assume by way of contradiction that $\mathcal{A}$ has an equivalent DDA $\mathcal{D}$ and infinitely many recoverable gaps. For every $q \in Q$, let

$$G_q = \{w \mid \mathcal{A} \text{ has a recoverable gap of the form } (w, q, p) \text{ for some } p\}$$

Since Q is finite and $\mathcal{A}$ has infinitely many recoverable gaps, there exists $q \in Q$ such that G_q is infinite. By Lemma 9, there is a finite collection Z of words such that every recoverable gap is recoverable with respect to some word in Z. Therefore there exist $z \in Z$ and an infinite subset $G'_q \subseteq G_q$ such that for every $w \in G'_q$, the gap (w, q, p) is recoverable with respect to z for some p. By Lemma 10, for every two words $w, w' \in G'_q$, the runs of $\mathcal{D}$ on w and w' end in different states, in contradiction to the fact that $\mathcal{D}$ has finitely many states.   □

Consider an NDA $\mathcal{A}$. By Lemmas 1, 6 to 9 and 11 we have that $\mathcal{A}$ has an equivalent DDA if and only if for every w, z such that $|w| \leq (\frac{\lambda}{\lambda-1}(\mathcal{N}|Q|+2m_\mathcal{A})+2)^{|Q|}$ and $|z| \leq 2^{2|Q|}$, it holds that w does not have the $\mathcal{N}$-separation property with respect to (U, L, q_u, z) for every U, L, q_u. Since the latter condition can be checked by traversing finitely many words and simulating the runs of $\mathcal{A}$ on each of them, we can conclude our main result.

Theorem 1. *The problem of whether an integral NDA has a deterministic equivalent is decidable.*

Remark 2 (Complexity of Determinization). Using the bounds on w, z, one can guess w, z on-the-fly, while keeping track of the weights of minimal runs to all states, discarding those that go above C as per Lemma 8, to check whether $\mathcal{A}$ has the $\mathcal{N}$-separation property. Since $\mathcal{N}$ is double exponential in the size of $\mathcal{A}$, this procedure can be done in NEXPSPACE = EXPSPACE. Thus, determinizability is in EXPSPACE. For a lower bound, determinizability is also PSPACE − hard by a standard reduction from NFA universality. Tightening this gap is left open. Note that for lowering the upper bound, we would need a refined application of the pigeonhole principle in Lemma 6, which seems somewhat out of reach for the pumping argument. Conversely, for increasing the lower bound, we would need to show that using discounting we can somehow force a double-exponential blowup in determinization. While this might be within reach, no such example are known for e.g., tropical weighted automata, suggesting that this may be very difficult.

Acknowledgments The authors thank Guy Raveh for fruitful discussion regarding Lemma 3.

References

1. de Alfaro, L., Henzinger, T.A., Majumdar, R.: Discounting the future in systems theory. In: Automata, Languages and Programming. pp. 1022–1037 (2003)
2. Almagor, S., Yeshurun, A.: Determinization of one-counter nets. In: 33rd International Conference on Concurrency Theory, CONCUR 2022 (Sep 2022)
3. Andersson, D.: An improved algorithm for discounted payoff games. In: ESSLLI Student Session. pp. 91–98. Citeseer (2006)
4. Boker, U., Hefetz, G.: Discounted-Sum Automata with Multiple Discount Factors. In: 29th EACSL Annual Conference on Computer Science Logic (CSL 2021). vol. 183, pp. 12:1–12:23 (2021)
5. Boker, U., Hefetz, G.: On the comparison of discounted-sum automata with multiple discount factors. Foundations of Software Science and Computation Structures LNCS 13992 p. 371 (2023)
6. Boker, U., Henzinger, T.A.: Determinizing Discounted-Sum Automata. In: Computer Science Logic (CSL'11) - 25th International Workshop/20th Annual Conference of the EACSL. Leibniz International Proceedings in Informatics (LIPIcs), vol. 12, pp. 82–96 (2011)

7. Boker, U., Henzinger, T.A.: Exact and approximate determinization of discounted-sum automata. Logical Methods in Computer Science **10** (2014)
8. Boker, U., Henzinger, T.A., Otop, J.: The target discounted-sum problem. In: 2015 30th Annual ACM/IEEE Symposium on Logic in Computer Science. pp. 750–761. IEEE (2015)
9. Broome, J.: Discounting the future. Philosophy & Public Affairs **23**(2), 128–156 (1994)
10. Cadilhac, M., Pérez, G.A., Van Den Bogaard, M.: The impatient may use limited optimism to minimize regret. In: Foundations of Software Science and Computation Structures: 22nd International Conference, FOSSACS 2019. pp. 133–149. Springer (2019)
11. Chatterjee, K., Doyen, L., Henzinger, T.A.: Quantitative languages. In: Computer Science Logic. pp. 385–400 (2008)
12. Chatterjee, K., Doyen, L., Henzinger, T.A.: Alternating weighted automata. In: Fundamentals of Computation Theory. pp. 3–13 (2009)
13. Chatterjee, K., Doyen, L., Henzinger, T.A.: Expressiveness and closure properties for quantitative languages. In: 2009 24th Annual IEEE Symposium on Logic In Computer Science. pp. 199–208 (2009). https://doi.org/10.1109/LICS.2009.16
14. Chatterjee, K., Forejt, V., Wojtczak, D.: Multi-objective discounted reward verification in graphs and mdps. In: Logic for Programming, Artificial Intelligence, and Reasoning. pp. 228–242 (2013)
15. Droste, M., Kuske, D.: Skew and infinitary formal power series. Theoretical Computer Science **366**(3), 199–227 (2006)
16. Filiot, E., Jecker, I., Lhote, N., Pérez, G.A., Raskin, J.F.: On delay and regret determinization of max-plus automata. In: 2017 32nd Annual ACM/IEEE Symposium on Logic in Computer Science (LICS). pp. 1–12. IEEE (2017)
17. Gimbert, H., Zielonka, W.: Limits of multi-discounted markov decision processes. In: 22nd Annual IEEE Symposium on Logic in Computer Science (LICS 2007). pp. 89–98 (2007). https://doi.org/10.1109/LICS.2007.28
18. Kirsten, D.: A burnside approach to the termination of mohri's algorithm for polynomially ambiguous min-plus-automata. RAIRO - Theoretical Informatics and Applications **42**(3), 553–581 (2008). https://doi.org/10.1051/ita:2008017
19. Madani, O., Thorup, M., Zwick, U.: Discounted deterministic markov decision processes and discounted all-pairs shortest paths. ACM Trans. Algorithms **6**(2) (apr 2010)
20. Mohri, M.: Finite-state transducers in language and speech processing. Computational Linguistics **23**(2), 269–311 (1997)
21. Zwick, U., Paterson, M.: The complexity of mean payoff games on graphs. Theoretical Computer Science **158**(1), 343–359 (1996)

Checking History-Determinism is NP-hard for Parity Automata

Aditya Prakash[(✉)] [iD]

University of Warwick, Coventry UK
aditya.prakash@warwick.ac.uk

Abstract. We show that the problem of checking if a given nondeterministic parity automaton simulates another given nondeterministic parity automaton is NP-hard. We then adapt the techniques used for this result to show that the problem of checking history-determinism for a given parity automaton is NP-hard. This is an improvement from Kuperberg and Skrzypczak's previous lower bound of solving parity games from 2015. We also show that deciding if Eve wins the one-token game or the two-token game of a given parity automaton is NP-hard. Finally, we show that the problem of deciding if the language of a nondeterministic parity automaton is contained in the language of a history-deterministic parity automaton can be solved in quasi-polynomial time.

1 Introduction

Deciding language inclusion between two automata is a fundamental problem in verification, wherein we ask whether all executions of an implementation satisfy a given specification. Unfortunately, the problem of checking language inclusion is often computationally hard. For parity automata—which are the focus of this paper—it is **PSPACE**-complete, with **PSPACE**-hardness already occurring for finite state automata [39].

On the other hand, simulation is a fundamental behavioural relation between two automata [33,23], which is a finer relation than language inclusion and is easier to check. For parity automata, simulation can be decided in polynomial time if the parity indices are fixed; otherwise it is in **NP** [13]. Note that while simulation between two automata is sufficient to guarantee language inclusion, it is not necessary.

For history-deterministic automata, however, the relation of language inclusion is equivalent to simulation [9,8], thus making them suitable for verification. These are nondeterministic automata where the nondeterminism can be resolved 'on-the-fly', just based on the prefix of the word read so far. The definition we use here was introduced by Henzinger and Piterman in 2006, where they dubbed it 'good-for-games' automata, while the term 'history-determinism' was coined by Colcombet [15] in the context of regular cost automata.

History-deterministic parity automata are more succinct than their deterministic counterparts [28] whilst still maintaining tractability for the problems of

N. Kobayashi and J. Worrell (Eds.): FoSSaCS 2024, LNCS 14574, pp. 212–233, 2024.
https://doi.org/10.1007/978-3-031-57228-9_11

verification and synthesis on them [24,28,8]. Consequently, history-deterministic parity automata have been the subject of extensive research [28,5,3,37,2,29], and has garnered significant attention over the recent years beyond parity automata as well, extending to quantitative automata [6,7], infinite state systems [21,31,35,10,20], and timed automata [9].

Despite these recent research efforts, a significant gap remains in understanding the complexity of checking whether a given parity automaton is history-deterministic. While Henzinger and Piterman have shown an **EXPTIME** upper bound [24], the best lower bound known so far is by Kuperberg and Skrzypczak since 2015 [28], who showed that checking for history-determinism is at least as hard as finding the winner of a parity game [28]—a problem that can be solved in quasi-polynomial time and is in **NP ∩ coNP** (and even in **UP ∩ coUP** [26]).

Kuperberg and Skrzypczak also gave a polynomial-time algorithm to check for history-determinism of co-Büchi automata in their work [28]. This was followed by a polynomial time algorithm to check for history-determinism of Büchi automata in 2018 by Bagnol and Kuperberg [3], who showed that in order to check if a Büchi automaton is history-deterministic, it suffices to find the winner of the so-called 'two-token game' of the automaton. This connection between history-determinism and two-token games was extended in 2020 to co-Büchi automata by Boker, Kuperberg, Lehtinen, and Skrzypczak [4]. It is conjectured that the winner of the two-token game of a parity automaton characterises its history-determinism. While the two-token conjecture is open to date, showing this conjecture would imply that one can check history-determinism of a given parity automata with a fixed parity index in polynomial time.

Our contributions. We show that checking for simulation between two parity automata is **NP**-hard when the parity index is not fixed. Since simulation is known to be in **NP**, this establishes the problem to be **NP**-complete (Theorem 11).

An adaptation of our proof of Theorem 11 gives us that checking history-determinism for a parity automata is also **NP**-hard (Theorem 15), when the parity index is not fixed. This is an improvement on Kuperberg and Skrzypczak's result from 2015, which shows that checking history-determinism for parity automata is at least as hard as solving parity games [28]. We also show, using the same reduction, that checking whether Eve wins the 2-token game (of a given parity automaton) is **NP**-hard, while checking whether Eve wins the 1-token game is **NP**-complete (Theorem 15).

As remarked earlier, for history-deterministic parity automata, the relation of language inclusion is equivalent to simulation. This gives us an immediate **NP** upper bound for checking language inclusion of a nondeterministic parity automaton in an HD-parity automata, as was observed by Schewe [37]. We show that we can do better, by showing the problem to be decidable in quasi-polynomial time (Theorem 20).

Overview of the paper: one reduction for all. The central problem used in our reduction is of checking whether Eve wins a 2-D parity game, which is known to be **NP**-complete due to Chatterjee, Henzinger and Piterman [13]. In Section 3, we give a reduction from this problem to checking for simulation between two parity automata, thus establishing its **NP**-hardness (Theorem 11). We then show, in Section 4.1, that the problem of checking whether Eve wins a good 2-D parity games—a technical subclass of 2-D parity games—is also **NP**-hard. In Section 4.2, we show that modifying the reduction in proof of Theorem 11 to take as inputs good 2-D parity games yields **NP**-hardness for the problems of checking history-determinism (Lemma 14) and of checking if Eve wins the 1-token game or the 2-token game (Theorem 15). Finally, in Section 5, we give a quasi-polynomial algorithm to check whether the language of a nondeterministic parity automaton is contained in the language of a history-deterministic parity automaton (Theorem 20), by reducing to finding the winner in a parity game.

2 Preliminaries

We let $\mathbb{N} = \{0, 1, 2, \cdots\}$ to be the set of natural numbers, and ω to be the cardinality of $\mathbb{N}$. We will use $[i, j]$ to denote the set of integers in the interval $\{i, i+1, \ldots, j\}$ for two natural numbers i, j with $i < j$, and $[j]$ for the interval $[0, j]$. An *alphabet* Σ is a finite set of *letters*. We use Σ^* and Σ^ω to denote the set of words with finite and ω length over Σ respectively. We also let ε denote the unique word of length 0.

2.1 Parity conditions

Let $G = (V, E)$ be a (finite or infinite) directed graph equipped with a *priority function* $\chi : E \to \mathbb{N}$ that assigns each edge with a natural number, called its *priority*. We say that an infinite path ρ in G satisfies the χ-*parity condition* if the highest priority occurring infinitely often in the path is even. When clear from the context, we will drop 'parity condition' and instead say that ρ satisfies χ.

A parity condition is easily *dualised*. Given a priority function χ as above, consider the priority function $\chi' := \chi + 1$ that is obtained by increasing all the labels by 1. Then, an infinite path satisfies χ' if and only if it does not satisfy χ.

2.2 Parity automata

A *nondeterministic parity automaton* $\mathcal{A} = (Q, \Sigma, q_0, \Delta, \Omega)$ contains a finite directed graph with edges labelled by letters in Σ. These edges are called *transitions*, which are elements of the set $\Delta \subseteq Q \times \Sigma \times Q$, and the vertices of this graph are called *states*, which are elements of the set Q.

Each automaton has a designated *initial state* $q_0 \in Q$, and a priority function $\Omega : \Delta \to [i, j]$ which assigns each transition a *priority* in $[i, j]$, for $i < j$ two natural numbers. For states p, q and an alphabet $a \in \Sigma$, we use $p \xrightarrow{a:c} q$ to denote a transition from p to q on the letter a that has the priority c.

A *run* on an infinite word w in Σ^ω is an infinite path in the automaton, starting at the initial state and following transitions that correspond to the letters of w in sequence. We say that such a run is *accepting* if it satisfies the Ω-parity condition, and a word w in Σ^ω is accepting if the automaton has an accepting run on w. The *language* of an automaton $\mathcal{A}$, denoted by $L(\mathcal{A})$, is the set of words that it accepts. We say that the automaton $\mathcal{A}$ *recognises* a language $\mathcal{L}$ if $L(\mathcal{A}) = \mathcal{L}$. A parity automaton $\mathcal{A}$ is said to be *deterministic* if for any given state in $\mathcal{A}$ and any given letter in Σ, there is at most one transition from the given state on the given letter.

If $\mathcal{A}$'s priorities are in $[i, j]$, we say that $(j - i + 1)$ is the *number of priorities* of $\mathcal{A}$. Since decreasing (or increasing) all of these priorities in the automaton by 2 does not change the acceptance of a run— and hence a word—in the automaton, we will often assume i to be 0 or 1. With this assumption, the interval $[i, j]$ is then said to be the *parity index* of $\mathcal{A}$. A *Büchi (resp. co-Büchi)* automaton is a parity automaton whose parity index is $[1, 2]$ (resp. $[0, 1]$).

Remark 1. We note that we allow an automatonto be *incomplete*, i.e. there might be letter and state pairs in an automaton such that there are no transitions on that letter from that state.

2.3   Game arenas

An *arena* is a directed graph $G = (V, E)$ with vertices partitioned as $V_\forall$ and $V_\exists$ between two players Adam and Eve respectively. Additionally, a vertex $v_0 \in V_\forall$ is designated as the initial vertex. We say that the set of vertices $V_\exists$ is owned by Eve while the set of vertices $V_\forall$ is owned by Adam. Additionally, we assume that the edges E don't have both its start and end vertex in $V_\exists$ or $V_\forall$.

Given an arena as above, a *play* of this arena is an infinite path starting at v_0, and is formed as follows. A play starts with a token at the start vertex v_0, and proceeds for countably infinite rounds. At each round, the player who owns the vertex on which the token is currently placed chooses an outgoing edge, and the token is moved along this edge to the next vertex for another round of play. This creates an infinite path in the arena, which we call a play of G.

A *game* $\mathcal{G}$ consists of an arena $G = (V, E)$ and a winning condition given by a language $L \subseteq E^\omega$. We say that Eve *wins a play* ρ in G if ρ is in L, and Adam wins otherwise. A *strategy* for Eve in such a game $\mathcal{G}$ is a function from the set of plays that end at an Eve's vertex to an outgoing edge from that vertex. Such an Eve strategy is said to be a *winning strategy* for Eve if any play that can be produced when she plays according to her strategy is winning for Eve. We say that Eve *wins the game* if she has a winning strategy. Winning strategies are defined for Adam analogously, and we say that Adam wins the game if he has a winning strategy.

In this paper we will deal with *ω-regular games*. These are games where the languages specifying the winning condition are recognised by a parity automata.

Such games are known to be determined [32,22], i.e. each game has a winner. Two games are *equivalent* if they have the same winner.

2.4   Parity games

A *parity game* $\mathcal{G}$ is played over a finite game arena $G = (V, E)$, with the edges of G labelled by a priority function $\chi : E \to \{0, 1, 2, \cdots, d\}$. A play ρ in the arena of $\mathcal{G}$ is winning for Eve if and only if ρ satisfies the χ-parity condition.

2.5   Muller conditions and Zielonka trees

A $(C, \mathcal{F})$-Muller conditions consists of a finite set of colours C, and a set $\mathcal{F}$ consisting of subsets of C. An infinite sequence in C^ω satisfies the $(C, \mathcal{F})$-Muller condition if the set of colours seen infinitely often along the sequence is in $\mathcal{F}$.

A *Muller game* $\mathcal{G}$ consists of an arena $G = (V, E)$, a colouring function $\pi : E \to C$ and a Muller condition $(C, \mathcal{F})$. An infinite play ρ in $\mathcal{G}$ is winning for Eve if the set of colours seen infinitely often along the play is in $\mathcal{F}$, and Eve wins the Muller game $\mathcal{G}$ if she has a winning strategy.

Every Muller game can be converted to an equivalent parity game, as shown by Gurevich and Harrington [22]. We will use the conversion of Dziembowski, Jurdziński, and Walukiewikz that involve Zielonka trees [16,12], which we define below.

Definition 2 (Zielonka tree). *Given a Muller condition $(C, \mathcal{F})$, the Zielonka tree of a Muller condition, denoted $Z_{C,\mathcal{F}}$, is a tree whose nodes are labelled by subsets of C, and is defined inductively. The root of the tree is labelled by C. For a node that is already constructed and labelled with the set X, its children are nodes labelled by distinct maximal non-empty subsets $X' \subsetneq X$ such that $X \in \mathcal{F} \Leftrightarrow X' \notin \mathcal{F}$. If there are no such X', then the node labelled X is a leaf of $Z_{C,\mathcal{F}}$ and has no attached children.*

Given a $(C, \mathcal{F})$-Muller condition, consider the language $L \subseteq C^\omega$ consisting of words w that satisfy the $(C, \mathcal{F})$-Muller condition. The language L is then said to be the language of the $(C, \mathcal{F})$-Muller condition, and can be recognised by a deterministic parity automaton, whose size depends on the size of the Zielonka tree [12].

Lemma 3 ([12]). *Let $(C, \mathcal{F})$ be a Muller condition with the Zielonka tree $Z_{C,\mathcal{F}}$ that has n leaves and height h. Then there is a deterministic parity automaton $\mathcal{D}_{C,\mathcal{F}}$ that can be constructed in polynomial time such that $\mathcal{D}_{C,\mathcal{F}}$ has n states and $(h + 1)$ priorities, and accepts the language of the the $(C, \mathcal{F})$-Muller condition.*

Consider a Muller game $\mathcal{G}$ on the arena $G = (V, E)$ with the colouring function $\pi : E \to C$ and the Muller condition $(C, \mathcal{F})$. We can then construct an equivalent parity game $\mathcal{G}'$ by taking the product of $\mathcal{G}$ with the automaton $\mathcal{D}_{C,\mathcal{F}}$ from Lemma 3. In more details, the set of vertices V' of $\mathcal{G}'$ consists of vertices of the form $v' = (v, q)$, where v is a vertex in $\mathcal{G}$ and q is a state in $\mathcal{D}_{C,\mathcal{F}}$. The owner of the vertex (v, q) is the owner of the vertex v, and the initial vertex is (ι, q_0), where ι is the initial vertex in $\mathcal{G}$ and q_0 is the initial state in $\mathcal{D}_{C,\mathcal{F}}$. We have the edge $e' = (v, q) \to (v', q')$ in $\mathcal{G}'$ if $e = v \to v'$ is an edge in $\mathcal{G}$ with the colour $\pi(e) = c$, and $\delta = q \xrightarrow{c} q'$ is a transition in $\mathcal{D}_{C,\mathcal{F}}$. The edge e' is assigned the priority $\Omega(\delta)$ in $\mathcal{G}'$, where Ω is the priority function of the automaton $\mathcal{D}_{C,\mathcal{F}}$. The game $\mathcal{G}'$ then is such that Eve wins $\mathcal{G}$ if and only if Eve wins $\mathcal{G}'$.

Lemma 4. *Let $\mathcal{G}$ be a Muller game on an arena consisting of m vertices with a Muller condition $(C, \mathcal{F})$ whose Zielonka tree $Z_{C,\mathcal{F}}$ has n leaves and height h. Then, $\mathcal{G}$ can be converted to an equivalent parity game $\mathcal{G}'$ which has mn many vertices and $h + 1$ priorities.*

2.6   2-dimensional parity game

Multi-dimensional parity games were introduced by Chatterjee, Henzinger and Piterman, where they called it generalised parity games [13]. For our purposes, it suffices to consider 2-dimensional (2-D) parity games, which is what we define now.

A *2-dimensional parity game* $\mathcal{G}$ is similar to a parity game, but we now have two priority functions $\pi_1 : E \to [0, d_1]$ and $\pi_2 : E \to [0, d_2]$ on E. Any infinite play in the game is winning for Eve if the following holds: *if the play satisfies π_1, then it satisfies π_2.*

We say that Adam wins the game otherwise. We call the problem of deciding whether Eve wins a 2-D parity game as 2-D PARITY GAME.

2-D PARITY GAME: Given a 2-D parity game $\mathcal{G}$, does Eve win $\mathcal{G}$?

If Eve has a strategy to win a 2-D parity game, then Eve has a positional winning strategy to do so, i.e. she can win by always choosing the same edge from each vertex in $V_\exists$, which is given by a function $\sigma : V_\exists \to E$. This can be inferred directly from seeing the 2-D parity game as a Rabin game, which are known to have positional strategies for Eve [17]. Furthermore, given a positional strategy σ for Eve in a 2-D parity game (or a Rabin game), one can check in polynomial time if σ is a winning strategy [17]. This gives us a nondeterministic polynomial time procedure to decide if Eve wins a given 2-D parity game. In 1988, Emerson and Jutla established **NP**-hardness for Rabin games [18,19]. This was later extended by Chatterjee, Henzinger, and Piterman in 2007 to show **NP**-hardness for 2-D parity games as well [13].

Theorem 5 ([13]). *The problem of deciding whether Eve wins a given 2-D parity game is **NP**-complete.*

Remark 6. Chatterjee, Henzinger and Piterman give a slightly different and a more natural definition of 2-D parity games [13], where the winning condition for Eve requires every play to satisfy either of two given parity conditions. It is easy to see, however, that both definitions are log-space inter-reducible to each other, by dualising the first parity condition. Our definition, although less natural, makes the connection to simulation games and our reductions in Sections 3 and 4 more transparent.

2.7   Simulation

We say a parity automaton $\mathcal{A}$ *simulates* another parity automaton $\mathcal{B}$ if for any (finite or infinite) run on $\mathcal{B}$, there is a corresponding run on $\mathcal{A}$ on the same word that can be constructed on-the-fly such that if the run in $\mathcal{B}$ is accepting, so is the corresponding run in $\mathcal{A}$. This is made more formal by the following *simulation game*.

Definition 7 (Simulation game). *Given nondeterministic parity automata* $\mathcal{A} = (Q, \Sigma, q_0, \Delta_A, \Omega_A)$ *and* $\mathcal{B} = (P, \Sigma, p_0, \Delta_B, \Omega_B)$, *the simulation game between* $\mathcal{A}$ *and* $\mathcal{B}$, *denoted* $Sim(\mathcal{A}, \mathcal{B})$, *is defined as a two player game between Adam and Eve as follows, with positions in* $P \times Q$. *A play of the simulation game starts at the position* (p_0, q_0), *and has* ω *many rounds. For each* $i \in \mathbb{N}$, *the* $(i + 1)^{th}$ *round starts at a position* $(p_i, q_i) \in P \times Q$, *and proceeds as follows:*

- *Adam selects a letter* $a \in \Sigma$, *and a transition* $p_i \xrightarrow{a} p_{i+1}$ *in* $\mathcal{B}$.
- *Eve selects a transition* $q_i \xrightarrow{a} q_{i+1}$ *on the same letter in* $\mathcal{A}$.

The new position is (p_{i+1}, q_{i+1}), *for another round of the play.*

The player Eve wins the above play if either her constructed run in $\mathcal{A}$ *is accepting, or Adam's constructed run in* $\mathcal{B}$ *is rejecting. If Eve has a winning strategy in* $Sim(\mathcal{A}, \mathcal{B})$, *then we say that* $\mathcal{A}$ *simulates* $\mathcal{B}$, *and denote it by* $\mathcal{B} \lesssim \mathcal{A}$.

We call the problem of checking whether a parity automaton simulates another as SIMULATION:

SIMULATION: Given two parity automata $\mathcal{A}$ and $\mathcal{B}$, does $\mathcal{A}$ simulate $\mathcal{B}$?

The simulation game $Sim(\mathcal{A}, \mathcal{B})$ can naturally be seen as a 2-D parity game, where the arena is the product of two automata with Adam selecting letters and transitions in $\mathcal{A}$ and Eve transitions in $\mathcal{B}$, and the priority functions χ_1 and χ_2 based on corresponding priorities of transitions in $\mathcal{A}$ and $\mathcal{B}$ respectively. Since 2-D PARITY GAME can be solved in **NP**, SIMULATION can be solved in **NP** as well.

2.8   History-determinism

A *history-deterministic* (HD) parity automaton is a nondeterministic parity automaton in which the nondeterminism can be resolved 'on-the-fly' just based on the prefix read so far, without knowing the rest of the word. The history-determinism of a parity automaton can be characterised by the letter game, which is a 2-player turn-based game between Adam and Eve, who take alternating turns to select a letter and a transition in the automaton (on that letter), respectively. After the game ends, the sequence of Adam's choices of letters is an infinite word, and the sequence of Eve's choices of transitions is a run on that word. Eve wins the game if her run is accepting or Adam's word is rejecting, and we say that an automaton is history-deterministic if Eve has a winning strategy in the history-determinism game.

Definition 8 (Letter game). *Given a parity automaton* $\mathcal{A} = (Q, \Sigma, q_0, \Delta, \Omega)$, *the* letter game *of* $\mathcal{A}$ *is defined between the two players Adam and Eve as follows, with positions in* $Q \times \Sigma^*$. *The game starts at* (q_0, ε) *and proceeds in* ω *many rounds. For each* $i \in \mathbb{N}$, *the* $(i+1)^{th}$ *round starts at a position* $(q_i, w_i) \in Q \times \Sigma^i$, *and proceeds as follows:*

- *Adam selects a letter* $a_i \in \Sigma$
- *Eve selects a transition* $q_i \xrightarrow{a_i} q_{i+1} \in \Delta$

The new position is (q_{i+1}, w_{i+1}), *where* $w_{i+1} = w_i a_i$.

Thus, the play of a letter game can be seen as Adam constructing a word letter-by-letter, and Eve constructing a run transition-by-transition on the same word. Eve wins such a play if the following holds: if Adam's word is in $L(\mathcal{A})$, *then Eve's run is accepting.*

We note that the letter game is an ω-regular game: the set of winning plays $\mathcal{P}$ for Eve are sequences of alternating letters and transitions, so that the word formed by just the letters is accepting in $\mathcal{A}$, while the run formed by just the transitions is rejecting. Since parity automata can be determinised, it is clear that $\mathcal{P}$ is an ω-regular language, hence the letter game is an ω-regular game, and therefore the letter game is determined [32,22].

If Eve has a winning strategy on the letter game of $\mathcal{A}$, then $\mathcal{A}$ is said to be *history-deterministic*. We are interested in the problem of checking whether a given parity automaton is history-deterministic, which we shall denote by HISTORY-DETERMINISTIC.

HISTORY-DETERMINISTIC: Given a parity automaton $\mathcal{A}$, is $\mathcal{A}$ history-deterministic?

2.9 Token games

Token games, or k-token games are defined on an automaton and are similar to letter games. Similar to as in a letter game, Adam constructs a word letter-by-letter and Eve constructs a run transition-by-transition on the same word over ω many rounds. But additionally, Adam also constructs k runs transition-by-transition on that word. The winning objective of Eve requires her to construct an accepting run if one of k Adam's runs is accepting.

Definition 9 (k-token game). *Given a nondeterministic parity automaton* $\mathcal{A} = (Q, \Sigma, , q_0, \Delta, \Omega)$, *the* k-token game *of* $\mathcal{A}$ *is defined between the two players Adam and Eve as follows, with positions in* $Q \times Q^k$. *The game starts at* $(q_0, (q_0)^k)$ *and proceeds in* ω *many rounds. For each* $i \in \mathbb{N}$, *the* $(i+1)^{th}$ *round starts at a position* $(q_i, (p_i^1, p_i^2, \cdots, p_i^k)) \in Q \times Q^k$, *and proceeds as follows:*

- *Adam selects a letter* $a_i \in \Sigma$
- *Eve selects a transition* $q_i \xrightarrow{a_i} q_{i+1} \in \Delta$
- *Adam selects* k *transitions* $p_i^1 \xrightarrow{a_i} p_{i+1}^1, p_i^2 \xrightarrow{a_i} p_{i+1}^2, \cdots p_i^k \xrightarrow{a_i} p_{i+1}^k$,

The new position is $(q_{i+1}, (p_{i+1}^1, p_{i+1}^2, \cdots, p_{i+1}^k))$, from where the $(i+2)^{th}$ round begins.

Thus, in a play of the k-token game, Eve constructs a run and Adam k runs, all on the same word. Eve wins such a play if the following holds: if one of Adam's k runs is accepting, then Eve's run is accepting.

Bagnol and Kuperberg have shown that for any parity automaton $\mathcal{A}$, the 2-token game of $\mathcal{A}$, and the k-token game of $\mathcal{A}$ for any $k \geq 2$, are equivalent.

Lemma 10 ([3]). *Given a parity automaton $\mathcal{A}$, Eve wins 2-token game of $\mathcal{A}$ if and only if Eve wins the k-token game of $\mathcal{A}$ for all $k \geq 2$.*

If $\mathcal{A}$ is a nondeterministic Büchi or co-Büchi automaton, then Eve wins the 2-token game of $\mathcal{A}$ if and only if $\mathcal{A}$ is history-deterministic[3,4], and it is conjectured that this result extends to all parity automata.

> TWO-TOKEN CONJECTURE: Given a nondeterministic parity automaton $\mathcal{A}$, Eve wins the 2-token game of $\mathcal{A}$ if and only if $\mathcal{A}$ is history-deterministic.

3   Simulation is NP-hard

In this section, we show that the problem of deciding if a parity automaton simulates another is **NP**-hard, by giving a reduction from the problem of deciding whether Eve wins a 2-D parity game, which was shown to be **NP**-complete by Chatterjee, Henzinger and Piterman [13]. Since a simulation game can be solved in **NP** (see Section 2.7), we obtain **NP**-completeness.

Theorem 11. *Given two parity automata $\mathcal{A}$ and $\mathcal{B}$, deciding if $\mathcal{A}$ simulates $\mathcal{B}$ is **NP**-complete.*

Since $\mathcal{A}$ simulates $\mathcal{B}$ if and only if Eve wins the simulation game, which is a 2-dimensional parity game (see Section 2.7), and deciding if Eve wins a 2-D parity game is in **NP** [13], we get that the problem of checking for simulation is in **NP**. Hence, we show that SIMULATION is **NP**-hard in the rest of this section, by giving a reduction from 2-D PARITY GAME.

Let $\mathcal{G}$ be a two-dimensional parity game played on the arena $G = (V, E)$, with two priority functions χ_1 and χ_2. We recall that the winning condition for Eve in such a game requires a play to satisfy the χ_2-parity condition if the χ_1-parity condition is satisfied (see Section 2.6).

Overview of the reduction. We shall construct two parity automata $\mathcal{H}$ and $\mathcal{D}$ such that $\mathcal{H}$ simulates $\mathcal{D}$ if and only if Eve wins $\mathcal{G}$. The automata $\mathcal{H}$ and $\mathcal{D}$ are over the alphabet $E \cup \{\$\}$, where $\$$ is a letter added for padding. The automaton $\mathcal{D}$ is deterministic, while the automaton $\mathcal{H}$ has nondeterminism on the letter $\$$ and contains a copy of $\mathcal{D}$.

Adam, by his choice of letter in $Sim(\mathcal{H}, \mathcal{D})$, captures his moves from Adam vertices in $\mathcal{G}$. Similarly, Eve, by means of choosing her transition on $ in $\mathcal{H}$, captures her moves from Eve vertices in $\mathcal{G}$. After each $-round in $Sim(\mathcal{H}, \mathcal{D})$, we require Adam to 'replay' Eve's choice as the next letter. Otherwise, Eve can take a transition to the same state as Adam (recall that $\mathcal{H}$ contains a copy of $\mathcal{D}$), from where she wins the play in $Sim(\mathcal{H}, \mathcal{D})$ by copying Adam's transitions in each round from here on-wards. The priorities of $\mathcal{D}$ are based on χ_1, while the priorities of $\mathcal{H}$ are based on χ_2. This way $\mathcal{D}$ and $\mathcal{H}$ roughly accept words that correspond to plays in $\mathcal{G}$ satisfying χ_1 and χ_2 respectively.

We first present our reduction on an example 2-D parity game whose sub-game consists of vertices u, v, v', w, w' with edges between them as shown in Fig. 1. For Adam's vertex u, we have corresponding states u_D in $\mathcal{D}$ and u_H in $\mathcal{H}$. An Adam move from u in $\mathcal{G}$ corresponds to one round of $Sim(\mathcal{H}, \mathcal{D})$ from the position (u_D, u_H). In $\mathcal{G}$, Adam chooses an outgoing edge, say $e = (u, v)$ from u such that $\chi_1(u) = c_1$ and $\chi_2(u) = c_2$. This corresponds to Adam choosing the letter e in $Sim(\mathcal{H}, \mathcal{D})$. We then have the corresponding unique transitions $u_D \xrightarrow{e:c_1} v_\$$ in $\mathcal{D}$ and $u_H \xrightarrow{e:c_2} v_H$ in $\mathcal{H}$, and hence the simulation game goes to $(v_\$, v_H)$. An Eve move from v in $\mathcal{G}$ corresponds to two rounds of the simulation game from $(v_\$, v_H)$. In $Sim(\mathcal{H}, \mathcal{D})$, Adam must select a letter $ and the unique $ transition $v_\$ \xrightarrow{\$:0} v_D$ on $\mathcal{D}$, since $ is the only letter on which there is an outgoing transition from $v_\$$. Eve must now select a transition on $ from v_H. Suppose she picks $v_H \xrightarrow{\$:0} (v_H, f)$ where $f = (v, w)$ is an outgoing edge from v in $\mathcal{G}$ with $\chi_1(f) = c_5$ and $\chi_2(f) = c_6$. This corresponds to Eve selecting the edge f from her vertex v in $\mathcal{G}$. The simulation game goes to the position $(v_D, (v_H, f))$. From here, Adam may select any outgoing edge from v as the letter. If he picks $f' = (v, w')$ and the transition $v_D \xrightarrow{f':c_7} w'_D$, then Eve can pick the transition $(v_H, f) \xrightarrow{f':c_8} w'_D$ and move to the same state as Adam: such transitions are indicated by dashed edges in Fig. 1. From here, Eve can win $Sim(\mathcal{H}, \mathcal{D})$ by simply copying Adam's transitions. Otherwise, Adam picks the edge f as the letter, same as Eve's 'choice' in the previous round, resulting in the transition $v_D \xrightarrow{f:c_5} w_D$ in $\mathcal{D}$ and $(v_H, f) \xrightarrow{f:c_6} w_H$ in $\mathcal{H}$, and the simulation game goes to the position (w_D, w_H), from where the game continues similarly.

The reduction. We now give formal descriptions of the two parity automata $\mathcal{D}$ and $\mathcal{H}$ such that $\mathcal{H}$ simulates $\mathcal{D}$ if and only if Eve wins $\mathcal{G}$. We encourage the reader to refer to Fig. 1 while reading the construction of the automata described below.

Both automata $\mathcal{D}$ and $\mathcal{H}$ are over the alphabet $\Sigma = E \cup \{\$\}$. The automaton $\mathcal{D}$ is given by $\mathcal{D} = (P, \Sigma, p_0, \Delta_D, \Omega_D)$, where the set P consists of the following states:

- states u_D for each Adam vertex $u \in V_\forall$,
- states $v_\$$ and v_D for each Eve vertex $v \in V_\exists$.

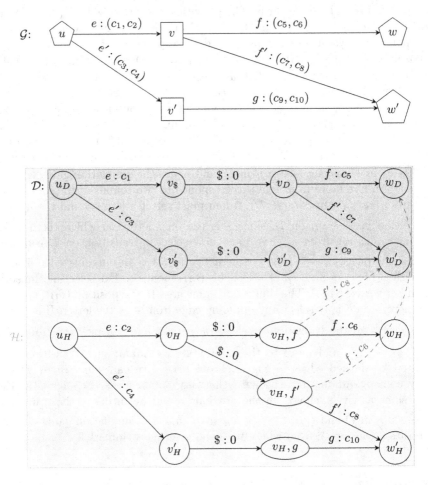

Fig. 1. A snippet of a game $\mathcal{G}$, and the corresponding automata $\mathcal{D}$ and $\mathcal{H}$ constructed in the reduction. The Adam vertices are represented by pentagons and Eve vertices by squares. The automaton $\mathcal{D}$ is deterministic, and $\mathcal{H}$ contains a copy of $\mathcal{D}$.

The state $p_0 = \iota_D$ is the initial vertex, where ι is the initial vertex of the game $\mathcal{G}$. The set Δ_D consists of the following transitions with their priorities (given by Ω_D) as indicated:

- transitions $u_D \xrightarrow{e:\chi_1(e)} v_\$$ for every edge $e = (u, v)$ in $\mathcal{G}$ such that $u \in V_\forall$ is an Adam-vertex in $\mathcal{G}$,
- transitions $v_\$ \xrightarrow{\$:0} v_D$ for every $v \in V_\exists$ that is an Eve-vertex in $\mathcal{G}$,
- transitions $v_D \xrightarrow{f:\chi_1(f)} w_D$ for every edge $f = (v, w)$ in $\mathcal{G}$ such that $v \in V_\exists$ is an Eve vertex in $\mathcal{G}$

The automaton $\mathcal{H}$ is given by $\mathcal{H} = (Q, \Sigma, q_0, \Delta_H, \Omega_H)$, where the set Q consists of the following states:

- states u_H for each Adam vertex $u \in V_\forall$,
- states v_H for each Eve vertex $v \in V_\exists$,
- states (v_H, f) for each edge $f = (v, w)$ in $\mathcal{G}$ such that $v \in V_\exists$ is an Eve vertex,
- all states in P, the set of states of $\mathcal{D}$.

The state $q_0 = \iota_H$ is the initial vertex. The set Δ_H consists of the following transitions with their priorities (given by Ω_H) as indicated:

- transitions $u_H \xrightarrow{e:\chi_2(e)} v_H$ for every edge $e = (u, v)$ in $\mathcal{G}$ such that $u \in V_\forall$ is an Adam-vertex in $\mathcal{G}$,
- transitions $v_H \xrightarrow{\$:0} (v_H, f)$ for every edge $f = (v, w)$ in $\mathcal{G}$ that is outgoing from an Eve-vertex $v \in V_\exists$,
- transitions $(v_H, f) \xrightarrow{f:\chi_2(f)} w_H$ for every edge $f = (v, w)$ in $\mathcal{G}$ outgoing from an Eve-vertex $v \in V_\exists$,
- transitions $(v_H, f) \xrightarrow{f':\chi_2(f')} w'_D$ for every edge $f' = (v, w') \neq f$ in $\mathcal{G}$ outgoing from an Eve-vertex $v \in V_\exists$,
- all transitions of $\mathcal{D}$.

Note that, by construction, $\mathcal{H}$ contains a copy of $\mathcal{D}$ as a sub-automaton.

Correctness of the reduction. We now show that Eve wins the simulation game $Sim(\mathcal{H}, \mathcal{D})$ if and only if Eve wins the game $\mathcal{G}$. Call any play of the simulation game *uncorrupted* if the following holds: whenever Eve's state in $\mathcal{H}$ is at (v_H, f) at the start of a round of $Sim(\mathcal{H}, \mathcal{D})$, Adam plays the letter f. If Adam plays a letter $f' \neq f$, then we call such a move *corrupted*. Any play consisting of a corrupted move is called a corrupted play.

It is clear that Eve wins any play of $Sim(\mathcal{H}, \mathcal{D})$ that is corrupted, since a corrupted move causes Eve's state in $\mathcal{H}$ and Adam's state in $\mathcal{D}$ to be the same in $Sim(\mathcal{H}, \mathcal{D})$. Then, both Eve's and Adam's runs are identical and determined by the choices of Adam's letters. In particular, Eve's run is accepting if Adam's run is.

Thus, it suffices to consider only uncorrupted plays. We first observe an invariant that is preserved throughout any uncorrupted play of $Sim(\mathcal{H}, \mathcal{D})$.

Invariant: At the start of any round of the simulation game $Sim(\mathcal{H}, \mathcal{D})$ following an uncorrupted play:

- Adam's state is at u_D for some $u \in V_\forall$ if and only if Eve's state is at u_H
- Adam's state is at $v_\$$ for some $v \in V_\exists$ if and only if Eve's state is at v_H
- Adam's state is at v_D for some $v \in V_\exists$ if and only if Eve's state is at (v_H, f) for some edge f that is outgoing from v.

This invariant is easy to observe from the construction, and can be shown by a routine inductive argument.

Note that if Adam constructs the word $w = e_0\$f_0e_1\$f_1 \ldots$—which we denote by $(e_i\$f_i)_{i \geq 0}$ for succinctness—in an uncorrupted play of $Sim(\mathcal{H}, \mathcal{D})$, then Eve's run on $\mathcal{H}$ is uniquely determined, since the letter f_i indicates how nondeterminism on $\mathcal{H}$ was resolved by Eve on the ith occurrence of $\$$ in $Sim(\mathcal{H}, \mathcal{D})$. Thus, any uncorrupted play in the simulation can be thought of as Adam selecting the e_i's and Eve selecting the f_i's, resulting in the word $w = (e_i\$f_i)_{i \geq 0}$ being constructed in the simulation game. Note that then, by construction, $(e_if_i)_{i \geq 0}$ is a play in $\mathcal{G}$. Conversely, if $(e_if_i)_{i \geq 0}$ is a play in $\mathcal{G}$, then there is an uncorrupted play of $Sim(\mathcal{H}, \mathcal{D})$ whose word is $w = (e_i\$f_i)_{i \geq 0}$.

Furthermore, observe that the transitions on a letter $e \in E$ in $\mathcal{D}$ and $\mathcal{H}$ in any uncorrupted play have the priorities $\chi_1(e)$ and $\chi_2(e)$ respectively, while transitions on $\$$ have priority 0. Thus, in a uncorrupted play of $Sim(\mathcal{H}, \mathcal{D})$ whose word is $(e_i\$f_i)_{i \geq 0}$, the highest priorities occurring infinitely often in the run on $\mathcal{D}$ and $\mathcal{H}$ are the same as the highest χ_1-priority and χ_2-priority occurring infinitely often in the play $(e_if_i)_{i \geq 0}$ respectively.

Thus, an uncorrupted play in $Sim(\mathcal{H}, \mathcal{D})$ whose word is $w = (e_i\$f_i)_{i \geq 0}$ is winning for Eve if and only if the play $(e_if_i)_{i \geq 0}$ in $\mathcal{G}$ is winning for Eve. Since Eve wins any corrupted play, the equivalence of the games $\mathcal{G}$ and $Sim(\mathcal{H}, \mathcal{D})$ follows easily now. If Eve has a winning strategy in $\mathcal{G}$, she can use her strategy to select transitions so that the word $w = (e_i\$f_i)_{i \geq 0}$ that is constructed in any uncorrupted play ρ of $Sim(\mathcal{H}, \mathcal{D})$ corresponds to a winning play for her in $\mathcal{G}$, and hence ρ is winning in $Sim(\mathcal{H}, \mathcal{D})$. If Adam ever makes a corrupted move, she wins trivially.

Conversely, if she has a winning strategy in $Sim(\mathcal{H}, \mathcal{D})$, then she can use her strategy to choose moves in $\mathcal{G}$ so that the play $(e_if_i)_{i \geq 0}$ corresponds to a winning uncorrupted play of $Sim(\mathcal{H}, \mathcal{D})$ in which the word $(e_i\$f_i)_{i \geq 0}$ is constructed, thus resulting in the play $(e_if_i)_{i \geq 0}$ to also be winning for Eve.

4   Checking History-Determinism is NP-hard

In this section, we show that the problem of deciding whether a given nondeterministic parity automaton is history-deterministic is **NP**-hard, as is the problem of deciding whether Eve wins the 1-token game or the 2-token game of a given parity automaton. To show this, we reduce from deciding whether Eve wins a 2-D parity game with priority functions χ_1 and χ_2 that satisfies the following property: any play satisfying the χ_2-parity condition also satisfies the χ_1-parity condition. We call such games 'good 2-D parity games'. We first show in Section 4.1 that deciding whether Eve wins a good 2-D parity game is **NP**-hard, and then use this to show **NP**-hardness for the problems mentioned above in Section 4.2.

4.1   Good 2-D parity games

Definition 12 (Good 2-D parity game). *A 2-D parity game $\mathcal{G}$ with the priority functions χ_1 and χ_2 is called* good *if any play in $\mathcal{G}$ that satisfies χ_2 also satisfies χ_1.*

We call the problem of deciding whether Eve wins a 2-D parity game as GOOD 2-D PARITY GAME. Chatterjee, Henzinger and Piterman's reduction from SAT to 2-D PARITY GAME [13] can also be seen as a reduction to GOOD 2-D PARITY GAME, as we show below.

Lemma 13. *Deciding whether Eve wins a good 2-D parity game is **NP**-hard.*

Proof. We reduce from the problem of SAT. Let ϕ be a Boolean formula over the variables $X = \{x_1, x_2, \cdots, x_M\}$ that is a conjunction of terms t_i for each $i \in [1, N]$, where each term t_i is a finite disjunction of *literals*—elements of the set $L = \{x_1, x_2, \cdots, x_M, \neg x_1, \neg x_2, \cdots, \neg x_M\}$. We shall construct a good 2-D parity game $\mathcal{G}_\phi$ such that Eve wins $\mathcal{G}_\phi$ if and only if ϕ has a satisfying assignment.

Let $T = \{t_1, t_2, \cdots, t_N\}$ be the set of all terms in ϕ. The game $\mathcal{G}_\phi$ has the set $T \cup L$ as its set of vertices. The elements of L are Adam vertices, while the elements of T are Eve vertices. We set the element x_1 in L to be the initial vertex. Each Adam vertex l in L has an outgoing edge $e = (l, t)$ to every term t in T, and every Eve vertex $t \in T$ has an outgoing edge $f = (t, l)$ to a literal l if l is a literal in t. Thus, each play in the game $\mathcal{G}_\phi$ can be seen as Adam and Eve choosing a term and a literal in that term in alternation respectively.

The game $\mathcal{G}_\phi$ has priority functions χ_1 and χ_2. To every edge $e = (l, t)$ that is outgoing from an Adam vertex, both priority functions χ_1 and χ_2 assign e the priority 0, i.e., $\chi_1(e) = \chi_2(e) = 0$. Every edge $e = (t, l)$ that is outgoing from an Eve vertex is assigned priorities as follows:

$$\chi_1(e) = \begin{cases} 2j + 2 & \text{if } l = x_j \\ 2j + 1 & \text{if } l = \neg x_j \end{cases} \qquad \chi_2(e) = \begin{cases} 2j & \text{if } l = x_j \\ 2j + 1 & \text{if } l = \neg x_j \end{cases}$$

This concludes our description of the game $\mathcal{G}_\phi$. We now show that $\mathcal{G}_\phi$ is a good 2-D parity game, which Eve wins if and only if ϕ is satisfiable.

$\mathcal{G}_\phi$ is a good 2-D parity game. Let ρ be a play in $\mathcal{G}_\phi$ that satisfies the χ_2 parity condition. If $2c$ is the largest χ_2-priority occurring infinitely often in ρ, then by construction, $2c + 2$ is the largest χ_1-priority occurring infinitely often in the ρ, which is also even. Thus, ρ satisfies the χ_1 parity condition.

If ϕ is satisfiable, then Eve wins $\mathcal{G}_\phi$. Let $f : \{x_1, x_2, \cdots, x_M\} \to \{\top, \bot\}$ be a satisfying assignment of ϕ. Let σ be a function which assigns, to each term t_i, a literal $l \in t_i$ that is assigned $\top$ in f. Consider the Eve-strategy $\sigma_\exists$ in $\mathcal{G}_\phi$ defined by $\sigma_\exists(t) = (t, \sigma(l))$. We claim that $\sigma_\exists$ is a winning strategy. Indeed, let ρ be a play in $\mathcal{G}_\phi$ following $\sigma_\exists$, and consider the largest i such that x_i or $\neg x_i$ appear infinitely often in ρ. Since $\sigma_\exists$ is obtained from a satisfying assignment, we know that either only x_i appears infinitely often, or only $\neg x_i$ appears infinitely often. In the former case, the highest χ_2 priority appearing infinitely often is $2i$, which is even, and hence ρ is winning for Eve. In the latter case, the highest χ_1 priority appearing infinitely often is $2i + 1$, which is odd, and hence the χ_1-parity condition is not satisfied, implying ρ is winning for Eve.

If Eve wins $\mathcal{G}_\phi$, then ϕ is satisfiable. If Eve wins $\mathcal{G}_\phi$, then we know she can win using a positional strategy since $\mathcal{G}_\phi$ is a 2-dimensional parity game. Let $\sigma_\exists : T \to L$ be such a strategy, where Eve chooses the edge $(t, \sigma_\exists(t))$ at a vertex t. If there are no two terms t, t' such that $\sigma_\exists(t) = x_i$ and $\sigma_\exists(t') = \neg x_i$ for some x_i, then consider the assignment σ defined as follows. The assignment σ maps all variables x that are in the image of $\sigma_\exists$ to $\top$, while any terms x_j such that neither x_j or $\neg x_j$ appear in the image are assigned $\top$ and $\bot$ respectively. It is clear then that σ is a satisfying assignment, since each term t in ϕ evaluates to $\top$.

Otherwise, if there are terms t, t' with $\sigma_\exists(t) = x_i$ and $\sigma_\exists(t') = \neg x_i$, we claim that Adam wins the game $\mathcal{G}_\phi$. Adam can alternate between picking t and t', and then the highest χ_1 priority appearing infinitely often is $2i + 2$ while the highest χ_2 priority appearing infinitely often is $2i + 1$. This implies that the play is winning for Adam, which is a contradiction since $\sigma_\exists$ is a winning strategy for Eve. □

4.2   NP-hardness of checking history-determinism

We now show that deciding the history-determinism, whether Eve wins the 1-token game, and whether Eve wins the 2-token game of a given parity automaton is **NP**-hard (Theorem 15). Much of the work towards this has already been done in the reduction from 2-D PARITY GAME to SIMULATION given in Section 3. We show that the automaton $\mathcal{H}$ that is constructed when using this reduction from a good 2-D parity game $\mathcal{G}$ is such that Eve wins $\mathcal{G}$ if and only if $\mathcal{H}$ is history-deterministic. Since GOOD 2-D PARITY GAME is **NP**-hard (Lemma 13), we get that HISTORY-DETERMINISTIC is **NP**-hard as well.

Lemma 14. *Checking whether a given nondeterministic parity automaton is history-deterministic is **NP**-hard.*

Proof. Let us consider a good 2-D parity game $\mathcal{G}$. Recall the construction of the automata $\mathcal{H}$ and $\mathcal{D}$ in Section 3, which is such that Eve wins $\mathcal{G}$ if and only if $\mathcal{H}$ simulates $\mathcal{D}$. We will show that if $\mathcal{G}$ is a good 2-D parity game, then the following statements are equivalent.

1. Eve wins $\mathcal{G}$.
2. $\mathcal{H}$ simulates $\mathcal{D}$.
3. $\mathcal{H}$ is history-deterministic.

The equivalence of 1 and 3 would then conclude the proof. The equivalence of 1 and 2 has already been shown in the proof of Theorem 11, and we now focus on showing that 2 and 3 are equivalent.

Towards this, let $\Sigma = E \cup \{\$\}$, and consider the languages $\mathcal{L}_j$ over Σ consisting of the words $(e_i \$ f_i)_{i \geq 0}$ such that $(e_i f_i)_{i \geq 0}$ is a play in $\mathcal{G}$ that satisfies χ_j, for $j = 1, 2$. By construction, we know $L(\mathcal{D}) = \mathcal{L}_1$, and $L(\mathcal{H}) = \mathcal{L}_1 \cup \mathcal{L}_2$. Furthermore, since $\mathcal{G}$ is good, we know that $\mathcal{L}_1 \supseteq \mathcal{L}_2$ and hence $L(\mathcal{D}) = L(\mathcal{H})$. Observe that by construction, $\mathcal{D}$ is deterministic.

If $\mathcal{H}$ is history-deterministic, then since $L(\mathcal{D}) = L(\mathcal{H})$, Eve wins the simulation game between $\mathcal{H}$ and $\mathcal{D}$: she can use her strategy in the letter game of $\mathcal{H}$ to play in $Sim(\mathcal{H}, \mathcal{D})$, ignoring Adam's transitions in $\mathcal{D}$.

The converse direction follows from [24, Theorem 4.1], where Henzinger and Piterman show that if a nondeterministic parity automaton $\mathcal{N}$ simulates a language-equivalent deterministic parity automaton, then $\mathcal{N}$ is history-deterministic. We include a proof nevertheless, for self-containment. Supposing $\mathcal{H}$ simulates $\mathcal{D}$, Eve can use her winning strategy in $Sim(\mathcal{H}, \mathcal{D})$ to win the letter game of $\mathcal{H}$ as follows. Eve, during the letter game of $\mathcal{H}$, will keep in her memory, a play of the game $Sim(\mathcal{H}, \mathcal{D})$. On each round in the letter game of $\mathcal{H}$, Adam gives a letter, and Eve, in the game $Sim(\mathcal{H}, \mathcal{D})$, lets Adam pick the same letter and the unique transition on that letter in $\mathcal{D}$. She then uses her strategy in $Sim(\mathcal{H}, \mathcal{D})$ to pick a transition in $\mathcal{H}$, and she plays the same transition in the letter game of $\mathcal{H}$. We claim that any resulting play of the letter game of $\mathcal{H}$ if Eve plays as above is winning for Eve. Indeed, if Adam constructs an accepting word in $\mathcal{H}$, then it is accepting in $\mathcal{D}$ as well. Hence, since $\mathcal{D}$ is deterministic, Adam's run on $\mathcal{D}$ in the simulation game between $\mathcal{H}$ and $\mathcal{D}$ that is stored in Eve's memory is accepting. Since Eve is playing according to a winning strategy in $Sim(\mathcal{H}, \mathcal{D})$, Eve's run in $\mathcal{H}$, which is the same in $Sim(\mathcal{H}, \mathcal{D})$ and the letter game of $\mathcal{H}$, is accepting as well. Hence, Eve wins the letter game of $\mathcal{H}$, and thus $\mathcal{H}$ is history-deterministic. □

We also argue in the full version of the paper [34] that the automaton $\mathcal{H}$ in proof of Lemma 14 above is such that Eve wins the 1-token game of $\mathcal{H}$ *if and only if* Eve wins the 2-token game of $\mathcal{H}$ *if and only if* $\mathcal{H}$ is history-deterministic. This gives us that checking whether Eve wins the 1-token game or the 2-token game of a parity automaton is **NP**-hard. Since 1-token games can naturally be seen as a 2-D parity game, we get that deciding whether Eve wins the 1-token game of a given parity automaton is in **NP**, and hence the problem is **NP**-complete.

Theorem 15. *The following problems are* **NP***-hard:*

1. *Given a parity automaton $\mathcal{A}$, is $\mathcal{A}$ history-deterministic?*
2. *Given a parity automaton $\mathcal{A}$, does Eve win the 2-token game of $\mathcal{A}$?*

Additionally, the following problem is **NP***-complete: Given a parity automaton $\mathcal{A}$, does Eve win the 1-token game of $\mathcal{A}$?*

5   Language Containment

In this section, we consider the following problem:

HD-AUTOMATON CONTAINMENT: Given two parity automata $\mathcal{A}$ and $\mathcal{B}$ such that $\mathcal{B}$ is history-deterministic, is $L(\mathcal{A}) \subseteq L(\mathcal{B})$?

While the problem of checking language inclusion between two non-deterministic parity automata is **PSPACE**-complete (regardless of whether the parity index is fixed or not) [30,1], the same for deterministic parity automata is **NL**-complete [36, Theorem 1]. For history-deterministic parity automata with fixed parity indices, however, the problem of language inclusion reduces to checking for simulation (Lemma 16), which can be solved in polynomial time when the parity indices are fixed [13]. This gives us that checking for language inclusion between two history-deterministic parity automata with fixed parity index can be done in polynomial time (Corollary 17). This observation has been treated as folklore, and we prove it here for completeness.

Lemma 16 ([37,9]). *Given a nondeterministic parity automaton $\mathcal{A}$ and a history-deterministic parity automaton $\mathcal{B}$, the following are equivalent:*

1. *$\mathcal{B}$ simulates $\mathcal{A}$*
2. *$L(\mathcal{A}) \subseteq L(\mathcal{B})$*

Proof. (1) $\Rightarrow$ (2): Fix $\sigma_\exists$ to be a winning strategy for Eve in $Sim(\mathcal{B}, \mathcal{A})$. Let w be a word accepted by $\mathcal{A}$ via an accepting run ρ. Consider a play of $Sim(\mathcal{B}, \mathcal{A})$ where Adam constructs the run ρ on the word w, and Eve plays according to $\sigma_\exists$. Then, the run in $\mathcal{B}$ that Eve constructs must be accepting, and hence w is accepted by $\mathcal{B}$.

(2) $\Rightarrow$ (1): Let σ_B be a winning strategy for Eve in the letter game of $\mathcal{B}$. Consider the strategy for Eve in $Sim(\mathcal{B}, \mathcal{A})$ where Eve chooses the transitions on $\mathcal{B}$ according to σ_B, ignoring Adam's transitions in $\mathcal{A}$. If Adam constructs an accepting run in $\mathcal{A}$ on a word w in $Sim(\mathcal{B}, \mathcal{A})$, then $w \in L(\mathcal{A}) \subseteq L(\mathcal{B})$. Hence σ_B would have constructed an accepting run in $\mathcal{B}$ in $Sim(\mathcal{B}, \mathcal{A})$. It follows that Eve wins $Sim(\mathcal{B}, \mathcal{A})$, and hence $\mathcal{B}$ simulates $\mathcal{A}$. $\square$

Corollary 17. *Given a nondeterministic parity automaton $\mathcal{A}$ and a history-deterministic parity automaton $\mathcal{B}$ such that both $\mathcal{A}$ and $\mathcal{B}$ have priorities in $[d]$ for a fixed d, the problem of whether $L(\mathcal{A}) \subseteq L(\mathcal{B})$ can be decided in polynomial time.*

We now focus on the problem HD-AUTOMATON CONTAINMENT when the parity index is not fixed. From Lemma 16, we know that this can be reduced to SIMULATION. Since SIMULATION is in **NP** [13], we get an immediate **NP**-upper bound for HD-AUTOMATON CONTAINMENT [37, Lemma 3]. We show that we can do better, in quasi-polynomial time, by giving a polynomial time reduction to finding the winner in a parity game[11,27].

Towards this, let us fix a nondeterministic parity automaton $\mathcal{A}$ and a history-deterministic parity automaton $\mathcal{B}$ over the alphabet Σ throughout the rest of this section, for which we want to decide if $L(\mathcal{A}) \subseteq L(\mathcal{B})$. Suppose that $\mathcal{A}$ has n_1 states and priorities in $[d_1]$, and $\mathcal{B}$ has n_2 states and priorities in $[d_2]$.

It is well known that every such parity automaton $\mathcal{A}$ can be converted efficiently to a language-equivalent nondeterministic Büchi automaton $\mathcal{A}'$ that has at most $(n_1 \cdot d_1)$ states [14,38]. Then, from Lemma 16, it suffices to check if Eve wins the game $Sim(\mathcal{B}, \mathcal{A}')$. Note that $Sim(\mathcal{B}, \mathcal{A}')$ is a 2-D parity game $\mathcal{G}$ with $(n_1 \cdot d_1 \cdot n_2 \cdot |\Sigma|)$-many vertices that has the priority functions $\chi_1 : V \to [1,2]$ and $\chi_2 : V \to [d_2]$, where V is the set of vertices of $\mathcal{G}$.

The game $\mathcal{G}$ can be viewed equivalently as a Muller game with the condition $(C, \mathcal{F})$, where $C = [1,2] \times [d_2]$ and $\mathcal{F}$ consists of sets $F \subseteq C$ such that if $\max(F|_1)$ is even, then $\max(F|_2)$ is even. Here, $F|_i$ for $i \in \{1,2\}$ indicates the projection of F onto the ith component. Call the Zielonka tree (Definition 2) of this Muller condition as Z_{d_2}. We shall show that the size of Z_{d_2} is polynomial in d_2.

Lemma 18. *The Zielonka tree Z_{d_2} has $(\lceil \frac{d_2}{2} \rceil)$ many leaves and its height is d_2.*

The proof of the lemma, obtained via an inductive argument, can be found in the full version of the paper [34]. Lemma 18 allows us to use Lemma 4 on $Sim(\mathcal{B}, \mathcal{A}')$ to obtain an equivalent Parity game $\mathcal{G}'$ with $(n_1 \cdot d_1 \cdot n_2 \cdot |\Sigma| \cdot \lceil \frac{d_2}{2} \rceil)$ vertices which has $d_2 + 1$ priorities, such that Eve wins $Sim(\mathcal{B}, \mathcal{A}')$ if and only if Eve wins $\mathcal{G}'$.

Lemma 19. *Given a nondeterministic parity automaton $\mathcal{A}$ with n_1 states and a history-deterministic parity automaton $\mathcal{B}$ with n_2 states whose priorities are in $[d_2]$ that are both over the alphabet Σ, the problem of deciding whether $L(\mathcal{A})$ is contained in $L(\mathcal{B})$ can be reduced in polynomial time to finding the winner of a parity game $\mathcal{G}$ which has $(n_1 \cdot d_1 \cdot n_2 \cdot |\Sigma| \cdot \lceil \frac{d_2}{2} \rceil)$ many vertices and $d_2 + 1$ priorities.*

Since parity games can be solved in quasi-polynomial time[11,27], Lemma 19 implies that the problem of language containment in a history-deterministic automaton can be solved in quasi-polynomial time as well.

Theorem 20. *Given a nondeterministic parity automaton $\mathcal{A}$ with n_1 states and priorities in $[d_1]$, and a history-deterministic parity automaton $\mathcal{B}$ with n_2 states whose priorities are in $[d_2]$, checking whether the language of $\mathcal{A}$ is contained in the language of $\mathcal{B}$ can be done in time*

$$(n_1 \cdot d_1 \cdot n_2 \cdot d_2 \cdot |\Sigma|)^{\mathcal{O}(\log d_2)}.$$

6   Discussion

We have shown **NP**-hardness for the problem of checking for simulation between two parity automata (when their parity indices are not fixed). We have also established upper and lower bounds of several decision problems relating to history-deterministic parity automata. The most significant amongst these, in our view, is the **NP**-hardness for the problem of deciding if a given parity automaton is history-deterministic, which is an improvement from the previous lower bound of solving a parity game [28].

There still remains a significant gap between the lower bound of **NP**-hardness and the upper bound of **EXPTIME** for checking history-determinism, however. Furthermore, note that even if one shows the two-token conjecture [3,4], this would only imply a **PSPACE**-upper bound (when the parity index is not fixed), since 2-token games can be seen as Emerson-Lei games [25]. Thus, a natural direction for future research is to try to show that the problem of checking for history-determinism is **PSPACE**-hard.

On the other hand, however, it is also plausible that checking whether Eve wins the 2-token game of a given parity automaton can be done in **NP**. A proof for this might show that if Eve wins a 2-token game, then she has a strategy that can be represented and verified polynomially. Such an approach, which would involve understanding the strategies for the players in the 2-token games better, could also yield crucial insights for proving or disproving the two-token conjecture (see Section 2.9).

Boker and Lehtinen showed in their recent survey that for a 'natural' class of automata T, checking history-determinism for T-automata is at least as hard as solving T-games [8]. Interestingly, the problem of checking history-determinism over T-automata also has the matching upper bound of solving T-games for all classes of automata T over finite words, and over infinite words with safety and reachability objectives on which the notion of history-determinism has been studied so far [7,21,35,9,20]. Our result of the problem of checking history-determinism being **NP**-hard for parity automata deviates from this trend (unless parity games are **NP**-hard, which would have the drastic and unlikely consequence of **NP** = **NP** ∩ **coNP**), and demonstrates the additional intricacy that parity conditions bring.

Acknowledgements We thank Marcin Jurdziński, Neha Rino, K. S. Thejaswini, and anonymous reviewers for their feedback and suggesting numerous improvements to the paper. Additionally, we are grateful to K. S. Thejaswini for several insightful discussions and pointing out a flaw in an earlier proof of Theorem 11.

References

1. Abdulla, P.A., Chen, Y., Clemente, L., Holík, L., Hong, C., Mayr, R., Vojnar, T.: Advanced ramsey-based büchi automata inclusion testing. In: CONCUR. Lecture Notes in Computer Science, vol. 6901, pp. 187–202. Springer (2011), https://doi.org/10.1007/978-3-642-23217-6_13

2. Abu Radi, B., Kupferman, O.: Minimization and canonization of GFG transition-based automata. Log. Methods Comput. Sci. **18**(3) (2022), https://doi.org/10.46298/lmcs-18(3:16)2022

3. Bagnol, M., Kuperberg, D.: Büchi good-for-games automata are efficiently recognizable. In: FSTTCS. LIPIcs, vol. 122, pp. 16:1–16:14. Schloss Dagstuhl - Leibniz-

Zentrum für Informatik (2018), https://doi.org/10.4230/LIPIcs.FSTTCS.2018.16

4. Boker, U., Kuperberg, D., Lehtinen, K., Skrzypczak, M.: On the succinctness of alternating parity good-for-games automata. CoRR **abs/2009.14437** (2020), https://arxiv.org/abs/2009.14437

5. Boker, U., Kupferman, O., Skrzypczak, M.: How deterministic are good-for-games automata? In: FSTTCS. LIPIcs, vol. 93, pp. 18:1–18:14. Schloss Dagstuhl - Leibniz-Zentrum für Informatik (2017), https://doi.org/10.4230/LIPIcs.FSTTCS.2017.18

6. Boker, U., Lehtinen, K.: History determinism vs. good for gameness in quantitative automata. In: FSTTCS. LIPIcs, vol. 213, pp. 38:1–38:20. Schloss Dagstuhl - Leibniz-Zentrum für Informatik (2021), https://doi.org/10.4230/LIPIcs.FSTTCS.2021.38

7. Boker, U., Lehtinen, K.: Token games and history-deterministic quantitative automata. In: FoSSaCS. Lecture Notes in Computer Science, vol. 13242, pp. 120–139. Springer (2022), https://doi.org/10.1007/978-3-030-99253-8_7

8. Boker, U., Lehtinen, K.: When a little nondeterminism goes a long way: An introduction to history-determinism. ACM SIGLOG News **10**(1), 24–51 (2023), https://doi.org/10.1145/3584676.3584682

9. Bose, S., Henzinger, T.A., Lehtinen, K., Schewe, S., Totzke, P.: History-deterministic timed automata. CoRR **abs/2304.03183** (2023), https://doi.org/10.48550/arXiv.2304.03183

10. Bose, S., Purser, D., Totzke, P.: History-deterministic vector addition systems. In: CONCUR. LIPIcs, vol. 279, pp. 18:1–18:17. Schloss Dagstuhl - Leibniz-Zentrum für Informatik (2023), https://doi.org/10.4230/LIPIcs.CONCUR.2023.18

11. Calude, C.S., Jain, S., Khoussainov, B., Li, W., Stephan, F.: Deciding parity games in quasi-polynomial time. SIAM J. Comput. **51**(2), 17–152 (2022), https://doi.org/10.1137/17m1145288

12. Casares, A., Colcombet, T., Fijalkow, N.: Optimal transformations of games and automata using muller conditions. In: ICALP. LIPIcs, vol. 198, pp. 123:1–123:14. Schloss Dagstuhl - Leibniz-Zentrum für Informatik (2021), https://doi.org/10.4230/LIPIcs.ICALP.2021.123

13. Chatterjee, K., Henzinger, T.A., Piterman, N.: Generalized parity games. In: FoSSaCS. Lecture Notes in Computer Science, vol. 4423, pp. 153–167. Springer (2007), https://doi.org/10.1007/978-3-540-71389-0_12

14. Choueka, Y.: Theories of automata on omega-tapes: A simplified approach. J. Comput. Syst. Sci. **8**(2), 117–141 (1974), https://doi.org/10.1016/S0022-0000(74)80051-6

15. Colcombet, T.: The theory of stabilisation monoids and regular cost functions. In: ICALP. Lecture Notes in Computer Science, vol. 5556, pp. 139–150. Springer (2009), https://doi.org/10.1007/978-3-642-02930-1_12

16. Dziembowski, S., Jurdzinski, M., Walukiewicz, I.: How much memory is needed to win infinite games? In: LICS. pp. 99–110. IEEE Computer Society (1997), https://doi.org/10.1109/LICS.1997.614939

17. Emerson, E.A.: Automata, tableaux and temporal logics (extended abstract). In: Logics of Programs, Conference, Brooklyn College, New York, NY, USA, June 17-19, 1985, Proceedings. Lecture Notes in Computer Science, vol. 193, pp. 79–88. Springer (1985), https://doi.org/10.1007/3-540-15648-8_7

18. Emerson, E.A., Jutla, C.S.: The complexity of tree automata and logics of programs (extended abstract). In: SFCS. pp. 328–337. IEEE Computer Society (1988), https://doi.org/10.1109/SFCS.1988.21949

19. Emerson, E.A., Jutla, C.S.: The complexity of tree automata and logics of programs. SIAM J. Comput. **29**(1), 132–158 (1999), https://doi.org/10.1137/S0097539793304741

20. Erlich, E., Guha, S., Jecker, I., Lehtinen, K., Zimmermann, M.: History-deterministic parikh automata. In: CONCUR. LIPIcs, vol. 279, pp. 31:1–31:16. Schloss Dagstuhl - Leibniz-Zentrum für Informatik (2023), https://doi.org/10.4230/LIPIcs.CONCUR.2023.31

21. Guha, S., Jecker, I., Lehtinen, K., Zimmermann, M.: A bit of nondeterminism makes pushdown automata expressive and succinct. In: MFCS. LIPIcs, vol. 202, pp. 53:1–53:20. Schloss Dagstuhl - Leibniz-Zentrum für Informatik (2021), https://doi.org/10.4230/LIPIcs.MFCS.2021.53

22. Gurevich, Y., Harrington, L.: Trees, automata, and games. In: STOC. pp. 60–65. ACM (1982), https://doi.org/10.1145/800070.802177

23. Henzinger, T.A., Kupferman, O., Rajamani, S.K.: Fair simulation. Inf. Comput. **173**(1), 64–81 (2002), https://doi.org/10.1006/inco.2001.3085

24. Henzinger, T.A., Piterman, N.: Solving games without determinization. In: CSL. Lecture Notes in Computer Science, vol. 4207, pp. 395–410. Springer (2006), https://doi.org/10.1007/11874683_26

25. Hunter, P., Dawar, A.: Complexity bounds for regular games. In: MFCS. Lecture Notes in Computer Science, vol. 3618, pp. 495–506. Springer (2005), https://doi.org/10.1007/11549345_43

26. Jurdzinski, M.: Deciding the Winner in Parity Games is in UP ∩ co-UP. Inf. Process. Lett. **68**(3), 119–124 (1998), https://doi.org/10.1016/S0020-0190(98)00150-1

27. Jurdzinski, M., Lazic, R.: Succinct progress measures for solving parity games. In: LICS. pp. 1–9. IEEE Computer Society (2017), https://doi.org/10.1109/LICS.2017.8005092

28. Kuperberg, D., Skrzypczak, M.: On determinisation of good-for-games automata. In: ICALP. Lecture Notes in Computer Science, vol. 9135, pp. 299–310. Springer (2015), https://doi.org/10.1007/978-3-662-47666-6_24

29. Kupferman, O.: Using the past for resolving the future. Frontiers Comput. Sci. **4** (2022), https://doi.org/10.3389/fcomp.2022.1114625

30. Kupferman, O., Vardi, M.Y.: Verification of fair transition systems. Chic. J. Theor. Comput. Sci. **1998** (1998), http://cjtcs.cs.uchicago.edu/articles/1998/2/contents.html

31. Lehtinen, K., Zimmermann, M.: Good-for-games ω-pushdown automata. Log. Methods Comput. Sci. **18**(1) (2022), https://doi.org/10.46298/lmcs-18(1:3)2022

32. Martin, D.A.: Borel determinacy. Annals of Mathematics **102**(2), 363–371 (1975), http://www.jstor.org/stable/1971035

33. Milner, R.: An algebraic definition of simulation between programs. In: IJCAI. pp. 481–489. William Kaufmann (1971), http://ijcai.org/Proceedings/71/Papers/044.pdf

34. Prakash, A.: Checking history-determinism is np-hard for parity automata. CoRR **abs/2310.13498** (2023), https://doi.org/10.48550/arXiv.2310.13498

35. Prakash, A., Thejaswini, K.S.: On history-deterministic one-counter nets. In: FoSSaCS 2023. Lecture Notes in Computer Science, vol. 13992, pp. 218–239. Springer (2023), https://doi.org/10.1007/978-3-031-30829-1_11

36. Schewe, S.: Beyond Hyper-Minimisation—Minimising DBAs and DPAs is NP-Complete. In: FSTTCS. LIPIcs, vol. 8, pp. 400–411. Schloss Dagstuhl - Leibniz-

Zentrum für Informatik (2010), https://doi.org/10.4230/LIPIcs.FSTTCS.2010.400

37. Schewe, S.: Minimising Good-For-Games Automata Is NP-Complete. In: FSTTCS. Leibniz International Proceedings in Informatics (LIPIcs), vol. 182, pp. 56:1–56:13. Schloss Dagstuhl–Leibniz-Zentrum für Informatik, Dagstuhl, Germany (2020), https://drops.dagstuhl.de/opus/volltexte/2020/13297

38. Seidl, H., Niwinski, D.: On distributive fixed-point expressions. RAIRO Theor. Informatics Appl. **33**(4/5), 427–446 (1999), https://doi.org/10.1051/ita:1999101

39. Stockmeyer, L.J., Meyer, A.R.: Word problems requiring exponential time: Preliminary report. In: STOC. pp. 1–9. ACM (1973), https://doi.org/10.1145/800125.804029

Tighter Construction of Tight Büchi Automata

Marek Jankola[1]* and Jan Strejček[2]

[1] LMU Munich, Munich, Germany
marek.jankola@sosy.ifi.lmu.de
[2] Masaryk University, Brno, Czechia
strejcek@fi.muni.cz

Abstract. Tight automata are useful in providing the shortest counterexample in LTL model checking and also in constructing a maximally
satisfying strategy in LTL strategy synthesis. There exists a translation of
LTL formulas to tight Büchi automata and several translations of Büchi
automata to equivalent tight Büchi automata. This paper presents another translation of Büchi automata to equivalent tight Büchi automata.
The translation is designed to produce smaller tight automata and it
asymptotically improves the best-known upper bound on the size of a
tight Büchi automaton equivalent to a given Büchi automaton. We also
provide a lower bound, which is more precise than the previously known
one. Further, we show that automata reduction methods based on quotienting preserve tightness. Our translation was implemented in a tool
called Tightener. Experimental evaluation shows that Tightener usually
produces smaller tight automata than the translation from LTL to tight
automata known as CGH.

1 Introduction

When a model checking algorithm decides that a given system violates a given
specification, a counterexample showing the undesired system behavior is produced. If the system has only finitely many states and it violates the specification
given by a formula of *Linear Temporal Logic (LTL)* or directly by a Büchi automaton accepting all erroneous behaviors, there exists a counterexample of the
form $u.v^\omega$ called *lasso-shaped* or *ultimately periodic*. A serious research effort
has been devoted to algorithms that produce short counterexamples, where the
length of a counterexample $u.v^\omega$ is given by $|uv|$ [7, 12, 13, 15, 19, 22, 24].

In 2005, Schuppan and Biere [24] defined *tight* Büchi automata, where each
lasso-shaped word accepted by such an automaton is accepted by a lasso-shaped
run of the same length. Hence, the product of a tight automaton $\mathcal{A}$ with an arbitrary transition system accepts the shortest lasso-shaped behavior of the system
that is in the language of $\mathcal{A}$ by the shortest lasso-shaped accepting run. This
property makes tight automata very useful for automata-based model checking

---

* This research was conducted partly during Master's studies of M. Jankola at Masaryk
University in Brno, Czechia.

© The Author(s) 2024
N. Kobayashi and J. Worrell (Eds.): FoSSaCS 2024, LNCS 14574, pp. 234–255, 2024.
https://doi.org/10.1007/978-3-031-57228-9_12

algorithms looking for shortest counterexamples, which was the original motivation for the definition. Tight automata found another application in autonomous robot action planning, where they are used in the algorithm synthesizing a maximally satisfying discrete control strategy while taking into account that the robot's action executions may fail [27].

There exist only few algorithms producing tight automata. The oldest is the translation of LTL formulas into generalized Büchi automata introduced by Clarke, Grumberg, and Hamaguchi [6] in 1994. The fact that this translation creates tight automata was shown about 10 years later by Schuppan and Biere [24], who named the translation CGH. They extended the translation to handle also past LTL operators and implemented it. The implementation produces automata in symbolic representation suitable for the model checker NuSMV [5].

There are also two constructions transforming Büchi automata into tight Büchi automata. The first was introduced by Schuppan [23] and it accepts even generalized Büchi automata as input. For a (non-generalized) Büchi automaton with n states, this construction creates a tight automaton with $\mathcal{O}((\sqrt{2}n)^{2n})$ states. The second (and completely different) construction was introduced by Ehlers [13] and it produces tight automata of size $2^{\mathcal{O}(n^2)}$ states. Kupferman and Vardi [20] provided the lower bound $2^{\Omega(n)}$ as a side result when analyzing counterexamples of safety properties. We are not aware of any implementation of these constructions.

This paper presents another construction transforming Büchi automata to tight Büchi automata. More precisely, our construction accepts (state-based) *Büchi automata (BA)* or *transition-based Büchi automata (TBA)* and produces tight BA or tight TBA. The construction is similar to the one of Schuppan [23], but it produces less states: while Schuppan's construction creates states that represent a sequence of up to $2n$ states of the original automata, our construction creates states representing at most n states of the original automaton and these n states are pairwise different (potentially with a single exception). The construction gives us an upper bound in $\mathcal{O}(n! \cdot n^3)$ which is strictly below both $\mathcal{O}((\sqrt{2}n)^{2n})$ and $2^{\mathcal{O}(n^2)}$. We also provide a lower bound in $\Omega(\frac{n-1}{2}!)$ for any transformation of BA into equivalent tight BA or TBA and a lower bound in $\Omega((n-1)!)$ for any transformation of TBA into equivalent tight BA or TBA. Note that the lower bound $\Omega(\frac{n-1}{2}!)$ is strictly above the previous lower bound $2^{\Omega(n)}$. Additionally, we show that tight automata can be reduced by quotienting with use of an arbitrary *good-for-quotienting (GFQ)* relation [8] and the resulting automaton is equivalent and tight.

Our paper also delivers some practical results. The tightening algorithm has been implemented in a tool called Tightener. The tool can be easily combined with other automata tools as it accepts and produces automata in the HOA format [2]. Furthermore, it also accepts LTL formulas on input. When Tightener receives an LTL formula, it calls the LTL to TBA translation of Spot [10] as the first step. We compare Tightener against the CGH translation as this is (as far as we know) the only other implemented algorithm producing tight automata. Our experimental evaluation shows that tight automata produced by CGH usually have more states than the ones by Tightener.

Contributions of the paper. The paper brings the following contributions:

- a construction transforming BA/TBA into tight BA/TBA with the lowest theoretical upper bound on the rise of the state space so far,
- lower bounds on any transformation of BA or TBA into equivalent tight BA/TBA that are currently the highest lower bounds,
- a proof that the automata reduction based on quotienting preserves tightness,
- a tool Tightener producing tight BA/TBA from LTL formulas or BA/TBA,
- an experimental comparison of tight automata by Tightener and CGH.

Structure of the paper. The following section introduces the basic terminology used in the paper. Section 3 formulates some observations crucial for our tightening construction, which is then presented in Section 4 together with the implied upper bound. Section 5 shows the lower bounds on the tightening process. The postprocessing of tight automata is discussed in Section 6. Section 7 describes the implementation of our tightening construction in Tightener and Section 8 compares it to the CGH translation in terms of the sizes of produced tight automata. Finally, Section 9 concludes the paper.

2   Preliminaries

A *transition-based Büchi automaton (TBA)* is a tuple $\mathcal{A} = (Q, \Sigma, \delta, I, \delta_F)$, where

- Q is a finite set of *states*,
- Σ is a finite *alphabet*,
- $\delta \subseteq Q \times \Sigma \times Q$ is a *transition relation*,
- $I \subseteq Q$ is a set of *initial states*, and
- $\delta_F \subseteq \delta$ is a set of *accepting transitions*.

A *run* of $\mathcal{A}$ over an infinite word $u = u_0 u_1 \ldots \in \Sigma^\omega$ is an infinite sequence $\rho = (q_0, u_0, q_1)(q_1, u_1, q_2) \ldots \in \delta^\omega$ of consecutive transitions starting in an initial state $q_0 \in I$. By ρ_i, we denote the transition (q_i, u_i, q_{i+1}) from ρ. A run ρ is *accepting* if $(q_i, u_i, q_{i+1}) \in \delta_F$ holds for infinitely many i. An automaton *accepts* a word u if there exists an accepting run over this word. A *language* of automaton $\mathcal{A}$ is the set $L(\mathcal{A})$ of all words in Σ^ω accepted by $\mathcal{A}$. Automata $\mathcal{A}, \mathcal{B}$ are *equivalent* if $L(\mathcal{A}) = L(\mathcal{B})$.

A transition $(p, a, q) \in \delta$ is also denoted as $p \xrightarrow{a} q$. In graphical representation, accepting transitions are these marked with the blue dot ●. In the following, *word* without any adjective refers to an infinite word. A *path* in $\mathcal{A}$ from a state q_0 to a state q_n over a finite word $r = r_0 r_1 \ldots r_{n-1} \in \Sigma^*$ is a finite sequence $\sigma = (q_0, r_0, q_1)(q_1, r_1, q_2) \ldots (q_{n-1}, r_{n-1}, q_n) \in \delta^n$ of consecutive transitions. We refer to a first state q_0 of a path as to *initial state* of the path. We naturally extend the notation for transitions and write that the path σ has the form $q_0 \xrightarrow{r} q_n$. If such a path exists, we say that q_n is *reachable* from q_0 over r. For a word or a run $u = u_0 u_1 \ldots$, by $u_{i..}$ we denote its suffix $u_i u_{i+1} \ldots$ and by $u_{i,j}$, for $i < j$, we denote its subpart $u_i u_{i+1} \ldots u_{j-1}$.

The paper intensively works with *lasso-shaped* words and runs, which are sequences of the form $s.l^\omega$, where s is called a *stem* and $l \neq \varepsilon$ is called a *loop*. Further, s is a *minimal stem* and l is a *minimal loop* of a lasso-shaped sequence $u = s.l^\omega$ if for each s', l' satisfying $u = s'.l'^\omega$ it holds $|s| + |l| \leq |s'| + |l'|$.

Lemma 1. *For each lasso-shaped sequence, there exist a unique minimal stem and a unique minimal loop.*

Proof. The existence of some minimal stem and loop for each lasso-shaped sequence u is obvious. We prove its uniqueness by contradiction. Assume that there are two different pairs s, l and s', l' of minimal stem and loop, which implies that $u = s.l^\omega = s'.l'^\omega$ and $|s| + |l| = |s'| + |l'|$. Without loss of generality, assume that $|s| < |s'|$ and $|l| > |l'|$. As $|s| + |l| = |s'| + |l'|$, we get $l^\omega = u_{|s|+|l|..} = u_{|s'|+|l'|..} = l'^\omega$ and thus $s.l^\omega = s.l'^\omega$. However, this is a contradiction with the minimality of s, l and s', l' as $|s| + |l'| < |s| + |l| = |s'| + |l'|$. $\qquad\square$

The minimal stem and the minimal loop of a lasso-shaped sequence u is denoted by $minS(u)$ and $minL(u)$, respectively. Moreover, we set $|minSL(u)| = |minS(u)| + |minL(u)|$ and call it the *size* of u.

If ρ is a lasso-shaped run over a word u, then u is a lasso-shaped word such that $|minS(u)| \leq |minS(\rho)|$ and $|minL(u)| \leq |minL(\rho)|$.

A TBA $\mathcal{A}$ is *tight* [24] iff for each lasso-shaped word $u \in L(\mathcal{A})$ there exists an accepting lasso-shaped run ρ satisfying $|minSL(u)| = |minSL(\rho)|$. We call such runs *tight*.

A *state-based Büchi automaton (BA)* is a tuple $\mathcal{A} = (Q, \Sigma, \delta, I, F)$, where Q, Σ, δ, I have the same meaning as in a TBA and $F \subseteq Q$ is a set of *accepting states*. The definition of all terms is the same as for TBA with the exception of accepting run. A run $\rho = (q_0, u_0, q_1)(q_1, u_1, q_2) \ldots \in \delta^\omega$ is *accepting* if $q_i \in F$ for infinitely many i. Note that BA can be seen as a special case of TBA as each BA can be easily transformed into an equivalent TBA only by replacing its accepting states F with the set of transitions δ_F leading to these states, i.e., $\delta_F = \{(p, a, q) \in \delta \mid q \in F\}$.

Finally, a *(state-based) generalized Büchi automaton (GBA)* is a tuple $\mathcal{A} = (Q, \Sigma, \delta, I, \mathcal{F})$, where Q, Σ, δ, I have the same meaning as in a TBA and $\mathcal{F} = \{F_1, \ldots, F_k\}$ is a finite set of sets $F_j \subseteq Q$. The definition of all terms is the same as for TBA, except for an accepting run. A run $\rho = (q_0, u_0, q_1)(q_1, u_1, q_2) \ldots \in \delta^\omega$ is *accepting* if for each $F_j \in \mathcal{F}$ there exist infinitely many i satisfying $q_i \in F_j$.

3 Observations

First of all, we explain why our definition of TBA considers multiple initial states. As every TBA can be transformed into an equivalent TBA with a single

Fig. 1. TBA with a single initial state (left) and an equivalent tight TBA with two initial states (right).

initial state, some definitions of TBA consider exactly one initial state. However, a tight TBA with one initial state would have only a restricted expressive power. Indeed, each TBA can be transformed to an equivalent tight TBA with multiple initial states (as we show in the following section), but there exist TBA that cannot be transformed into equivalent tight TBA with a single initial state.

Lemma 2. *There exists a TBA such that there is no equivalent tight TBA with a single initial state.*

Proof. Let $\mathcal{A}$ be the TBA in Figure 1 (left). For the sake of contradiction, assume that there is a tight TBA $\mathcal{B}$ with one initial state q_0 and equivalent to $\mathcal{A}$. Then $\mathcal{B}$ must accept a^ω and b^ω. Furthermore, since $|minSL(a^\omega)| = |minSL(b^\omega)| = 1$ and $\mathcal{B}$ is tight, there must exist accepting self-loops over a and b in q_0. However, $\mathcal{B}$ then accepts for instance $a.b^\omega \notin L(\mathcal{A})$, which is a contradiction. $\square$

As the (un)tightness of an automaton depends purely on lasso-shaped words accepted by the automaton and the corresponding accepting runs, we turn our attention to these words. We start with the definition of *significant positions* in a lasso-shaped word u as positions i where $u_{i..} = minL(u)^\omega$. Formally, we set

$$Sign(u) = \{k, k+o, k+2o, k+3o, \ldots\}$$

where $k = |minS(u)|$ and $o = |minL(u)|$. We first prove that for every lasso-shaped word u accepted by a TBA, there exists a lasso-shaped accepting run over u.

Lemma 3. *Let $\mathcal{A}$ be a TBA. For each lasso-shaped word $u \in L(\mathcal{A})$ there exists a lasso-shaped accepting run over u of the form $\tau.\pi^\omega$, where*

- *τ is a path over $minS(u).minL(u)^i$ for some $i \geq 0$ and*
- *π is a path over $minL(u)^k$ for some $k > 0$.*

Proof. Let $u \in L(\mathcal{A})$ be a lasso-shaped word and $\rho = (q_0, u_0, q_1)(q_1, u_1, q_2) \ldots$ be an accepting run of $\mathcal{A}$ over this word. We focus on states of this run at significant positions, i.e., states $q_k, q_{k+o}, q_{k+2o}, \ldots$ where $k = |minS(u)|$ and $o = |minL(u)|$. The run and its states at significant positions are depicted in Figure 2. Since $\mathcal{A}$ has finitely many states, there are positions $p_1, p_2 \in Sign(u)$ such that $p_1 < p_2$, $q_{p_1} = q_{p_2}$, and path ρ_{p_1,p_2} contains an accepting transition. We set $\tau = \rho_{0,p_1}$ and $\pi = \rho_{p_1,p_2}$. As p_1, p_2 are significant positions, τ is a path over $minS(u).minL(u)^i$ for some $i \geq 0$ and π is a path over $minL(u)^k$ for some $k > 0$. As $q_{p_1} = q_{p_2}$ and π contains an accepting transition, $\tau.\pi^\omega$ is an accepting run over u. $\square$

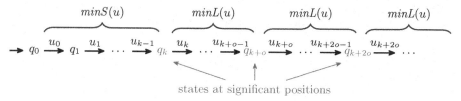

Fig. 2. A run over a lasso-shaped word $u = u_0 u_1 \ldots$ with states at significant positions typeset in red.

Once we know that each lasso-shaped word $u \in L(\mathcal{A})$ has a lasso-shaped accepting run, we also know that there exists at least one accepting lasso-shaped run ρ over u with the minimal size $|minSL(\rho)|$. We call such runs *minimal*. For example, consider the word $b.(abc)^\omega$ accepted by the automaton in Figure 3. The minimal run for this word is $\tau.\pi^\omega$ with the following minimal stem τ and minimal loop π.

$$\tau = p_0 \xrightarrow{b} p_1 \xrightarrow{a} p_2 \xrightarrow{b} p_3 \xrightarrow{c} p_4 \xrightarrow{a} r_0$$

$$\pi = r_0 \xrightarrow{b} r_1 \xrightarrow{c} r_2 \xrightarrow{a} r_2 \xrightarrow{b} r_3 \xrightarrow{c} r_4 \xrightarrow{a} r_5 \xrightarrow{b} r_4 \xrightarrow{c} \bullet \rightarrow r_6 \xrightarrow{a} r_0$$

Now we formulate and prove Lemma 4, which says that each minimal run ρ has a specific property regarding repetition of states. The property considers states of ρ at the positions at least $|minS(u)|$ and less than $|minSL(\rho)|$. The property says that there cannot be the same state twice on the considered positions from which the same suffix of u is read. It can be illustrated on the minimal run $\tau.\pi^\omega$ mentioned above. If we write the states of this run such that the states reading the same suffix of u are vertically aligned (see Table 1), the considered states in each column are pairwise different.

Lemma 4. *Let $\mathcal{A}$ be a TBA and $\rho = (q_0, u_0, q_1)(q_1, u_1, q_2) \ldots$ be a minimal run over a lasso-shaped word $u \in L(\mathcal{A})$. For each $|minS(u)| \leq m < l < |minSL(\rho)|$ satisfying $u_{m..} = u_{l..}$ it holds that the states q_m and q_l are different.*

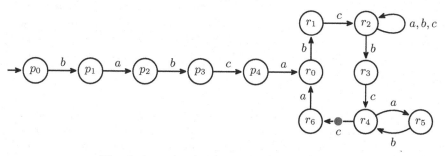

Fig. 3. An example of a TBA that is not tight.

Table 1. Illustration of the property formulated in Lemma 4. Unconsidered states are struck through and states at significant positions are typeset in red.

suffices of u:	$b.(abc)^\omega$	$(abc)^\omega$	$bc.(abc)^\omega$	$c.(abc)^\omega$
states of $\tau.\pi^\omega$:	~~p_0~~	p_1	p_2	p_3
		p_4	r_0	r_1
		r_2	r_2	r_3
		r_4	r_5	r_4
		r_6	~~p_0~~	~~p_1~~
		~~p_2~~	~~p_4~~	~~p_3~~
		~~p_4~~	$\cdots$	

Proof. Let $\mathcal{A}$ be a TBA and $\rho = (q_0, u_0, q_1)(q_1, u_1, q_2) \ldots$ be a minimal run over a lasso-shaped word $u \in L(\mathcal{A})$. For the sake of contradiction, assume that there are positions $|minS(u)| \leq m < l < |minSL(\rho)|$ such that $u_{m..} = u_{l..}$ and $q_m = q_l$. We will show that there exists another lasso-shaped accepting run ρ' over u of a smaller size than ρ. This will give us a contradiction with the minimality of ρ.

We start with the case that the path $\rho_{m,l}$ from q_m to q_l contains an accepting transition. The equation $u_{m..} = u_{l..}$ implies that $u_{m..} = u_{l..} = (u_{m,l})^\omega$. Hence, $\rho' = \rho_{0,m}.(\rho_{m,l})^\omega$ is a lasso-shaped accepting run over u. Moreover, the size of ρ' is smaller than the size of ρ as $|minSL(\rho')| \leq |\rho_{0,m}| + |\rho_{m,l}| = l < |minSL(\rho)|$.

Now we solve the case when there is no accepting transition in the path $\rho_{m,l}$. First, assume that $\rho_{m,l}$ is completely included in the minimal stem of ρ, i.e., $m < l \leq |minS(\rho)|$. Then we simply exclude $\rho_{m,l}$ from the stem and get an accepting lasso-shaped run ρ' over u, which has a shorter stem than ρ. Second, assume that $\rho_{m,l}$ is partly in the minimal stem and partly in the minimal loop of ρ, i.e., $m < |minS(\rho)| < l$. Let $\rho' = \rho_{0,m}.\rho_{l..}$ be the run ρ without the path $\rho_{m,l}$. Note that ρ' is again an accepting run over u as $u_{m..} = u_{l..}$. As ρ is lasso-shaped, we know that $\rho_{l..} = (\rho_{l,l+|minL(\rho)|})^\omega$. Hence, $\rho' = \rho_{0,m}.(\rho_{l,l+|minL(\rho)|})^\omega$ is also lasso-shaped. Moreover, the size of ρ' is smaller than the size of ρ as $|minSL(\rho')| \leq m + |minL(\rho)| < |minS(\rho)| + |minL(\rho)| = |minSL(\rho)|$. Finally, assume that $\rho_{m,l}$ is completely included in the minimal loop of ρ, i.e., $|minS(\rho)| \leq m < l$. Then we exclude $\rho_{m,l}$ from the minimal loop of ρ and get an accepting run $\rho' = \rho_{0,|minS(\rho)|}.(\rho_{|minS(\rho)|,m}.\rho_{l,|minSL(\rho)|})^\omega$ of a smaller size than ρ. We need to show that ρ' accepts u. The run ρ' accepts the word $u' = u_{0,|minS(\rho)|}.(u_{|minS(\rho)|,m}.u_{l,|minSL(\rho)|})^\omega = u_{0,m}.(u_{l,|minSL(\rho)|}.u_{|minS(\rho)|,m})^\omega$. As $u_{|minS(\rho)|,m} = u_{|minSL(\rho)|,m+|minL(\rho)|}$, we get $u' = u_{0,m}.(u_{l,m+|minL(\rho)|})^\omega$. Further, $u_{m..} = u_{l..} = u_{m+|minL(\rho)|..}$ implies $u_{m..} = u_{l..} = (u_{l,m+|minL(\rho)|})^\omega$ and thus $u' = u_{0,m}.u_{m..} = u$. $\square$

The next lemma shows that each minimal run over u can be denoted as a lasso-shaped structure build from one path over $minS(u)$ and at most n paths over $minL(u)$, where n is the number of states in the automaton. For example, the minimal run $\tau.\pi^\omega$ over $b.(abc)^\omega$ presented above can be also denoted as

$\pi_0\pi_1\pi_2.(\pi_3\pi_4\pi_5)^\omega$, where the paths π_i are defined as follows.

$$\pi_0 = p_0 \xrightarrow{b} p_1$$

$$\pi_1 = p_1 \xrightarrow{a} p_2 \xrightarrow{b} p_3 \xrightarrow{c} p_4$$

$$\pi_2 = p_4 \xrightarrow{a} r_0 \xrightarrow{b} r_1 \xrightarrow{c} r_2$$

$$\pi_3 = r_2 \xrightarrow{a} r_2 \xrightarrow{b} r_3 \xrightarrow{c} r_4$$

$$\pi_4 = r_4 \xrightarrow{a} r_5 \xrightarrow{b} r_4 \xrightarrow{c} \bullet\!\!\rightarrow r_6$$

$$\pi_5 = r_6 \xrightarrow{a} r_0 \xrightarrow{b} r_1 \xrightarrow{c} r_2$$

Note that the stem $\pi_0\pi_1\pi_2$ is not the minimal stem and $\pi_3\pi_4\pi_5$ is not the minimal loop of the minimal run $\tau.\pi^\omega$. Further, note that the paths $\pi_1, \ldots, \pi_5$ start in the considered states at significant positions, which are typeset in red and not struck through in Table 1.

Lemma 5. *Let $\mathcal{A}$ be a TBA with n states and ρ be a minimal run over a lasso-shaped word $u \in L(\mathcal{A})$. Then ρ can be denoted as $\pi_0\pi_1 \ldots \pi_i.(\pi_{i+1}\pi_{i+2} \ldots \pi_k)^\omega$, where π_0 is a path over $minS(u)$, $\pi_1, \pi_2, \ldots, \pi_k$ are paths over $minL(u)$, and $0 \le i < k \le n$. Moreover, $|minSL(\rho)| \le |\pi_0\pi_1 \ldots \pi_k| < |minSL(\rho)| + |minL(u)|$ and the last $|\pi_0\pi_1 \ldots \pi_k| - |minSL(\rho)|$ transitions of π_k and π_i are identical.*

Proof. Let $\mathcal{A}$ be a TBA with n states and $\rho = (q_0, u_0, q_1)(q_1, u_1, q_2) \ldots$ be a minimal run over a lasso-shaped word $u \in L(\mathcal{A})$. The lasso shape of ρ implies that $\rho_{|minS(\rho)|..} = \rho_{|minSL(\rho)|..}$ and thus also $u_{|minS(\rho)|..} = u_{|minSL(\rho)|..}$. This means that $|minL(\rho)| = j \cdot |minL(u)|$ for some $j > 0$.

Let $i \ge 0$ be the smallest number such that $minS(u).minL(u)^i$ is at least as long as $minS(\rho)$. As $|minL(\rho)| = j \cdot |minL(u)|$, then $k = i + j$ is the smallest number such that $minS(u).minL(u)^k$ is at least as long as $minS(\rho).minL(\rho)$. Let $p_1, p_2, \ldots, p_k, p_{k+1}$ be the first $k + 1$ significant positions in u. We set $\pi_0 = \rho_{0,p_1}$ to be the prefix of ρ over $minS(u)$ and, for each $1 \le l \le k$, we set $\pi_l = \rho_{p_l,p_{l+1}}$ to be the l-th successive subpart of ρ over $minL(u)$. The definition of k implies that $|minSL(\rho)| \le |\pi_0\pi_1 \ldots \pi_k| < |minSL(\rho)| + |minL(u)|$.

We have $\pi_0\pi_1 \ldots \pi_k = minS(\rho).minL(\rho).\pi'$, where π' is a prefix of $minL(\rho)$ such that $0 \le |\pi'| = |\pi_0\pi_1 \ldots \pi_k| - |minSL(\rho)| < |\pi_k|$. As $|\pi_{i+1}\pi_{i+2} \ldots \pi_k| = j \cdot |minL(u)| = |minL(\rho)|$, we get that $\pi_0\pi_1 \ldots \pi_i = minS(\rho).\pi'$ and this means that π_k and π_i have the same suffix π' of the length $|\pi'| = |\pi_0\pi_1 \ldots \pi_k| - |minSL(\rho)|$. Note that this holds also in the case when $i = 0$ because this situation implies that $\pi_0 = minS(\rho)$ and thus $\pi' = \varepsilon$.

As $\pi_0\pi_1 \ldots \pi_i = minS(\rho).\pi'$ and $\pi_0\pi_1 \ldots \pi_k = minS(\rho).minL(\rho).\pi'$, we get that there exists π'' such that $minL(\rho) = \pi'.\pi''$. Then

$$\pi_0\pi_1 \ldots \pi_i.(\pi_{i+1}\pi_{i+2} \ldots \pi_k)^\omega = minS(\rho).\pi'.(\pi''.\pi')^\omega = minS(\rho).(\pi'.\pi'')^\omega = \rho.$$

It remains to show that $k \le n$. For each significant position p_l such that $1 \le l \le k$, it holds that $|minS(u)| \le p_l < |minSL(\rho)|$ and $u_{p_l..} = minL(u)^\omega$. Lemma 4 says that states of the run ρ at positions $p_1, p_2, \ldots, p_k$ are pairwise different. Hence, $k \le n$. $\square$

4 Tightening construction and upper bound

Our tightening construction extends a given automaton $\mathcal{A}$ with new states and transitions to make it tight. Let n be the number of states in $\mathcal{A}$. Lemmata 3–5 imply that for each lasso-shaped word $u \in L(\mathcal{A})$, there exists an accepting run $\rho = \pi_0\pi_1 \ldots \pi_i.(\pi_{i+1}\pi_{i+2} \ldots \pi_k)^\omega$ over u where $0 \le i < k \le n$, π_0 is a path over $minS(u)$ and $\pi_1, \pi_2, \ldots, \pi_k$ are paths over $minL(u)$. Moreover, the states at an arbitrary but fixed position in $\pi_1, \pi_2, \ldots, \pi_k$ are pairwise different with the exception of the last x states of π_k for some $0 < x \le |minL(u)|$, which are identical to the corresponding states in π_i.

To accept a lasso-shaped word $u \in L(\mathcal{A})$ by a tight run, the extended automaton nondeterministically guesses the moment when $minS(u)$ is read and the path π_0 terminates. In this moment, it nondeterministically guesses the numbers i, k and the initial states of $\pi_2, \ldots, \pi_k$ and sets the initial state of π_1 to the current state of the original automaton. When reading $minL(u)$, it simultaneously tracks these paths and if there are more than one possible successors in a path, it chooses one nondeterministically. The extended automaton closes a cycle over $minL(u)$ via an accepting transition if the tracked paths $\pi_1, \pi_2, \ldots, \pi_k$ form together a path $\pi_1\pi_2 \ldots \pi_k$ leading to the first state of π_{i+1} and such that $\pi_{i+1}\pi_{i+2} \ldots \pi_k$ contains at least one accepting transition.

Note that our tightening construction considers only the cases when $k \ge 2$. If $k = 1$, then ρ can be denoted as $\pi_0.\pi_1^\omega$ where π_0 is a path over $minS(u)$ and π_1 is a path over $minL(u)$. This means that the run ρ of $\mathcal{A}$ is tight and we do not have to extend the automaton because of the corresponding word u.

Let $\mathcal{A}$ be a TBA with n states. The tightening construction adds to $\mathcal{A}$ so-called *macrostates*. Each macrostate $s_1 \ldots s_i[s_{i+1} \ldots s_k]_j^\star$ represents

- the current states $s_1, s_2, \ldots, s_k$ of paths $\pi_1, \pi_2, \ldots, \pi_k$ where $2 \le k \le n$,
- the number $0 \le i < k$ marking the beginning of the loop $\pi_{i+1}\pi_{i+2} \cdots \pi_k$,
- the number $i < j \le k$ such that π_j is the leftmost path in this loop containing an accepting transition, and
- the information $\star \in \{\circ, \bullet\}$ whether the accepting transition of π_j has been already passed ($\bullet$) or not ($\circ$).

As the paths $\pi_1, \pi_2, \ldots, \pi_k$ are tracked in a parallel and synchronous way, the states $s_1, s_2, \ldots, s_k$ of a macrostate have to be pairwise different with a possible exception of states $s_i = s_k$. Formally, we define the set of macrostates built from the set of states Q as

$$M_Q = \{ s_1 \ldots s_i[s_{i+1} \ldots s_k]_j^\star \mid 2 \le k \le |Q|,\ 0 \le i < j \le k,\ \star \in \{\circ, \bullet\},$$
$$s_1, \ldots, s_k \in Q \text{ where } s_m = s_l \text{ implies } m = l \text{ or } m, l \in \{i, k\}\}.$$

Now we are ready to define a tight automaton $\mathcal{A}^\dagger$ equivalent to $\mathcal{A}$.

Definition 1. *Let* $\mathcal{A} = (Q, \Sigma, \delta, I, \delta_F)$ *be a TBA. We define the TBA* $\mathcal{A}^\dagger$ *as* $\mathcal{A}^\dagger = (Q \cup M_Q, \Sigma, \delta \cup \delta', I \cup I', \delta_F \cup \delta'_F)$*, where*

- $\delta' = \delta_1 \cup \delta_2 \cup \delta_3$ *consists of three kinds of transitions,*
- $I' = \{s_1...s_i[s_{i+1}...s_k]_j^\circ \in M_Q \mid s_1 \in I,\ s_i \neq s_k\},$ *and*
- $\delta'_F = \delta_3.$

The transitions in $\delta_1 \cup \delta_2 \cup \delta_3$ *involve macrostates. They are defined as follows.*

$$\delta_1 = \{q \xrightarrow{a} s_1...s_i[s_{i+1}...s_k]_j^\circ \mid q \xrightarrow{a} s_1 \in \delta,\ s_i \neq s_k\}$$

These transitions are used to nondeterministically choose the numbers i, j, k *and the initial states of* $\pi_2, \pi_3, \ldots, \pi_k$ *when reading the last symbol of* $minS(u)$. *If* $minS(u) = \varepsilon$, *the nondeterministic choice is done by starting the computation in a macrostate of* I'.

$$\delta_2 = \{s_1...s_i[s_{i+1}...s_k]_j^\star \xrightarrow{a} r_1...r_i[r_{i+1}...r_k]_j^\star \mid *, \star \in \{\circ, \bullet\},$$
$$\forall\, 1 \leq l \leq k \,.\, if\ i < l < j\ then\ s_l \xrightarrow{a} r_l \in \delta \setminus \delta_F\ else\ s_l \xrightarrow{a} r_l \in \delta,$$
$$s_i = s_k\ implies\ r_i = r_k,\ if\ s_j \xrightarrow{a} r_j \in \delta_F\ then\ \star = \bullet\ else\ \star = *\}$$

These transitions simultaneously track the progress on the paths $\pi_1, \pi_2, \ldots, \pi_k$ *including the information whether* π_j *has already passed an accepting transition or not. The condition* $s_l \xrightarrow{a} r_l \notin \delta_F$ *for* $i < l < j$ *enforces that* π_j *is the leftmost path on the loop* $\pi_1\pi_2\ldots\pi_k$ *containing an accepting transition.*

$$\delta_3 = \{s_1...s_i[s_{i+1}...s_k]_j^\star \xrightarrow{a} r_1...r_i[r_{i+1}...r_k]_j^\circ \mid \star \in \{\circ, \bullet\},\ s_k \xrightarrow{a} r_{i+1} \in \delta,$$
$$\forall\, 1 \leq l < k \,.\, if\ i < l < j\ then\ s_l \xrightarrow{a} r_{l+1} \in \delta \setminus \delta_F\ else\ s_l \xrightarrow{a} r_{l+1} \in \delta,$$
$$r_i \neq r_k,\ \star = \bullet\ or\ (j < k \wedge s_j \xrightarrow{a} r_{j+1} \in \delta_F)\ or\ (j = k \wedge s_k \xrightarrow{a} r_{i+1} \in \delta_F)\}$$

These accepting transitions can enclose a cycle on macrostates if the last state of π_l *matches the first state of* π_{l+1} *for each* $1 \leq l < k$, *the last state of* π_k *matches the first state of* π_{i+1}, *and* π_j *has passed an accepting transition in the past or during this step.*

Theorem 1. *Let* $\mathcal{A} = (Q, \Sigma, \delta, I, \delta_F)$ *be a TBA. Then* $L(\mathcal{A}) = L(\mathcal{A}^\dagger)$.

Proof. The inclusion $L(\mathcal{A}) \subseteq L(\mathcal{A}^\dagger)$ is trivial as each accepting run of $\mathcal{A}$ is also an accepting run of $\mathcal{A}^\dagger$.

We show that $L(\mathcal{A}^\dagger) \subseteq L(\mathcal{A})$. Let σ be an accepting run of $\mathcal{A}^\dagger$ that involves some macrostates. Note that all macrostates in the run have to use the same numbers i, j, k. We construct an accepting run ρ of $\mathcal{A}$ over the same word as σ. Intuitively, ρ will consistently use the transitions of some element of the macrostates in σ, starting with the first element. Each time σ uses a transition of δ_3, ρ will switch to the next element and after the k-th element, it will switch back to the $(i + 1)$-st element.

First we define an auxiliary function g that determines for each $l \geq 0$ the element of the macrostate in σ that will be followed by the transition ρ_l.

$$g(0) = 1 \qquad g(l+1) = \begin{cases} g(l) & \text{if } \sigma_l \notin \delta_3 \\ g(l) + 1 & \text{if } \sigma_l \in \delta_3 \text{ and } g(l) < k \\ i + 1 & \text{if } \sigma_l \in \delta_3 \text{ and } g(l) = k \end{cases}$$

Now we construct ρ as follows.

$$\rho_l = \begin{cases} \sigma_l & \text{if } \sigma_l \in \delta \\ q \xrightarrow{a} s_1 & \text{if } \sigma_l = q \xrightarrow{a} s_1...s_i[s_{i+1}...s_k]_j^\circ \in \delta_1 \\ s_{g(l)} \xrightarrow{a} r_{g(l+1)} & \text{if } \sigma_l = s_1...s_i[s_{i+1}...s_k]_j^* \xrightarrow{a} r_1...r_i[r_{i+1}...r_k]_j^* \in \delta_2 \cup \delta_3 \end{cases}$$

One can easily check that ρ is a run of $\mathcal{A}$ over the same word as σ. Further, because σ is accepting, it contains infinitely many transitions of δ_3. Hence, there are infinitely many pairs m, l such that $0 < m < l$ and

$$g(m-1) \neq j = g(m) = g(m+1) = \ldots = g(l-1) \neq g(l).$$

The definition of g implies that $\sigma_{m-1,l} \in \delta_3.\delta_2^*.\delta_3$, which means that the j-th element of some macrostate in $\sigma_{m,l}$ takes an accepting transition in δ_F. The construction of ρ guarantees that $\rho_{m,l}$ contains the same transition in δ_F. Hence, ρ contains infinitely many accepting transitions and it is therefore accepting. □

Theorem 2. *Let* $\mathcal{A} = (Q, \Sigma, \delta, I, \delta_F)$ *be a TBA. Then* $\mathcal{A}^\dagger$ *is tight.*

Proof. Lemma 3 implies that for each lasso-shaped word $u \in L(\mathcal{A})$, there exists a minimal run of $\mathcal{A}$ over u. The validity of the statement then follows directly from the properties of minimal runs proven in Lemmata 4 and 5 and from the design of the tightening construction. □

4.1 State-based tight automata

While our tightening construction produces automata with transition-based acceptance, the previous tightening constructions [13, 23, 24] produce automata with state-based acceptance. Some applications [27] also work with tight state-based automata on the input. Therefore, we present a transformation of a tight TBA to an equivalent BA preserving tightness.

Let $\mathcal{A} = (Q, \Sigma, \delta, I, \delta_F)$ be a tight TBA. An equivalent tight BA $\mathcal{B}$ can be constructed as follows. We define the set of accepting states as duplicates of states $q \in Q$ that have some accepting transition starting in q, i.e., $F = \{\overline{q} \mid q \xrightarrow{a} p \in \delta_F\}$. We extend the initial states and the transition relation in such a way that whenever the original automaton can use an accepting transition from a state q, the resulting state-based automaton can reach the corresponding state $\overline{q}$ and use an analogous transition from it. Formally, the tight BA $\mathcal{B}$ equivalent to $\mathcal{A}$ is constructed as $\mathcal{B} = (Q \cup F, \Sigma, \delta \cup \delta', I \cup I', F)$, where

- $I' = \{\overline{q} \mid q \in I\} \cap F$ and
- $\delta' = \{p \xrightarrow{a} \overline{q} \mid p \xrightarrow{a} q \in \delta, \ \overline{q} \in F\} \cup \{\overline{p} \xrightarrow{a} \overline{q} \mid p \xrightarrow{a} q \in \delta_F, \ \overline{p}, \overline{q} \in F\} \cup$
  $\cup \{\overline{p} \xrightarrow{a} q \mid p \xrightarrow{a} q \in \delta_F, \ \overline{p} \in F\}$.

Each accepting run σ of $\mathcal{B}$ can be transformed to an accepting run ρ of $\mathcal{A}$ over the same word simply by replacing each state $\overline{q} \in F$ by the corresponding state q. Thus we get $L(\mathcal{B}) \subseteq L(\mathcal{A})$.

Further, each accepting run ρ of $\mathcal{A}$ can be transformed into an accepting run σ of $\mathcal{B}$ over the same word simply by replacing each state q from which an accepting transition is taken with the corresponding state $\bar{q}$. This implies $L(\mathcal{A}) \subseteq L(\mathcal{B})$. Moreover, when we apply this transformation to a tight run ρ of $\mathcal{A}$, we obtain a tight run σ of $\mathcal{B}$. To sum up, the automata $\mathcal{A}$ and $\mathcal{B}$ are equivalent and if $\mathcal{A}$ is tight, then $\mathcal{B}$ is also tight.

4.2 Upper bound for tightening

Lemma 6. *Let $\mathcal{A}$ be a TBA with n states. The number of states in $\mathcal{A}^\dagger$ is at most*

$$n + 2 \cdot \sum_{k=2}^{n} \frac{n! \cdot k \cdot (k+1)}{(n-k)!} \in \mathcal{O}(n! \cdot n^3).$$

Proof. Let Q be the set of states of $\mathcal{A}$. First we bound the number of macrostates of the form $s_1 \ldots s_i [s_{i+1} \ldots s_k]_j^\star \in M_Q$ for a fixed i, j, k. There are $\frac{n!}{(n-k)!} \cdot 2$ cases where all states $s_1, s_2, \ldots, s_k$ are pairwise different and $\frac{n!}{(n-(k-1))!} \cdot 2$ cases where $s_1, s_2, \ldots, s_{k-1}$ are pairwise different and $s_k = s_i$. The factor 2 comes from $\star \in \{\circ, \bullet\}$. Altogether, M_Q contains at most $4 \cdot \frac{n!}{(n-k)!}$ macrostates for fixed i, j, k. Further, for a fixed $k \geq 2$, there are $\frac{k \cdot (k+1)}{2}$ possible pairs of values of i, j satisfying $0 \leq i < j \leq k$. Altogether, the number of macrostates in M_Q can be bounded by $\sum_{k=2}^{n} 4 \cdot \frac{n!}{(n-k)!} \cdot \frac{k \cdot (k+1)}{2} = 2 \cdot \sum_{k=2}^{n} \frac{n! \cdot k \cdot (k+1)}{(n-k)!}$. When we add the number $n = |Q|$ of the original states, we get the statement. $\square$

Recall that each BA can be seen as a special case of a TBA. Further, note that the transformation of tight TBA to tight BA presented in Section 4.1 only doubles the state space in the worst case. Hence, we also proved that each BA or TBA with n states can be transformed into an equivalent tight BA with at most $\mathcal{O}(n! \cdot n^3)$ states. This upper bound is tighter (i.e., asymptotically smaller) than the upper bound $2^{\mathcal{O}(n^2)}$ by Ehlers [13] and than the upper bound $\mathcal{O}((\sqrt{2}n)^{2n})$ derived from the Schuppan's construction [23] as

$$\lim_{n \to \infty} \frac{n! \cdot n^3}{2^{n^2}} = \lim_{n \to \infty} \frac{n! \cdot n^3}{(\sqrt{2}n)^{2n}} = 0.$$

5 Lower bound for tightening

We present a lower bounds for any transformation of a TBA or a BA to an equivalent tight TBA or BA.

Lemma 7. *For each $n > 0$, there is a TBA $\mathcal{A}$ with $n+1$ states and an equivalent BA $\mathcal{A}'$ with $2n+1$ states such that for every equivalent tight TBA $\mathcal{B}$ with the set of states Q it holds that*

$$|Q| \geq \sum_{k=1}^{n} \frac{n!}{(n-k)!}.$$

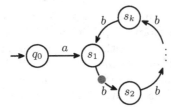

Fig. 4. The construction of the transitions for a given $[s_1...s_k]$.

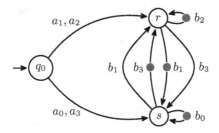

Fig. 5. The automaton $\mathcal{A}$ for $n = 2$. The construction considers 4 sequences and each sequence induced transitions that accept the following words: $[s]$ relates to the word $a_0.b_0^\omega$, $[rs]$ to $a_1.b_1^\omega$, $[r]$ to $a_2.b_2^\omega$, and $[sr]$ to $a_3.b_3^\omega$.

Proof. Let us fix some $n > 0$. We construct the TBA $\mathcal{A}$ with $n{+}1$ states gradually as follows. The automaton uses states $\{q_0\} \cup Q'$ where q_0 is the only initial state and Q' contains another n states. The construction works with nonempty sequences $[s_1...s_k]$ of pairwise different states from Q'. For each $[s_1...s_k]$, we add fresh symbols a, b to the alphabet of $\mathcal{A}$ and the transitions depicted in Figure 4 to the transition relation of $\mathcal{A}$. The automaton accepts $a.b^\omega$ with these transitions. The constructed automaton for $n = 2$ is in Figure 5. The equivalent BA $\mathcal{A}'$ is constructed from $\mathcal{A}$ by the transformation given in Section 4.1.

Now we assume that $\mathcal{B} = (Q, \Sigma, \delta, I, \delta_F)$ is a tight TBA equivalent to $\mathcal{A}$. Each $[s_1...s_k]$ induces the acceptance of a new word $a.b^\omega \in L(\mathcal{A}) = L(\mathcal{B})$. As $\mathcal{B}$ is tight and $|minSL(a.b^\omega)| = 2$, there have to be transitions $p \xrightarrow{a} q \in \delta$ and $q \xrightarrow{b} q \in \delta_F$ for some states $p \in I$ and $q \in Q$. We prove by contradiction that the state q has to be different for each $[s_1...s_k]$.

Let us assume that $[s_1...s_k]$ and $[r_1...r_{k'}]$ are different sequences inducing the acceptance of $a.b^\omega$ and $a'.b'^\omega$, respectively, and $\mathcal{B}$ accepts these two words using transitions $p \xrightarrow{a} q, p' \xrightarrow{a'} q \in \delta$ and $q \xrightarrow{b} q, q \xrightarrow{b'} q \in \delta_F$. The situation is depicted in Figure 6. We distinguish two cases.

1. $\{s_1, \ldots, s_k\} \neq \{r_1, \ldots, r_{k'}\}$: Without loss of generality, we assume that $r_j \notin \{s_1, \ldots, s_k\}$. As $\mathcal{B}$ accepts all words in $\{a'\}.\{b, b'\}^\omega$, it also accepts the word $a'.b'^{j-1}.b^\omega$. However, this word is not in $L(\mathcal{A})$ as $\mathcal{A}$ is deterministic and after reading $a'.b'^{j-1}$ it gets to state r_j which has no transition over b since

Fig. 6. Illustration of the assumption. States p and p' does not have to be different.

since $r_j \notin \{s_1, \ldots, s_k\}$. Hence, $L(\mathcal{B}) \neq L(\mathcal{A})$ and this is a contradiction with our assumptions.

2. $\{s_1, \ldots, s_k\} = \{r_1, \ldots, r_{k'}\}$: As states in each sequence are not repeating, we get $k = k'$. We use the fact that the only accepting transitions over b and b' in $\mathcal{A}$ are those from s_1 and r_1, respectively. We distinguish two subcases:

   (a) $s_1 = r_1$: Let j be the smallest number such that $s_j \neq r_j$. As the sets of states are equal, there exist $j < m, m' \leq k$, such that $s_m = r_j$ and $r_{m'} = s_j$. Consider the run $\tau.\pi^\omega$ of $\mathcal{A}$, where

$$\tau = q_0 \xrightarrow{a} s_1 \xrightarrow{b} s_2 \xrightarrow{b} \ldots \xrightarrow{b} s_j \quad \text{and}$$
$$\pi = s_j \xrightarrow{b} s_{j+1} \ldots \xrightarrow{b} (s_m = r_j) \xrightarrow{b'} r_{j+1} \xrightarrow{b'} \ldots \xrightarrow{b'} (r_{m'} = s_j).$$

   As π contains no accepting transition, the run is not accepting. Since $\mathcal{A}$ is deterministic, it is the only run of $\mathcal{A}$ over $a.b^{j-1}.(b^{m-j}b'^{m'-j})^\omega$. As the word is accepted by $\mathcal{B}$, we get a contradiction with $L(\mathcal{A}) = L(\mathcal{B})$.

   (b) $s_1 \neq r_1$: Since the sequences contain the same states, there are some $1 \leq m, l \leq k$ such that $s_1 \xrightarrow{b'^m} r_1$ and $r_1 \xrightarrow{b^l} s_1$. Consider the run $\tau.\pi^\omega$ of $\mathcal{A}$, where

$$\tau = q_0 \xrightarrow{a} s_1 \quad \text{and} \quad \pi = s_1 \xrightarrow{b'^m} r_1 \xrightarrow{b^l} s_1.$$

   The run is not accepting as the only accepting transitions over b or b' starting in s_1 and r_1, respectively, are never taken. Since $\mathcal{A}$ is deterministic, it is the only run of $\mathcal{A}$ over $a.(b'^m b^l)^\omega$. As the word is accepted by $\mathcal{B}$, we get a contradiction with $L(\mathcal{A}) = L(\mathcal{B})$.

To sum up, we proved that every tight TBA $\mathcal{B}$ satisfying $L(\mathcal{B}) = L(\mathcal{A})$ must have at least one state for every nonempty sequence $[s_1 \ldots s_k]$. This directly implies that the number of its states is at least $\sum_{k=1}^{n} \frac{n!}{(n-k)!}$.   □

The previous lemma says that for each $n \geq 2$, there exists a TBA with n states such that the smallest equivalent tight TBA (and thus also the smallest equivalent tight BA) has at least $\sum_{k=1}^{n-1} \frac{(n-1)!}{(n-1-k)!}$ states. This function is clearly in $\Omega((n-1)!)$ as

$$\lim_{n \to \infty} \frac{\sum_{k=1}^{n-1} \frac{(n-1)!}{(n-1-k)!}}{(n-1)!} = \lim_{n \to \infty} \sum_{k=1}^{n-1} \frac{1}{(n-1-k)!} = \sum_{k=0}^{\infty} \frac{1}{k!} = e.$$

Note that the difference between the upper bound $\mathcal{O}(n! \cdot n^3) = \mathcal{O}((n-1)! \cdot n^4)$ given in Lemma 6 and the lower bound $\Omega((n-1)!)$ is only the factor n^4.

Analogous arguments lead to the statement that for each odd $n \geq 3$ there exists a BA with n states such that the smallest equivalent tight TBA (and thus also the smallest equivalent tight BA) has at least $\Omega(\frac{n-1}{2}!)$ states. This lower bound is above the previously known lower bound $2^{\Omega(n)}$ as for each c it holds

$$\lim_{n \to \infty} \frac{2^{cn}}{\frac{n-1}{2}!} = 0.$$

6 Postprocessing of tight automata

This section shows that a standard automata reduction technique called *quotienting* [8] preserves tightness. Hence, it can be applied to reduce tight automata before they are further processed.

Consider an automaton with the set of states Q. A *preorder* $\sqsubseteq \subseteq Q \times Q$ is a reflexive and transitive relation. Every preorder defines an *induced equivalence* $\approx = \sqsubseteq \cap \sqsupseteq$. Given a state q, we denote by $[q]$ the equivalence class of q with respect to a fixed equivalence $\approx$. Furthermore, for every $P \subseteq Q$, by $[P]$ we denote the set $[P] = \{[q] \mid q \in P\}$ of all equivalence classes of states in P.

Given a TBA $\mathcal{A} = (Q, \Sigma, \delta, I, \delta_F)$ and a preorder $\sqsubseteq$ on Q with its induced equivalence $\approx$, the *quotient* of $\mathcal{A}$ is the TBA $\mathcal{A}/\sqsubseteq = ([Q], \Sigma, \delta', [I], \delta'_F)$, where $\delta' = \{[q] \xrightarrow{a} [p] \mid q \xrightarrow{a} p \in \delta\}$ and $\delta'_F = \{[q] \xrightarrow{a} [p] \mid q \xrightarrow{a} p \in \delta_F\}$.

A preorder $\sqsubseteq$ is *good for quotienting (GFQ)* [8] if $L(\mathcal{A}) = L(\mathcal{A}/\sqsubseteq)$ for each TBA $\mathcal{A}$. There exist many preorders that are GFQ, for example various kinds of forward or backward simulation or trace inclusion. For their definition and more information about automata reduction techniques we refer to the comprehensive paper by Clemente and Mayr [8].

Lemma 8. *Let $\mathcal{A}$ be a tight TBA and let $\sqsubseteq$ be a GFQ preorder. The automaton $\mathcal{A}/\sqsubseteq$ is tight and $L(\mathcal{A}) = L(\mathcal{A}/\sqsubseteq)$.*

Proof. The language equivalence trivially follows from the definition of *GFQ*. Let us consider an arbitrary lasso-shaped word $u \in L(\mathcal{A})$. As $\mathcal{A}$ is tight, it has an accepting run $\rho = \tau.\pi^\omega$ where τ has the form $q_0 \xrightarrow{minS(u)} l$ and π has the form $l \xrightarrow{minL(u)} l$. The definition of quotient implies that for each accepting run of $\mathcal{A}$ there exists an accepting run over the same word through the corresponding equivalence classes in $\mathcal{A}/\sqsubseteq$. Hence, $\mathcal{A}/\sqsubseteq$ has an accepting run $\rho' = \tau'.\pi'^\omega$ where τ' has the form $[q_0] \xrightarrow{minS(u)} [l]$ and π' has the form $[l] \xrightarrow{minL(u)} [l]$. It is easy to see that $|minSL(\rho')| \leq |minSL(\rho)|$ and thus ρ' is tight. Therefore, the automaton $\mathcal{A}/\sqsubseteq$ is tight. $\square$

7 Implementation

We have implemented our tightening construction in a tool called *Tightener*. The tool is written in Python 3.8.15 and it is built upon the library for LTL and

ω-automata called Spot [10] in version 2.11.4. Spot provides state-of-the-art LTL to automata translations, efficient transformations of arbitrary automata in the HOA format [2] to equivalent TBA, and some automata reduction techniques, in particular *direct simulation* [8] that is good for quotienting.

Tightener can take as an input either an LTL formula or an automaton in the HOA format. The input is internally translated into an equivalent TBA using the functionality provided by the Spot library. The TBA is then transformed into a tight TBA or tight state-based BA using the construction presented in this paper. The tight automaton is then optionally reduced using Spot's function `reduce_direct_sim` which performs quotienting by direct simulation. The resulting tight automaton is encoded in DOT or in the HOA format.

Tightener is available in an artifact at Zenodo[3] and at the project repository[4] under the GNU Public License, version 3 [1]. The tool can be run in the directory `Tightener_project` using the command `python Tightener.py [flags] "input"`. The tool supports the following flags.

- `-h` or `--help` describes the basic usage of the tool.
- `-f` or `--formula` says that the `"input"` is an LTL formula (e.g., `"Fp1 | Fp2"`) on the command line. Tightener uses the same syntax for LTL formulas as Spot, see `https://spot.lre.epita.fr/ioltl.html`.
- `-F` or `--file` says that the `"input"` is a path to a text file containing an LTL formula in the format mentioned above.
- `-a` or `--HOA` says that the `"input"` is a path to a file containing an automaton in the HOA format.
- `-s` or `--sbacc` asks to produce a state-based tight automaton. The tool produces tight TBA by default.
- `-r` or `--reduces` applies reductions preserving tightness before the tight automaton is returned. These reductions are not applied by default.
- `-o` or `--outputHOA` outputs the tight automaton in the HOA format. By default, the tool returns a tight automaton in DOT format, which can be easily visualized, for example at `https://dreampuf.github.io/GraphvizOnline/`. Note that the DOT format does not support multiple initial states. Hence, if the returned automaton has multiple initial states, one of them is marked as initial and the others are identified by an auxiliary incoming edge labeled with *init*.

8   Experimental results

We compare Tightener against the translation of LTL to state-based generalized Büchi automata introduced by Clarke, Grumberg, and Hamaguchi [6] and called CGH. Schuppan and Biere [24] proved that the automata produced by CGH are tight. As far as we know, this is the only existing implementation besides Tightener that produces tight automata. Still, the comparison is not entirely

---

[3] `https://zenodo.org/records/10512677`
[4] `https://gitlab.com/mjankola/tightener/-/tree/main?ref_type=heads`

Table 2. We compare the tight TBA and BA produced by Tightener against the GBA constructed by CGH. For both datasets, the table shows the number [#] and the percentage [%] of cases where the corresponding tool provided a tight automaton with fewer states than the other tool. Columns *avg. size* represent the average number of states of the automata constructed by the corresponding tool. Columns *TO* indicate the number of timeouts. Cases where Tightener timed out are counted in the CGH winning columns, but these cases are excluded from the computation of average size.

tool	642 random formulas				219 formulas from literature			
	[#]	[%]	avg. size	TO	[#]	[%]	avg. size	TO
Tightener (TBA)	482	75.1%	20.03	44	179	81.7%	37.00	28
CGH (GBA)	149	23.2%	73.9	0	39	17.8%	161.51	0
Tightener (BA)	381	59.3%	32.54	44	141	64.4%	60.44	28
CGH (GBA)	243	37.8%	73.9	0	72	32.8%	161.51	0

fair as Tightener and CGH have different input and different output: Tightener can transform any LTL formula or automaton in the HOA format to tight TBA or BA, CGH accepts only an LTL formula and produces a tight GBA. BA can be seen as a special case of both TBA and GBA, but the opposite does not hold. We provided a transformation of tight TBA into equivalent tight BA in Section 4.1. Each GBA can be transformed into an equivalent BA (this so-called *degeneralization* process has been recently significantly improved [3]), but the transformation increases the number of states and it does not guarantee to preserve tightness. We therefore compare the size of tight GBA produced by CGH against the size of tight TBA and tight BA produced by Tightener.

Since CGH produces tight GBA in symbolic representation, we implemented a process that enumerates automata states from this symbolic representation and uses the SMT solver Z3 [9] to prune unreachable and contradictory states. In the end, we count the number of reachable states. This implementation can be also found in our repository in script `Tightener_project/CGH_implementation.py`.

We compare CGH and Tightener on two sets of LTL formulas. The first dataset contains 642 formulas produced by random formulas generator `rand_ltl` of Spot's. These formulas are stored in file `ltlDataSet_random.txt` in our repository. The second dataset consists of 219 formulas taken from literature [4,11,14,16–18,21,25,26]. We obtained these formulas from the tool `gen_ltl` of Spot and they are stored in file `ltlDataSet_pattern.txt` in our repository.

We ran the experiments on a machine with an AMD Ryzen 7 PRO 4750U processor and 32 GB of RAM. We set 15 minutes timeout limit per task with no explicit memory limit.

Each formula has been translated by Tightener to a tight TBA and to a tight BA with reduction switched on in both cases, and by CGH to a tight GBA. Table 2 summarizes the cummulative results for the two datasets. One can see that Tightener constructs smaller automata in substantially more cases than

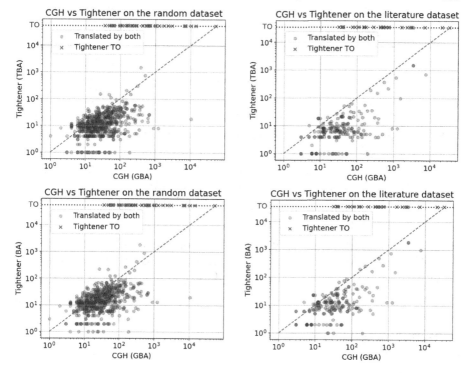

Fig. 7. The comparison of the number of states of the tight automata produced by Tightener and CGH for individual LTL formulas of each dataset. In the top row, Tightener produces tight TBA. In the bottom row, it produces tight BA. CGH always produces GBA. The red crosses display the cases where Tightener reaches a time limit.

CGH in both considered datasets and with both settings. However, Tightener run out of time in some cases.

The scatter plots in Figure 7 compare the number of states of the tight automata constructed by CGH and Tightener for individual LTL formulas in each dataset. Since some of the produced automata are rather large, we use logarithmic scale in all of the scatter plots. The graphs clearly show that Tightener often produces dramatically smaller tight automata than CGH.

8.1   Experiments on formulas for robot action planning

Tumova et al. [27] introduced a technique that generates control strategies for a robot planning problem. They represent the strategies as lasso-shaped words, where alphabet is a set of locations and possible actions in the respective location. Their approach takes advantage of tight BA to obtain the strategies with the shortest length of the stem and the loop.

The paper contains three LTL formulas representing meaningful properties. Table 3 compares the sizes of tight BA obtained from Tightener and tight GBA

Table 3. Sizes of tight automata constructed from LTL formulas taken from the study of Tumova et al. [27]. *TO* indicates a timeout.

	number of states	
formula	Tightener (BA)	CGH (GBA)
$GF(R_4 \wedge grab \wedge F(R_2 \wedge drop)) \wedge GF\,light\_up$	38	224
$GF(((R_4 \wedge grab) \vee (R_5 \wedge grab)) \wedge F(R_2 \wedge drop)) \wedge GF\,light\_up$	43	317
$G(R_1 \rightarrow \bigwedge_{i \neq 1} \neg R_i \ U \ R_2 \wedge (\bigwedge_{i \neq 2} \neg R_i \ U \ R_3 \wedge$ $(\bigwedge_{i \neq 3} \neg R_i \ U \ (R_6 \wedge drop) \wedge \bigwedge_{i \neq 6} \neg R_i \ U \ R_5 \wedge$ $(\bigwedge_{i \neq 5} \neg R_i \ U \ (R_4 \wedge drop) \wedge (\bigwedge_{i \neq 4} \neg R_i \ U \ R_1))))) \wedge GF\,light\_up$	2	TO

from CGH on these formulas. For two of the formulas, Tightener constructed dramatically smaller automaton than CGH. On the third formula, Tightener produced a tight BA with 2 states while CGH ran out of time.

9   Conclusions

In this paper, we presented a new approach for converting TBA or BA to tight TBA or BA. We proved that the asymptotical rise of the state space is $O(n! \cdot n^3)$, which is the smallest upper bound so far reached. Further, we proved the highest lower bounds on the rise of the state-space of tight automata so far reached, making the theoretical construction of tight automata significantly tighter. We also showed that the good-for-quotienting simulations can be used to reduce automata while preserving tightness.

Our tool Tightener opens new ways to construct tight automata as it is the first tool that can create tight automata from arbitrary automata in the HOA format or from LTL formulas. We compared Tightener against the LTL to tight automata translation CGH on two datasets of LTL formulas. Experiments show that Tightener constructs smaller tight automata in substantially more cases. Moreover, we compared the two tools on three formulas for which a tight automaton was explicitly desired before. In all three cases, Tightener provided a dramatically better result.

Funding Statement. Until June 2023, M. Jankola was supported by the European Union's Horizon Europe program under the grant agreement No. 101087529 and since July 2023 by the Deutsche Forschungsgemeinschaft (DFG) – 378803395 (ConVeY). J. Strejček was supported by the Czech Science Foundation grant GA23-06506S.

References

1. GNU general public license, version 3. `http://www.gnu.org/licenses/gpl.html`, June 2007. Last retrieved 2020-01-01.
2. Tomáš Babiak, František Blahoudek, Alexandre Duret-Lutz, Joachim Klein, Jan Křetínský, David Müller, David Parker, and Jan Strejček. The Hanoi omega-automata format. In Daniel Kroening and Corina S. Pasareanu, editors, *Computer Aided Verification - 27th International Conference, CAV 2015, San Francisco, CA, USA, July 18-24, 2015, Proceedings, Part I*, volume 9206 of *Lecture Notes in Computer Science*, pages 479–486. Springer, 2015. See also `http://adl.github.io/hoaf/`.
3. Antonio Casares, Alexandre Duret-Lutz, Klara J. Meyer, Florian Renkin, and Salomon Sickert. Practical applications of the alternating cycle decomposition. In Dana Fisman and Grigore Rosu, editors, *Tools and Algorithms for the Construction and Analysis of Systems - 28th International Conference, TACAS 2022, Held as Part of the European Joint Conferences on Theory and Practice of Software, ETAPS 2022, Munich, Germany, April 2-7, 2022, Proceedings, Part II*, volume 13244 of *Lecture Notes in Computer Science*, pages 99–117. Springer, 2022.
4. Jacek Cichoń, Adam Czubak, and Andrzej Jasiński. Minimal Büchi automata for certain classes of LTL formulas. In *Proceedings of the Fourth International Conference on Dependability of Computer Systems (DEPCOS'09)*, pages 17–24. IEEE Computer Society, 2009.
5. Alessandro Cimatti, Edmund Clarke, Enrico Giunchuglia, Fausto Giunchiglia, Marco Pistore, Macro Roveri, Roberto Sebastiani, and Armando Tacchella. Nusmv 2: An opensource tool for symbolic model checking. In E. Brinksma and K. Guldstrand Larsen, editors, *Proceedings of the 14th International Conference on Computer Aided Verification (CAV'02)*, volume 2404 of *Lecture Notes in Computer Science*, pages 359–364, Copenhagen, Denmark, July 2002. Springer-Verlag.
6. Edmund M. Clarke, Orna Grumberg, and Kiyoharu Hamaguchi. Another look at LTL model checking. In David L. Dill, editor, *Computer Aided Verification, 6th International Conference, CAV '94, Stanford, California, USA, June 21-23, 1994, Proceedings*, volume 818 of *Lecture Notes in Computer Science*, pages 415–427. Springer, 1994.
7. Edmund M. Clarke, Orna Grumberg, Kenneth L. McMillan, and Xudong Zhao. Efficient generation of counterexamples and witness in symbolic model checking. In *Proceedings of the 32nd ACM/IEEE Design Automation Conference (DAC'95)*, pages 427–432, San Francisco, California, USA, June 1995. ACM Press.
8. Lorenzo Clemente and Richard Mayr. Efficient reduction of nondeterministic automata with application to language inclusion testing. *Log. Methods Comput. Sci.*, 15(1), 2019.
9. Leonardo Mendonça de Moura and Nikolaj S. Bjørner. Z3: an efficient SMT solver. In C. R. Ramakrishnan and Jakob Rehof, editors, *Tools and Algorithms for the Construction and Analysis of Systems, 14th International Conference, TACAS 2008, Held as Part of the Joint European Conferences on Theory and Practice of Software, ETAPS 2008, Budapest, Hungary, March 29-April 6, 2008. Proceedings*, volume 4963 of *Lecture Notes in Computer Science*, pages 337–340. Springer, 2008.
10. Alexandre Duret-Lutz, Etienne Renault, Maximilien Colange, Florian Renkin, Alexandre Gbaguidi Aisse, Philipp Schlehuber-Caissier, Thomas Medioni, Antoine Martin, Jérôme Dubois, Clément Gillard, and Henrich Lauko. From Spot 2.0 to

Spot 2.10: What's new? In Sharon Shoham and Yakir Vizel, editors, *Computer Aided Verification - 34th International Conference, CAV 2022, Haifa, Israel, August 7-10, 2022, Proceedings, Part II*, volume 13372 of *Lecture Notes in Computer Science*, pages 174–187. Springer, 2022.

11. Matthew B. Dwyer, George S. Avrunin, and James C. Corbett. Property specification patterns for finite-state verification. In Mark Ardis, editor, *Proceedings of the 2nd Workshop on Formal Methods in Software Practice (FMSP'98)*, pages 7–15, New York, March 1998. ACM Press.

12. Rüdiger Ehlers. Short witnesses and accepting lassos in ω-automata. In Adrian-Horia Dediu, Henning Fernau, and Carlos Martín-Vide, editors, *Language and Automata Theory and Applications, 4th International Conference, LATA 2010, Trier, Germany, May 24-28, 2010. Proceedings*, volume 6031 of *Lecture Notes in Computer Science*, pages 261–272. Springer, 2010.

13. Rüdiger Ehlers. How hard is finding shortest counter-example lassos in model checking? In Maurice H. ter Beek, Annabelle McIver, and José N. Oliveira, editors, *Formal Methods - The Next 30 Years - Third World Congress, FM 2019, Porto, Portugal, October 7-11, 2019, Proceedings*, volume 11800 of *Lecture Notes in Computer Science*, pages 245–261. Springer, 2019.

14. Kousha Etessami and Gerard J. Holzmann. Optimizing Büchi automata. In C. Palamidessi, editor, *Proceedings of the 11th International Conference on Concurrency Theory (Concur'00)*, volume 1877 of *Lecture Notes in Computer Science*, pages 153–167, Pennsylvania, USA, 2000. Springer-Verlag.

15. Paul Gastin, Pierre Moro, and Marc Zeitoun. Minimization of counterexamples in SPIN. In S. Graf and L. Mounier, editors, *Proceedings of the 11th International SPIN Workshop on Model Checking of Software (SPIN'04)*, volume 2989 of *Lecture Notes in Computer Science*, pages 92–108, April 2004.

16. Paul Gastin and Denis Oddoux. Fast LTL to Büchi automata translation. In G. Berry, H. Comon, and A. Finkel, editors, *Proceedings of the 13th International Conference on Computer Aided Verification (CAV'01)*, volume 2102 of *Lecture Notes in Computer Science*, pages 53–65, Paris, France, 2001. Springer-Verlag.

17. Jaco Geldenhuys and Henri Hansen. Larger automata and less work for LTL model checking. In *Proceedings of the 13th International SPIN Workshop (SPIN'06)*, volume 3925 of *Lecture Notes in Computer Science*, pages 53–70. Springer, 2006.

18. Jan Holeček, Tomáš Kratochvíla, Vojtěch Řehák, David Šafránek, and Pavel Šimeček. Verification results in Liberouter project. Technical Report 03/2004, CESNET Technical Report, 2004.

19. Orna Kupferman and Sarai Sheinvald-Faragy. Finding shortest witnesses to the nonemptiness of automata on infinite words. In Christel Baier and Holger Hermanns, editors, *CONCUR 2006 - Concurrency Theory, 17th International Conference, CONCUR 2006, Bonn, Germany, August 27-30, 2006, Proceedings*, volume 4137 of *Lecture Notes in Computer Science*, pages 492–508. Springer, 2006.

20. Orna Kupferman and Moshe Y. Vardi. Model checking of safety properties. In N. Halbwachs and D. Peled, editors, *Proceedinfs of the 11th International Conference on Computer Aided Verification (CAV'99)*, volume 1633 of *Lecture Notes in Computer Science*, pages 172–183. Springer-Verlag, 1999.

21. Radek Pelánek. BEEM: benchmarks for explicit model checkers. In *Proceedings of the 14th international SPIN conference on Model checking software*, Lecture Notes in Computer Science, pages 263–267. Springer-Verlag, 2007.

22. Kavita Ravi, Roderick Bloem, and Fabio Somenzi. A comparative study of symbolic algorithms for the computation of fair cycles. In J. W. O'Leary M. D. Aagaard, editor, *Proceedings of the 4th International Conference on Formal Methods*

in *Computer Aided Design (FMCAD'00)*, volume 2517 of *Lecture Notes in Computer Science*, pages 143–160. Springer-Verlag, 2000.

23. Viktor Schuppan. *Liveness checking as safety checking to find shortest counterexamples to linear time properties*. PhD thesis, ETH Zurich, 2006.

24. Viktor Schuppan and Armin Biere. Shortest counterexamples for symbolic model checking of LTL with past. In Nicolas Halbwachs and Lenore D. Zuck, editors, *Tools and Algorithms for the Construction and Analysis of Systems, 11th International Conference, TACAS 2005, Held as Part of the Joint European Conferences on Theory and Practice of Software, ETAPS 2005, Edinburgh, UK, April 4-8, 2005, Proceedings*, volume 3440 of *Lecture Notes in Computer Science*, pages 493–509. Springer, 2005.

25. Fabio Somenzi and Roderick Bloem. Efficient Büchi automata for LTL formulæ. In *Proceedings of the 12th International Conference on Computer Aided Verification (CAV'00)*, volume 1855 of *Lecture Notes in Computer Science*, pages 247–263, Chicago, Illinois, USA, 2000. Springer-Verlag.

26. Deian Tabakov and Moshe Y. Vardi. Optimized temporal monitors for SystemC. In *Proceedings of the 1st International Conference on Runtime Verification (RV'10)*, volume 6418 of *Lecture Notes in Computer Science*, pages 436–451. Springer, November 2010.

27. Jana Tumova, Alejandro Marzinotto, Dimos V. Dimarogonas, and Danica Kragic. Maximally satisfying LTL action planning. In *2014 IEEE/RSJ International Conference on Intelligent Robots and Systems, Chicago, IL, USA, September 14-18, 2014*, pages 1503–1510. IEEE, 2014.

Synthesis with Privacy Against an Observer*

Orna Kupferman[iD], Ofer Leshkowitz[(⊠)][iD],
and Naama Shamash Halevy[(⊠)][iD]

School of Computer Science and Engineering, The Hebrew University, Jerusalem,
Israel
`ofer.leshkowitz@mail.huji.ac.il`

Abstract. We study automatic *synthesis* of systems that interact with their environment and maintain *privacy* against an observer to the interaction. The system and the environment interact via sets I and O of input and output signals. The input to the synthesis problem contains, in addition to a specification, also a list of *secrets*, a function $\mathsf{cost} : I \cup O \to \mathbb{N}$, which maps each signal to the cost of hiding it, and a bound $b \in \mathbb{N}$ on the budget that the system may use for hiding of signals. The desired output is an (I/O)-transducer $\mathcal{T}$ and a set $\mathcal{H} \subseteq I \cup O$ of signals that respects the bound on the budget, thus $\sum_{s \in \mathcal{H}} \mathsf{cost}(s) \leq b$, such that for every possible interaction of $\mathcal{T}$, the generated computation satisfies the specification, yet an observer from which the signals in $\mathcal{H}$ are hidden, cannot evaluate the secrets.

We first show that the complexity of the problem is 2EXPTIME-complete for specifications and secrets in LTL, thus it is not harder than synthesis with no privacy requirements. We then analyze the complexity of the problem more carefully, isolating the two aspects that do not exist in traditional synthesis, namely the need to hide the value of the secrets and the need to choose the set $\mathcal{H}$. We do this by studying settings in which traditional synthesis can be solved in polynomial time – when the specification formalism is deterministic automata and when the system is closed, and show that each of the two aspects involves an exponential blow-up in the complexity. We continue and study *bounded synthesis with privacy*, where the input also includes a bound on the size of the synthesized transducer, as well as a variant of the problem in which the observer has *knowledge about the specification*, which can be helpful in evaluating the secrets. We study the effect of both variants on the different aspects of the problem and provide algorithms with a tight complexity.

1 Introduction

Synthesis is the automated construction of correct systems from their specifications [2]. While synthesized systems are correct, there is no guarantee about their *quality*. Since designers will be willing to give up manual design only after being convinced that the automatic process replacing it generates systems of comparable quality, it is extremely important to develop and study quality measures for automatically-synthesized systems. An important quality measure is *privacy*:

---

* This research is supported by the Israel Science Foundation, Grant 2357/19, and the European Research Council, Advanced Grant ADVANSYNT.

N. Kobayashi and J. Worrell (Eds.): FoSSaCS 2024, LNCS 14574, pp. 256–277, 2024.
https://doi.org/10.1007/978-3-031-57228-9_13

making sure that the system and its environment do not reveal information they prefer to keep private. Privacy is a vivid research area in Theoretical Computer Science. There, the notion of *differential privacy* is used for formalizing when an algorithm maintains privacy. Essentially, an algorithm is differentially private if by observing its output, one cannot tell if a particular individual's information is used in the computation [9,11]. Another related notion is *obfuscation* in system development, where we aim to develop systems whose internal operation is hidden [1,15]. Obfuscation has been mainly studied in the context of software, where it has exciting connections with cryptography [1,15].

In the setting of automated synthesis in formal methods, a very basic notion of privacy has been studied by means of synthesis with *incomplete information* [27,21,7]. There, the system should satisfy its specification eventhough it only has a partial view of the environment. Lifting differential privacy to formal methods, researchers have introduced the temporal logic *HyperLTL*, which extends LTL with explicit trace quantification [8]. Such a quantification can relate computations that differ only in non-observable elements, and can be used for specifying that computations with the same observable input have the same observable output. The synthesis problem of HyperLTL is undecidable, yet is decidable for the fragment with a single existential quantifier, which can specify interesting properties [13]. In [18], the authors suggested a general framework for automated synthesis of privacy-preserving reactive systems. In their framework, the input to the synthesis problem includes, in addition to the specification, also *secrets*. During its interaction with the environment, the system may keep private some of the assignments to the output signals, and it directs the environment which assignments to the input signals it should keep private. Consequently, the satisfaction value of the specification and secrets may become unknown. The goal is to synthesize a system that satisfies the specification yet keeps the value of the secrets unknown. Finally, lifting obfuscation to formal methods, researchers have studied the synthesis of obfuscation policies for temporal specifications. In [32], an obfuscation mechanism is based on edit functions that alter the output of the system, aiming to make it impossible for an observer to distinguish between secret and non-secret behaviors. In [10], the goal is to synthesize a control function that directs the user which actions to disable, so that the observed sequence of actions would not disclose a secret behavior.

In this paper we continue to study privacy-preserving reactive synthesis. As in [18], our setting is based on augmenting the specification with secrets whose satisfaction value should remain unknown. Unlike [18], the system and the environment have complete information about the assignments to the input and output signals, and the goal is to hide the secrets from a third party, and to do so by hiding the assignment to some of the signals throughout the interaction. As an example, consider a system that directs a robot patrolling a warehouse storage. Typical specifications for the system require it to direct the robot so that it eventually reaches the shelves of requested items, it never runs out of energy, etc. An observer to the interaction between the system and the robot may infer properties we may want to keep private, like dependencies between customers and shelves visited, locations of battery docking stations, etc. If we want to prevent the observer from inferring these properties (a.k.a,. the secrets), we have to hide the interaction from it. Different effort should be made in order to hide different components of the interaction (alarm sound, content of shelves, etc.). Our framework synthesizes a system that realizes the specification without the secrets being revealed, subject to restrictions on hiding of signals. As another example, consider a scheduler that should grant access to a joint resource. The scheduler should maintain mutual exclusion (grants are not given to different users simultaneously) and non-starvation (all requests are granted), while hiding details like waiting time or priority to specific users. In Examples 1 and 2, we describe in detail the application of our framework for the synthesis of such a scheduler, as well as its application in the synthesis of a robot that paints parts of manufactured pieces. The robot should satisfy some requirements about the generated pattern of colors while hiding other features of the pattern.

Formally, we consider a reactive system that interacts with its environments via sets I and O of input and output signals. At each moment in time, the system reads a truth assignment, generated by the environment, to the signals in I, and it generates a truth assignment to the signals in O. The interaction between the system and its environment generates a *computation*. The system *realizes* a specification φ if all its computations satisfy φ [25]. We introduce and study the problem of *synthesis with privacy in the presence of an observer*. Given a specification φ, and secrets $\psi_1, \ldots, \psi_k$ over $I \cup O$, our goal is to return, in addition to a system that realizes the specification φ, also a set $\mathcal{H} \subseteq I \cup O$ of *hidden signals*, such that the satisfaction value of the secrets $\psi_1, \ldots, \psi_k$ is unknown to an observer that does not know the truth values of the signals in $\mathcal{H}$. Thus, secrets are evaluated according to a *three-valued semantics*. The use of secrets enables us to hide *behaviors*, rather than just signals. [1] Obviously, hiding all signals guarantees that the satisfaction value of every secret is unknown. Hiding of signals, however, is not always possible or involves some cost. We formalize this by adding to the setting a function $\mathsf{cost} : I \cup O \to \mathbb{N}$, which maps each signal to the cost of hiding its value, and a bound $b \in \mathbb{N}$ on the budget that the system may use for hiding of signals. The set $\mathcal{H}$ of hidden signals has to respect the bound, thus $\sum_{s \in \mathcal{H}} \mathsf{cost}(s) \leq b$.

In some cases, it is desirable to hide the truth value of a secret only when some condition holds. For example, we may require to hide the content of selves only in some sections of the warehouse. We extend our framework to *conditional secrets*: pairs of the form $\langle \theta, \psi \rangle$, where the satisfaction value of the secret ψ should be hidden from the observer only when the trigger θ holds. In particular, when $\theta = \psi$, we require to hide the secret only when it holds. For example, we may require to hide an unfair scheduling policy only when it is applied. Note that a conditional secret $\langle \theta, \psi \rangle$ is not equivalent to a secret $\theta \to \psi$ or $\theta \to \neg \psi$, and that the synthesized system may violate the trigger, circumventing the need

---

[1] Hiding of signals is a special case of our framework. Specifically, hiding of a signal p can be done with the secrets Fp and $F \neg p$.

to hide the secret. For example, by synthesizing a fair scheduler, the designer circumvents the need to hide an unfair policy.

We show that synthesis with privacy is 2EXPTIME-complete for specifications and secrets in LTL. Essentially, once the set $\mathcal{H}$ of hidden signals is determined, we can compose an automaton for the specification with automata that verify, for each secret, that the assignments to the signals in $(I \cup O) \setminus \mathcal{H}$ can be completed both in a way that satisfies the secret and in a way that does not satisfy it. A similar algorithm works for conditional secrets.

While the complexity of our algorithm is not higher than that of LTL synthesis with no privacy, it would be misleading to conclude that handling of privacy involves no increase in the complexity. The 2EXPTIME complexity follows from the need to translate LTL specifications to deterministic automata on infinite words. Such a translation involves a doubly-exponential blow-up [22,20], which possibly dominates other computational tasks of the algorithm. In particular, two aspects of synthesis with privacy that do not exist in usual synthesis are a need to go over all possible choices of signals to hide, and a need to go over all assignments to the hidden signals.

Our main technical contribution is a finer complexity analysis of the problem, which reveals that each of the two aspects above involves an exponential complexity: the first in the number of signals and the second in the size of the secret. We start with the need to go over all assignments of hidden signals and show that even when the specification is T, the set $\mathcal{H}$ of hidden signals is given, and there is only one secret, given by a deterministic Büchi automaton, synthesis with privacy is EXPTIME-complete. This is exponentially higher than synthesis of deterministic Büchi automata, which can be solved in polynomial time. We continue to the need to go over all possible choices of $\mathcal{H}$. For that, we focus on the closed setting, namely when $I = \varnothing$, and the case the specification and secrets are given by deterministic automata. We show that while synthesis with privacy can be then solved in polynomial time for a given set $\mathcal{H}$, it is NP-complete when $\mathcal{H}$ is not given, even when the function cost is uniform.

We continue and study two variants of the problem: *bounded synthesis* and *knowledgeable observer*. One way for coping with the 2EXPTIME complexity of LTL synthesis, which is carried over to a doubly-exponential lower bound on the size of the generated system [28], is bounded synthesis. There, the input to the problem includes also a bound on the size of the system [30,12,19]. In a setting with no privacy, the bound reduces the complexity of LTL synthesis to PSPACE, as one can go over all candidate systems. We study bounded synthesis with privacy and show that privacy makes the problem much harder: it is EXPSPACE-complete when the specification and secrets are given by LTL formulas, and is PSPACE-complete when they are given by deterministic parity (or Büchi) automata.

Finally, recall that a system keeps a secret ψ private if an observer cannot reveal the truth value of ψ: every observable computation can be completed both to a computation that satisfies ψ and to a computation that does not satisfy ψ. We study a setting in which the observer knows the specification φ of the system.

Consequently, the observer knows that only completions that satisfy φ should be taken into account. If, for example, $\varphi \to \psi$, then ψ cannot be kept private. We describe an algorithm for this variant of the problem and analyze the way knowledge of the specification influences the complexity. In particular, we show that the problem becomes EXPTIME-complete even when the specification is given by a deterministic Büchi automaton and the secrets are of a fixed size.

Due to the lack of space, some examples and proofs are omitted and can be found in the full version, in the authors' URLs.

2   Preliminaries

2.1   Synthesis

For a finite nonempty alphabet Σ, an infinite *word* $w = \sigma_0 \cdot \sigma_1 \cdots \in \Sigma^\omega$ is an infinite sequence of letters from Σ. A *language* $L \subseteq \Sigma^\omega$ is a set of infinite words.

Let I and O be disjoint finite sets of input and output signals, respectively. We consider the alphabet $2^{I \cup O}$ of truth assignments to the signals in $I \cup O$. Then, a languages $L \subseteq (2^{I \cup O})^\omega$ can be viewed as a *specification*, and the *truth value* of L in a computation $w \in (2^{I \cup O})^\omega$ is T if $w \in L$, and is F otherwise.

An (I/O)-*transducer* is a tuple $\mathcal{T} = \langle I, O, S, s_0, \eta, \tau \rangle$, where S is a finite set of states, $s_0 \in S$ is an initial state, $\eta : S \times 2^I \to S$ is a transition function, and $\tau : S \to 2^O$ is a labeling function. We extend the transition function η to words in $(2^I)^*$ in the expected way, thus $\eta^* : S \times (2^I)^* \to S$ is such that for all $s \in S$, $x_I \in (2^I)^*$, and $i \in 2^I$, we have that $\eta^*(s, \epsilon) = s$, and $\eta^*(s, x_I \cdot i) = \eta(\eta^*(s, x_I), i)$. For a word $w_I = i_0 \cdot i_1 \cdot i_2 \cdots \in (2^I)^\omega$, we define the *computation of* $\mathcal{T}$ *on* w_I to be the word $\mathcal{T}(w_I) = (i_0 \cup o_0) \cdot (i_1 \cup o_1) \cdots \in (2^{I \cup O})^\omega$, where for all $j \geq 0$, we have that $o_j = \tau(\eta^*(s_0, i_0 \cdots i_j))$. The *language* of $\mathcal{T}$, denoted $L(\mathcal{T})$, is the set of computations of $\mathcal{T}$, that is $L(\mathcal{T}) = \{ \mathcal{T}(w_I) : w_I \in (2^I)^\omega \}$.

We say that $\mathcal{T}$ *realizes* a language $L \subseteq (2^{I \cup O})^\omega$ if $L(\mathcal{T}) \subseteq L$. We say that a language $L \subseteq (2^{I \cup O})^\omega$ is *realizable* if there is an (I/O)-transducer that realizes it. In the *synthesis* problem, we are given a specification language $L \subseteq (2^{I \cup O})^\omega$ and we have to return an (I/O)-transducer that realizes L or decide that L is not realizable. The language L is given by an automaton over the alphabet $2^{I \cup O}$ or a temporal logic formula over $I \cup O$ (see definitions in Section 2.4).

2.2   Synthesis with privacy

In the *synthesis with privacy* problem, we are given, in addition to the specification language $L_\varphi \subseteq (2^{I \cup O})^\omega$, also a *secret* $L_\psi \subseteq (2^{I \cup O})^\omega$, which defines a behavior that we want to hide from an observer[2]. Thus, we seek an (I/O)-transducer that realizes L_φ without revealing the truth value of L_ψ in the generated computations. Keeping the truth value of L_ψ secret is done by hiding the truth value of some signals in $I \cup O$. Before we define synthesis with privacy formally, we first need some notations.

---

[2] See Remark 2.3 for an extension of the setting to multiple and conditional secrets.

Consider a set $\mathcal{H} \subseteq I \cup O$ of *hidden signals*. Let $\mathcal{V} = (I \cup O) \setminus \mathcal{H}$ denote the set of *visible signals*. For an assignment $\sigma \in 2^{I \cup O}$, let $\text{hide}_{\mathcal{H}}(\sigma) \in 2^{\mathcal{V}}$ be the restriction of σ to the visible signals. That is, $\text{hide}_{\mathcal{H}}(\sigma)(v) = \sigma(v)$ for all $v \in \mathcal{V}$. Also, let $\text{noise}_{\mathcal{H}}(\sigma) \subseteq 2^{I \cup O}$ be the set of assignments that differ from σ in assignments to the signals in $\mathcal{H}$. Thus, $\text{noise}_{\mathcal{H}}(\sigma) = \{\sigma' \in 2^{I \cup O} : \sigma \cap \mathcal{V} = \sigma' \cap \mathcal{V}\}$. Then, for an infinite computation $w = \sigma_0 \cdot \sigma_1 \cdots \in (2^{I \cup O})^{\omega}$, we have that $\text{hide}_{\mathcal{H}}(w) = \text{hide}_{\mathcal{H}}(\sigma_0) \cdot \text{hide}_{\mathcal{H}}(\sigma_1) \cdots \in (2^{\mathcal{V}})^{\omega}$ and $\text{noise}_{\mathcal{H}}(w)$ is the set of all computations that differ from w in assignments to the signals in $\mathcal{H}$. Formally, $\sigma_0' \cdot \sigma_1' \cdots \in \text{noise}_{\mathcal{H}}(w)$ iff $\sigma_i' \in \text{noise}_{\mathcal{H}}(\sigma_i)$ for all $i \geq 0$. Note that for all $w, w' \in (2^{I \cup O})^{\omega}$, it holds that $w' \in \text{noise}_{\mathcal{H}}(w)$ iff $w \in \text{noise}_{\mathcal{H}}(w')$ iff $\text{hide}_{\mathcal{H}}(w') = \text{hide}_{\mathcal{H}}(w)$, and that $w \in \text{noise}_{\mathcal{H}}(w)$ for all $w \in (2^{I \cup O})^{\omega}$ and $\mathcal{H} \subseteq I \cup O$. Intuitively, when the signals in $\mathcal{H}$ are hidden, then an observer of a computation $w \in (2^{I \cup O})^{\omega}$ only knows that the computation is in $\text{noise}_{\mathcal{H}}(w)$.

Consider a specification $L_{\varphi} \subseteq (2^{I \cup O})^{\omega}$ and a secret $L_{\psi} \subseteq (2^{I \cup O})^{\omega}$. For a set $\mathcal{H} \subseteq I \cup O$ of hidden signals, we say that an (I/O)-transducer $\mathcal{T}$ $\mathcal{H}$-*hides* L_{ψ} if for all words $w_I \in (2^I)^{\omega}$, the truth value of the secret L_{ψ} in the computation $\mathcal{T}(w_I)$ cannot be deduced from $\text{hide}_{\mathcal{H}}(\mathcal{T}(w_I))$. Formally, for every $w_I \in (2^I)^{\omega}$, there exist two computations $w^+, w^- \in \text{noise}_{\mathcal{H}}(\mathcal{T}(w_I))$, such that $w^+ \in L_{\psi}$ and $w^- \notin L_{\psi}$. We say that $\mathcal{T}$ *realizes* $\langle L_{\varphi}, L_{\psi}, \mathcal{H} \rangle$ *with privacy* if $\mathcal{T}$ realizes L_{φ} and $\mathcal{H}$-hides L_{ψ}. We say that $\langle L_{\varphi}, L_{\psi}, \mathcal{H} \rangle$ is *realizable with privacy* if there exists an (I/O)-transducer that realizes $\langle L_{\varphi}, L_{\psi}, \mathcal{H} \rangle$ with privacy.

Clearly, hiding is monotone with respect to $\mathcal{H}$, in the sense that the larger $\mathcal{H}$ is, the more likely it is for an (I/O)-transducer $\mathcal{T}$ to $\mathcal{H}$-hide L_{ψ}. Indeed, if $\mathcal{T}$ $\mathcal{H}$-hides L_{ψ}, then $\mathcal{T}$ $\mathcal{H}'$-hides L_{ψ} for all $\mathcal{H}'$ with $\mathcal{H} \subseteq \mathcal{H}'$. In particular, taking $\mathcal{H} = I \cup O$, we can hide all non-trivial secrets. Hiding of signals, however, is not always possible, and may sometimes involve a cost. Formally, we consider a *hiding cost function* $\text{cost} : I \cup O \to \mathbb{N}$, which maps each signal to the cost of hiding it, and a *hiding budget* $b \in \mathbb{N}$, which bounds the cost that the system may use for hiding of signals. The cost of hiding a set $\mathcal{H} \subseteq I \cup O$ of signals is then $\text{cost}(\mathcal{H}) = \sum_{p \in \mathcal{H}} \text{cost}(p)$, and we say that $\mathcal{H}$ *respects* b if $\text{cost}(\mathcal{H}) \leq b$. Note that if $\text{cost}(p) > b$, for $p \in I \cup O$, then p cannot be hidden. Also, when $\text{cost}(p) = 1$ for all $p \in I \cup O$, we say that cost is *uniform*. Note that then, b bounds the number of signals we may hide.

Now, we say that $\langle L_{\varphi}, L_{\psi}, \text{cost}, b \rangle$ is *realizable with privacy* if there exists a set $\mathcal{H} \subseteq I \cup O$ such that $\mathcal{H}$ respects b and $\langle L_{\varphi}, L_{\psi}, \mathcal{H} \rangle$ is realizable with privacy. Finally, in the *synthesis with privacy* problem, we are given L_{φ}, L_{ψ}, cost, and b, and we have to return a set $\mathcal{H} \subseteq I \cup O$ that $\mathcal{H}$ respects b and an (I/O)-transducer $\mathcal{T}$ that realizes $\langle L_{\varphi}, L_{\psi}, \mathcal{H} \rangle$ with privacy, or determine that $\langle L_{\varphi}, L_{\psi}, \text{cost}, b \rangle$ is not realizable with privacy.

2.3   Multiple and conditional secrets

In this section we discuss two natural extensions of our setting. First, often we need to hide form the observer more than one secret. We extend the definition of synthesis with privacy to a set of secrets $S = \{L_{\psi_1}, L_{\psi_2}, \ldots, L_{\psi_k}\}$ in the natural way. Thus, an (I/O)-transducer $\mathcal{T}$ realizes $\langle \varphi, S, \mathcal{H} \rangle$ with privacy if it realizes φ and $\mathcal{H}$-hides L_{ψ_i}, for all $i \in [k]$. Note that

Then, a *conditional secret* is a pair $\langle L_\theta, L_\psi \rangle$, consisting of a *trigger* and a *secret*. The truth value of the secret should be unknown only in computations that satisfy the trigger. Formally, for a set $\mathcal{H} \subseteq I \cup O$ of hidden signals, we say that an (I/O)-transducer $\mathcal{T}$ $\mathcal{H}$-*hides* $\langle L_\theta, L_\psi \rangle$ if for all input sequences $w_I \in (2^I)^\omega$ such that $\mathsf{noise}_{\mathcal{H}}(\mathcal{T}(w_I)) \subseteq L_\theta$, the truth value of L_ψ in the computation $\mathcal{T}(w_I)$ cannot be deduced from $\mathsf{hide}_{\mathcal{H}}(\mathcal{T}(w_I))$, thus there exist two computations $w^+, w^- \in \mathsf{noise}_{\mathcal{H}}(\mathcal{T}(w_I))$, such that $w^+ \in L_\psi$ and $w^- \notin L_\psi$. A useful special case of conditional secrets is when the trigger and the secret coincide, and so we have to hide the truth value of the secret only if there are computations where the value of secret is T. Formally, $\mathcal{T}$ $\mathcal{H}$-hides $\langle L_\psi, L_\psi \rangle$ if for all input sequences $w_I \in (2^I)^\omega$, there exists a computation $w^- \in \mathsf{noise}_{\mathcal{H}}(\mathcal{T}(w_I))$ such that $w^- \notin L_\psi$.

Note that unlike a collection of specifications, which can be conjuncted, hiding a set of secrets is not equivalent to hiding their conjunction. Likewise, hiding a conditional secret is not equivalent to hiding the implication of the secret by the trigger. Thus, the two variants require an extension of the solution for the case of a single or unconditional secret. In Remark 2, we describe such an extension.

2.4 Automata and LTL

An *automaton* on infinite words is $\mathcal{A} = \langle \Sigma, Q, q_0, \delta, \alpha \rangle$, where Σ is an alphabet, Q is a finite set of *states*, $q_0 \in Q$ is an *initial state*, $\delta : Q \times \Sigma \to 2^Q$ is a *transition function*, and α is an *acceptance condition*, to be defined below. For states $q, s \in Q$ and a letter $\sigma \in \Sigma$, we say that s is a σ-successor of q if $s \in \delta(q, \sigma)$. Note that we do not require the transition function to be *total*. That is, we allow that $\delta(q, \sigma) = \varnothing$. If $|\delta(q, \sigma)| \leq 1$ for every state $q \in Q$ and letter $\sigma \in \Sigma$, then $\mathcal{A}$ is *deterministic*. For a deterministic automaton $\mathcal{A}$ we view δ as a function $\delta : Q \times \Sigma \to Q \cup \{\bot\}$, where $\bot$ is a distinguished symbol, and instead of writing $\delta(q, \sigma) = \{q\}$ and $\delta(q, \sigma) = \varnothing$, we write $\delta(q, \sigma) = q$ and $\delta(q, \sigma) = \bot$, respectively.

A *run* of $\mathcal{A}$ on $w = \sigma_0 \cdot \sigma_1 \cdots \in \Sigma^\omega$ is an infinite sequence of states $r = r_0 \cdot r_1 \cdot r_2 \cdots \in Q^\omega$, such that $r_0 = q_0$, and for all $i \geq 0$, we have that $r_{i+1} \in \delta(r_i, \sigma_i)$. The acceptance condition α determines which runs are "good". We consider here the *Büchi*, *co-Büchi*, *generalized Büchi* and *parity* acceptance conditions. All conditions refer to the set $inf(r) \subseteq Q$ of states that r traverses infinitely often. Formally, $inf(r) = \{q \in Q : q = r_i \text{ for infinitely many } i\text{'s}\}$. In generalized Büchi the acceptance condition is of the form $\alpha = \{\alpha_1, \alpha_2, \ldots, \alpha_k\}$, for $k \geq 1$ and sets $\alpha_i \subseteq Q$. In a generalized Büchi automaton, a run r is accepting if for all $1 \leq i \leq k$, we have that $inf(r) \cap \alpha_i \neq \varnothing$. Thus, r visits each of the sets in α infinitely often. Büchi automata is a special case of its generalized form with $k = 1$. That is, a run r is accepting with respect to the Büchi condition $\alpha \subseteq Q$, if $inf(r) \cap \alpha \neq \varnothing$. Dually, in co-Büchi automata, a run r is accepting if $inf(r) \cap \alpha = \varnothing$. Finally, in a parity automaton, the acceptance condition $\alpha : Q \to \{1, \ldots, k\}$, for some $k \geq 1$, maps states to ranks, and a run r is accepting if the maximal rank of a state in $inf(r)$ is even. Formally, $\max_{q \in inf(r)} \{\alpha(q)\}$ is even. A run that is not accepting is *rejecting*. We refer to the number k in α as the *index* of the automaton. A word w is accepted by $\mathcal{A}$ if there is an accepting run of $\mathcal{A}$ on w. The language of $\mathcal{A}$, denoted $L(\mathcal{A})$, is the set of words that $\mathcal{A}$ accepts. Two automata are *equivalent* if their languages are equivalent.

We denote the different classes of automata by three-letter acronyms in $\{D,N\} \times \{B,C,GB,P\} \times \{W\}$. The first letter stands for the branching mode of the automaton (deterministic, nondeterministic); the second for the acceptance condition type (Büchi, co-Büchi, generalized Büchi, or parity); and the third indicates we consider automata on words. For example, NBWs are nondeterministic Büchi word automata.

LTL is a linear temporal logic used for specifying on-going behaviors of reactive systems [24]. Specifying the behavior of (I/O)-transducers, formulas of LTL are defined over the set $I \cup O$ of signals using the usual Boolean operators and the temporal operators G ("always") and F ("eventually"), X ("next time") and U ("until"). The semantics of LTL is defined with respect to infinite computations in $(2^{I \cup O})^\omega$. Thus, each LTL formula φ over $I \cup O$ induces a language $L_\varphi \subseteq (2^{I \cup O})^\omega$ of all computations that satisfy φ.

Recall that the input to the synthesis with privacy problem includes languages L_φ and L_ψ. We sometimes replace L_φ and L_ψ in the different notations with automata or LTL formulas that describe them, thus talk about realizability with privacy of $\langle \mathcal{A}_\varphi, \mathcal{A}_\psi, \mathcal{H} \rangle$ or $\langle \varphi, \psi, \mathcal{H} \rangle$, for automata $\mathcal{A}_\varphi$ and $\mathcal{A}_\psi$, or LTL formulas φ and ψ.

Example 1. Consider a scheduler that serves two users and grant them with access to a joint resource. The scheduler can be viewed as an open system with $I = \{\mathsf{req}_1, \mathsf{req}_2\}$, with req_i ($i \in \{1,2\}$) standing for a request form User i, and $O = \{\mathsf{grant}_1, \mathsf{grant}_2\}$, with grant_i standing for a grant to User i. The system should satisfy mutual exclusion and non-starvation. Formally, the specification for the system is $\varphi_1 \wedge \varphi_2 \wedge \varphi_3$, for $\varphi_1 = G((\neg\mathsf{grant}_1) \vee (\neg\mathsf{grant}_2))$, $\varphi_2 = G(\mathsf{req}_1 \rightarrow F\mathsf{grant}_1)$, and $\varphi_3 = G(\mathsf{req}_2 \rightarrow F\mathsf{grant}_2)$.

We may want to hide from an observer of the interaction the exact policy scheduling of the system. For example,[3] the secret $\psi_1 = ((\neg\mathsf{grant}_1)W\mathsf{req}_1) \wedge G(\mathsf{grant}_1 \rightarrow X((\neg\mathsf{grant}_1)W\mathsf{req}_1))$ reveals whether the system gives User 1 grants only after requests that have not been granted yet. Indeed, ψ_1 specifies that once a grant to User 1 is given, no more grants are given to her, unless a new request from her arrives. A similar secret can be specified for User 2. Note that in order to hide ψ_1, it is sufficient to hide only one of the signals req_1 or grant_1. In fact, this is true even when the observer knows the specification for the system. Then, the secret $\psi_2 = G((\mathsf{req}_1 \rightarrow \mathsf{grant}_1 \vee X\mathsf{grant}_1) \wedge (\mathsf{req}_2 \rightarrow \mathsf{grant}_2 \vee X\mathsf{grant}_2))$ reveals whether delays in grants are limited to one cycle. Here, unlike with ψ_2, it is not sufficient hiding only a single request or even both. Indeed, some policies disclose the satisfaction value of ψ_2 even when requests are hidden. For example, a system that simply alternates between grants, thus outputs $\{\mathsf{grant}_1\}, \{\mathsf{grant}_2\}, \{\mathsf{grant}_1\}, \{\mathsf{grant}_2\}, \ldots$, satisfies the specification and clearly satisfies ψ_2 regardless of the users' requests.

---

[3] The LTL operator W is "weak Until", thus $p_1 W p_2 = (p_1 U p_2) \vee G p_1$.

Consider now the secret $\psi_3 = FG(\text{req}_1 \rightarrow \text{grant}_1)$, which asserts that eventually, the requests of User 1 are always granted immediately. A system that satisfies ψ_3 is unfair to User 2. Aiming to hide this unfair behavior, we can use the conditional secret $\langle \psi_3, \psi_3 \rangle$, which requires a system that satisfies ψ_3 to hide its satisfaction.

Some computations that satisfy ψ_3, however, may still be fair to User 2. For example, if ψ_3 is satisfied vacuously or if only finitely many requests are sent from User 2, then the behavior specified in ψ_3 is fair, and we need not hide it. Accordingly, we can strengthen the trigger ψ_3 and restrict further the computations in which the satisfaction value of ψ_3 should be hidden. Formally, we replace the trigger ψ_3 by a trigger $\psi_3 \wedge \theta$, for a behavior θ in which a scheduling policy that satisfies ψ_3 is not fair (and hence, need to be hidden).

Let us consider possible behaviors θ for the conditional secret $\langle \psi_3 \wedge \theta, \psi_3 \rangle$. As discussed above, behaviors that make ψ_3 unfair are $GF\text{req}_1$, implying that ψ_3 is not satisfied vacuously, and $GF\text{req}_2$, implying that the immediate grants to User 1 are not due to no requests from User 2. Taking $\theta = (GF\text{req}_1) \wedge (GF\text{req}_2)$ results in a more precise conditional secret.

The trigger θ can be made more precise: taking $\theta = GF(\text{req}_1 \wedge \text{req}_2)$ still guarantees no vacuous satisfaction and also asserts that immediate grants to User 1 are given even when the requests of User 1 arrive together with those of User 2. In fact, $\theta = GF(\text{req}_2 \wedge (\neg\text{grant}_2)U\text{req}_1))$ is even more precise, as it excludes the possibility that the requests of User 1 arrive before those of User 2. Note that the secret can be made less restrictive too, for example with $\psi_3' = FG(\text{req}_1 \rightarrow ((\neg\text{grant}_2)U\text{grant}_1))$, which specifies that eventually, grants to User 1 are always given before grants to User 2. $\square$

Example 2. As a different example, consider a paint robot that paints parts of manufactured pieces. The set O includes 3 signals c_0, c_1, c_2 that encode 8 colors. The encoding corresponds to the composition of the color from paint in three different containers. For example, color 101 stands for the robot mixing paints from containers 0 and 2. The observer does not see the generated pattern, but, unless we hide it, may see the arm of the robot when it reaches a container. Accordingly, hiding of signals in O involve different costs.

The user instructs the robot whether to stay with the current color or change it, thus $I = \{\text{change}\}$. We seek a system that directs the robot which color to chose, in a way that satisfies requirements about the generated pattern. For example, in addition to the requirement to respect the changing instructions ($\xi_{respect}$), the specification φ may require the pattern to start with color 000, and if there are infinitely many changes, then all colors are used (ξ_{all}), yet color 000 repeats between each two colors (ξ_{repeat}). Formally,

- $\xi_{respect} = G((X\neg\text{change}) \leftrightarrow ((c_0 \leftrightarrow Xc_0) \wedge (c_1 \leftrightarrow Xc_1) \wedge (c_2 \leftrightarrow Xc_2)))$,
- $\xi_{all} = GF(\bar{c}_0 \wedge \bar{c}_1 \wedge \bar{c}_2) \wedge GF(\bar{c}_0 \wedge \bar{c}_1 \wedge c_2) \wedge \cdots \wedge GF(c_0 \wedge c_1 \wedge c_2)$,
- $\xi_{repeat} = G((c_0 \vee c_1 \vee c_2) \rightarrow X(\text{change} \rightarrow (\bar{c}_0 \wedge \bar{c}_1 \wedge \bar{c}_2)))$, and
- $\varphi = (\bar{c}_0 \wedge \bar{c}_1 \wedge \bar{c}_2) \wedge \xi_{respect} \wedge ((GF\text{change}) \rightarrow (\xi_{all} \wedge \xi_{repeat}))$.

We may want to hide from an observer certain patterns that the robot may produce. For example, the fact color 111 is used only after color 110 (with color

000 between them), the fact there are colors other than 000 that repeat without a color different from 000 between them, and more. Note that not all the signals in O need to be hidden, and that the choice of signals to hide depends on the secrets as well as the cost of hiding. $\qquad\square$

3 Solving Synthesis with Privacy

In this section we describe a solution to the problem of synthesis with privacy for LTL specifications and show that it is 2EXPTIME-complete, thus not harder than LTL synthesis. The solution is based on replacing the specification by one that guarantees the hiding of the secret. For this, we need the following two constructions.

Lemma 1. *Consider a nondeterministic automaton $\mathcal{A} = \langle 2^{I \cup O}, Q, q_0, \delta, \alpha \rangle$. Given a set $\mathcal{H} \subseteq I \cup O$, there is a transition function $\delta^{\mathcal{H}} : Q \times 2^{I \cup O} \to 2^Q$ such that the nondeterministic automaton $\mathcal{A}^{\mathcal{H}} = \langle 2^{I \cup O}, Q, q_0, \delta^{\mathcal{H}}, \alpha \rangle$ is such that $L(\mathcal{A}^{\mathcal{H}}) = \mathsf{noise}_{\mathcal{H}}(L(\mathcal{A}))$.*

Proof. Intuitively, the transition function $\delta^{\mathcal{H}}$ increases the nondeterminism of δ by guessing an assignment to the signals in $\mathcal{H}$. Formally, for $q \in Q$ and $\sigma \in 2^{I \cup O}$, we define $\delta^{\mathcal{H}}(q, \sigma) = \bigcup_{\sigma' \in \mathsf{noise}_{\mathcal{H}}(\sigma)} \delta(q, \sigma')$. It is easy to see that a word w' is accepted by $\mathcal{A}^{\mathcal{H}}$ iff there is a word w accepted by $\mathcal{A}$ such that $w' \in \mathsf{noise}_{\mathcal{H}}(w)$. $\qquad\square$

Note that while $\mathcal{A}^{\mathcal{H}}$ maintains the state space and acceptance condition of $\mathcal{A}$, it does not preserve determinism. Indeed, unless $\mathcal{H} = \varnothing$, we have that $\mathcal{A}^{\mathcal{H}}$ is nondeterministic even when $\mathcal{A}$ is deterministic. Next, in Lemma 2 we construct automata that accept computations that satisfy the specification and hide the secret when a given set of signals is hidden.

Lemma 2. *Consider two disjoint finite sets I and O, a subset $\mathcal{H} \subseteq I \cup O$, and ω-regular languages L_φ and L_ψ over the alphabet $2^{I \cup O}$. There exists a DPW $\mathcal{D}^{\mathcal{H}}_{\varphi,\psi}$ with alphabet $2^{I \cup O}$ that accepts a computation $w \in (2^{I \cup O})^\omega$ iff $w \in L_\varphi$ and there exist $w^+, w^- \in \mathsf{noise}_{\mathcal{H}}(w)$ such that $w^+ \in L_\psi$ and $w^- \notin L_\psi$.*

1. *If L_φ and L_ψ are given by LTL formulas φ and ψ, then $\mathcal{D}^{\mathcal{H}}_{\varphi,\psi}$ has $2^{2^{O(|\varphi|+|\psi|)}}$ states and index $2^{O(|\varphi|+|\psi|)}$.*
2. *If L_φ and L_ψ are given by DPWs $\mathcal{D}_\varphi$ and $\mathcal{D}_\psi$ with n_φ and n_ψ states, and of indices k_φ and k_ψ, then $\mathcal{D}^{\mathcal{H}}_{\varphi,\psi}$ has $2^{O(n_\varphi \cdot k_\varphi \cdot (n_\psi \cdot k_\psi)^2 \log(n_\varphi \cdot k_\varphi \cdot n_\psi \cdot k_\psi))}$ states and index $O(n_\varphi \cdot k_\varphi \cdot (n_\psi \cdot k_\psi)^2)$.*

Proof. We start with the case L_φ and L_ψ are given by LTL formulas φ and ψ. Let $\mathcal{A}_\varphi$, $\mathcal{A}_\psi$ and $\mathcal{A}_{\neg\psi}$ be NGBWs for L_φ, L_ψ, and $L_{\neg\psi}$. By [31], such NGBWs exist, and are of size exponential in the corresponding LTL formulas. Let $\mathcal{A}^{\mathcal{H}}_\psi$ and $\mathcal{A}^{\mathcal{H}}_{\neg\psi}$ be the NGBWs for $\mathsf{noise}_{\mathcal{H}}(L(\mathcal{A}_\psi))$ and $\mathsf{noise}_{\mathcal{H}}(L(\mathcal{A}_{\neg\psi}))$, respectively, constructed as in Lemma 1.

Now, let $\mathcal{N}^{\mathcal{H}}_{\varphi,\psi}$ be an NGBW for the intersection of the three automata $\mathcal{A}_\varphi$, $\mathcal{A}^{\mathcal{H}}_\psi$, and $\mathcal{A}^{\mathcal{H}}_{\neg\psi}$. The NGBW $\mathcal{N}$ can be easily defined on top of the product of the three automata, and hence is of size $2^{O(|\varphi|+|\psi|)}$ and index $O(|\varphi|+|\psi|)$. Observe that indeed, a word $w \in (2^{I \cup O})^\omega$ is accepted by $\mathcal{N}^{\mathcal{H}}_{\varphi,\psi}$ iff $w \models \varphi$ and there exist $w^+, w^- \in \mathsf{noise}_{\mathcal{H}}(w)$ such that $w^+ \models \psi$ and $w^- \not\models \psi$. By [29,23], determinizing $\mathcal{N}^{\mathcal{H}}_{\varphi,\psi}$ results in a DPW $\mathcal{D}^{\mathcal{H}}_{\varphi,\psi}$ with $2^{2^{O(|\varphi|+|\psi|)}}$ states and index $2^{O(|\varphi|+|\psi|)}$, and we are done.

We continue with the case L_φ and L_ψ are given by DPWs $\mathcal{D}_\varphi$ and $\mathcal{D}_\psi$. We first obtain from $\mathcal{D}_\psi$ two NBWs, $\mathcal{A}_\psi$ and $\mathcal{A}_{\neg\psi}$ for L_ψ and $L_{\neg\psi} = (2^{I \cup O})^\omega \setminus L_\psi$ respectively, and also we translate $\mathcal{D}_\varphi$ into an equivalent NBW $\mathcal{A}_\varphi$. Note that the NBWs $\mathcal{A}_\psi$ and $\mathcal{A}_{\neg\psi}$ can be defined with $O(n_\psi \cdot k_\psi)$ states, and that $\mathcal{A}_\varphi$ can be defined with $O(n_\varphi \cdot k_\varphi)$ states. We then obtain the NBWs $\mathcal{A}^{\mathcal{H}}_\psi$ and $\mathcal{A}^{\mathcal{H}}_{\neg\psi}$ by applying the construction in Lemma 1 on $\mathcal{A}_\psi$ and $\mathcal{A}_{\neg\psi}$, respectively. We then define an NBW of size $O(n_\varphi \cdot k_\varphi \cdot (n_\psi \cdot k_\psi)^2)$ for the intersection of the three NBWs $\mathcal{A}_\varphi$, $\mathcal{A}^{\mathcal{H}}_\psi$, and $\mathcal{A}^{\mathcal{H}}_{\neg\psi}$, and finally determinize it into a DPW $\mathcal{D}^{\mathcal{H}}_{\varphi,\psi}$ with $2^{O(n_\varphi \cdot k_\varphi \cdot (n_\psi k_\psi)^2 \log(n_\varphi \cdot k_\varphi \cdot n_\psi k_\psi))}$ states and index $O(n_\varphi \cdot k_\varphi \cdot (n_\psi k_\psi)^2)$. □

Remark 1. [**The size of** $\mathcal{D}^{\mathcal{H}}_{\varphi,\psi}$ **for specifications and secrets given by DBWs**] The exponential dependency of $\mathcal{D}^{\mathcal{H}}_{\varphi,\psi}$ in the DPW $\mathcal{D}_\varphi$ in the construction in Lemma 2 follows from the exponential blow up in DPW intersection [4]. When L_φ is given by a DBW $\mathcal{D}_\varphi$, we can first construct a DPW for the intersection of $\mathcal{A}^{\mathcal{H}}_\psi$ and $\mathcal{A}^{\mathcal{H}}_{\neg\psi}$, and only then take its intersection with $\mathcal{D}_\varphi$. This results in a DPW $\mathcal{D}^{\mathcal{H}}_{\varphi,\psi}$ of size exponential in $\mathcal{D}_\psi$, but only polynomial in $\mathcal{D}_\varphi$. □

We can now solve synthesis with privacy for LTL formulas.

Theorem 1. [**Synthesis with privacy, LTL**] *Given two disjoint finite sets I and O, LTL formulas φ and ψ over $I \cup O$, a cost function $\mathsf{cost} : I \cup O \to \mathbb{N}$, and a budget $b \in \mathbb{N}$, deciding whether $\langle \varphi, \psi, \mathsf{cost}, b \rangle$ is realizable with privacy is 2EXPTIME-complete.*

Proof. We start with the upper bound. Given φ, ψ, cost, and b, we go over all $\mathcal{H} \subseteq I \cup O$ such that $\mathsf{cost}(\mathcal{H}) \le b$, construct the DPW $\mathcal{D}^{\mathcal{H}}_{\varphi,\psi}$ defined in Lemma 2, and check whether $L(\mathcal{D}^{\mathcal{H}}_{\varphi,\psi})$ is realizable. Since realizability of a DPW with n states and index k can be solved in time at most $O(n^k)$ [5], the 2EXPTIME upper bound follows from $\mathcal{D}^{\mathcal{H}}_{\varphi,\psi}$ having $2^{2^{O(|\varphi|+|\psi|)}}$ states and index $2^{O(|\varphi|+|\psi|)}$.

For the lower bound, we describe a reduction from LTL synthesis with no privacy. Note that adding to a specification φ a secret T or F does not work, as an observer knows its satisfaction value. It is easy, however, to add a secret that is independent of the specification. Specifically, given a specification φ over $I \cup O$, let $O' = O \cup \{p\}$, where p is a fresh signal not in $I \cup O$. Consider the secret $\psi = p$ and a cost function with $\mathsf{cost}(p) = 0$. Clearly, an (I/O)-transducer $\mathcal{T}$ realizes φ iff the (I/O')-transducer $\mathcal{T}'$ that agrees with $\mathcal{T}$ and always assigns F to p, realizes φ and $\{p\}$-hides ψ. Conversely, for an I/O'-transducer $\mathcal{T}'$, let $\mathcal{T}$ be the (I/O)-transducer obtained from $\mathcal{T}'$ by ignoring the assignments to p.

Clearly, $\mathcal{T}'$ $\{p\}$-hides ψ. In addition, as φ does not refer to p, we have that $\mathcal{T}'$ realizes φ iff $\mathcal{T}$ realizes φ. Thus, φ is realizable iff $\langle\varphi,\psi,\mathsf{cost},0\rangle$ is realizable with privacy. $\qquad\square$

Remark 2. [**Solving privacy with multiple and conditional secrets**] Recall that for a set of secrets $S = \{\psi_1,\psi_2,\ldots,\psi_k\}$, an (I/O)-transducer $\mathcal{T}$ realizes $\langle\varphi,S,\mathcal{H}\rangle$ with privacy if it realizes φ and $\mathcal{H}$-hides ψ_i, for all $i \in [k]$. It is easy to extend Theorem 1 to the setting of multiple secrets by replacing the DPW $\mathcal{D}^{\mathcal{H}}_{\varphi,\psi}$ by a DPW obtained by determinizing the product of $\mathcal{A}_\varphi$ with automata $\mathcal{A}^{\mathcal{H}}_{\psi_i}$ and $\mathcal{A}^{\mathcal{H}}_{\neg\psi_i}$, for all $1 \le i \le k$.

As for conditional secrets, recall that a computation should satisfy the specification, and from the point of view of an observer, either the trigger is not triggered, thus $\pi \in L(\mathcal{A}^{\mathcal{H}}_{\neg\theta})$, or the secret is hidden, thus $\pi \in L(\mathcal{A}^{\mathcal{H}}_\psi) \cap L(\mathcal{A}^{\mathcal{H}}_{\neg\psi})$. Accordingly, we need to construct a deterministic automaton for $L(\mathcal{A}_\varphi) \cap (L(\mathcal{A}^{\mathcal{H}}_{\neg\theta}) \cup L(\mathcal{A}^{\mathcal{H}}_\psi) \cup L(\mathcal{A}^{\mathcal{H}}_{\neg\psi}))$. This can done by determinizing an NBW that is defined on top of the product of $\mathcal{A}_\varphi, \mathcal{A}^{\mathcal{H}}_{\neg\theta}, \mathcal{A}^{\mathcal{H}}_\psi$, and $\mathcal{A}^{\mathcal{H}}_{\neg\psi}$. $\qquad\square$

While the complexity of our algorithm is not higher than that of LTL synthesis with no privacy, it would be misleading to state that handling of privacy involves no increase in the complexity. Indeed, the algorithm involved two components whose complexity may have been dominated by the doubly exponential translation of the LTL formulas to deterministic automata:

1. A need to go over all candidate sets $\mathcal{H} \subseteq I \cup O$.
2. A need to check that the generated transducer $\mathcal{H}$-hides the secret.

In the next two sections, we isolate these two components of synthesis with privacy and show that each of them involves an exponential complexity: the first in the number of signals and the second in the size of the secret.

3.1   Hiding secrets is hard

The synthesis problem for DBWs can be solved in polynomial time. Indeed, the problem can be reduced to solving a Büchi game played on top of the specification automaton. In this section we show that synthesis with privacy is EXPTIME hard even for a given set $\mathcal{H}$ of hidden signals (in fact, even a singleton set $\mathcal{H} \subseteq I$), a trivial specification, and a secret given by a DBW.

We start by showing that $\mathcal{H}$-hiding is hard even for secrets given by DBWs.

Theorem 2. *Given two disjoint finite sets I and O, a DBW $\mathcal{D}_\psi$ over $2^{I\cup O}$, and a set $\mathcal{H} \subseteq I \cup O$ of hidden signals, deciding whether there exists an (I/O)-transducer that $\mathcal{H}$-hides $\mathcal{D}_\psi$ is EXPTIME-hard. The problem is EXPTIME-hard already when $\mathcal{H} \subseteq I$.*

Proof. We describe a polynomial-time reduction from NBW realizability, which is EXPTIME-hard [26,17]. Given an NBW $\mathcal{A}$ over $2^{I\cup O}$, we define a set of signals $\mathcal{H}$ and a DBW $\mathcal{D}_\psi$ over $2^{I\cup O\cup\mathcal{H}}$, such that $L(\mathcal{A})$ is realizable iff there exists an $((I \cup \mathcal{H})/O)$-transducer that $\mathcal{H}$-hides $L(\mathcal{D}_\psi)$.

Let $\mathcal{A} = \langle 2^{I \cup O}, Q, q_0, \delta, \alpha \rangle$. W.l.o.g, we assume that $\mathcal{A}$ has a single initial state and that every word in $(2^{I \cup O})^\omega$ has at least one rejecting run in $\mathcal{A}$. The latter can be achieved, for example, by adding a nondeterministic transition from the initial state to a rejecting sink upon any assignment $i \cup o \in 2^{I \cup O}$. Let $\mathcal{H}$ be a set of signals that encode Q. Thus, each assignment $s \in 2^{\mathcal{H}}$ is associated with a single state in Q. We refer to a letter in $2^{I \cup O \cup \mathcal{H}}$ as a pair $\langle \sigma, q \rangle \in 2^{I \cup O} \times Q$, and we view a word in $(2^{I \cup O \cup \mathcal{H}})^\omega$ as the combination $w \oplus r$, of a word $w \in (2^{I \cup O})^\omega$ with a word $r \in Q^\omega$. Formally, for $w = \sigma_0 \cdot \sigma_1 \cdots \in (2^{I \cup O})^\omega$ and $r = r_1 \cdot r_2 \cdots \in Q^\omega$, let $w \oplus r = \langle \sigma_0, r_1 \rangle \cdot \langle \sigma_1, r_2 \rangle \cdots \in (2^{I \cup O \cup \mathcal{H}})^\omega$. Then, we define $\mathcal{D}_\psi$ so that $L(\mathcal{D}_\psi) = \{w \oplus r \in (2^{I \cup O \cup \mathcal{H}})^\omega :$ the sequence $q_0 \cdot r$ is an accepting run of $\mathcal{A}$ on $w\}$. Note that since every word in $(2^{I \cup O})^\omega$ has at least one rejecting run in $\mathcal{A}$, then every word $w \in L(\mathcal{A})$ has at least one word $r^+ \in Q^\omega$ such that $w \oplus r^+ \in L(\mathcal{D}_\psi)$ and at least one word $r^- \in Q^\omega$ such that $w \oplus r^- \notin L(\mathcal{D}_\psi)$.

Formally, $\mathcal{D}_\psi = \langle 2^{I \cup O \cup \mathcal{H}}, Q, q_0, \delta', \alpha \rangle$ has the same state space and acceptance condition as $\mathcal{A}$, and it uses the Q-component of each letter in order to resolve the nondeterministic choices in $\mathcal{A}$. Thus, the transitions function $\delta' : Q \times 2^{I \cup O \cup \mathcal{H}} \to Q$ is defined as follows. For every state $q \in Q$ and letter $\langle \sigma, s \rangle \in 2^{I \cup O \cup \mathcal{H}}$, we have that $\delta'(q, \langle \sigma, s \rangle) = s$ if $s \in \delta(q, \sigma)$, and otherwise $\delta'(q, \langle \sigma, s \rangle) = \bot$. We prove that indeed $L(\mathcal{D}_\psi)$ accepts exactly all words $w \oplus r$ such that $q_0 \cdot r$ is an accepting run of $\mathcal{A}$ on w. By definition of δ', it holds that r' is a run of $\mathcal{D}_\psi$ over $w \oplus r$ iff $r' = q_0 \cdot r$, and $q_0 \cdot r$ is a run of $\mathcal{A}$ over w. Hence, a run $r' = q_0 \cdot r$ of $\mathcal{D}_\psi$ over $w \oplus r$ is accepting, iff $inf(r') \cap \alpha \neq \varnothing$, iff $r' = q_0 \cdot r$ is an accepting run of $\mathcal{A}$ over w, and we are done. $\square$

Note that in the proof of Theorem 2, we could have defined $\mathcal{H}$ so that it resolves the nondeterminism in $\mathcal{A}$ in a more concise way. In particular, if we assume that the nondeterminism degree in $\mathcal{A}$ is at most 2, then a set $\mathcal{H}$ of size 1 can resolve the nondeterminizm of δ. Hence, as NBW synthesis is EXPTIME-hard already for NBWs with branching degree 2 (this follows from the fact that a bigger branching degree can be decomposed along several transitions), EXPTIME hardness holds already when hiding a single input signal.

Theorem 3. [Synthesis with privacy, DPWs] *Given two disjoint finite sets I and O, DPWs $\mathcal{D}_\varphi$ and $\mathcal{D}_\psi$ over $2^{I \cup O}$, and a set $\mathcal{H} \subseteq I \cup O$ of hidden signals, deciding whether $\langle \mathcal{D}_\varphi, \mathcal{D}_\psi, \mathcal{H} \rangle$ is realizable with privacy is EXPTIME-complete. Moreover, hardness holds already when the specification is trivial and the secret is given by a DBW.*

Proof. For the upper bound, we solve the synthesis problem for the DPW $\mathcal{D}^{\mathcal{H}}_{\varphi, \psi}$ defined in Lemma 2. As specified there, the size of $\mathcal{D}^{\mathcal{H}}_{\varphi, \psi}$ is exponential in the size and index of both $\mathcal{D}_\varphi$ and $\mathcal{D}_\psi$, and its index is polynomial in the size and index of $\mathcal{D}_\varphi$ and $\mathcal{D}_\psi$. Membership in EXPTIME then follows from the complexity of the synthesis problem for DPWs [2].

For the lower bound, fix a DBW $\mathcal{D}_T$ such that $L(\mathcal{D}_T) = (2^{I \cup O})^\omega$. Then, it is easy to see that $\langle \mathcal{D}_T, \mathcal{D}_\psi, \mathcal{H} \rangle$ is realizable with privacy iff there is an (I/O)-transducer that $\mathcal{H}$-hides $\mathcal{D}_\psi$. Thus, hardness in EXPTIME follows from Theorem 2. $\square$

3.2   Searching for a set of signals to hide is hard

Another component in the algorithm that is dominated by the doubly-exponential translation of LTL to DPWs is the need to go over all subsets of $I \cup O$ in a search for the set $\mathcal{H}$ of signals to hide. Trying to isolate the influence of this search, it is not enough to consider specifications and secrets that are given by DBWs, as the synthesis with privacy problem is EXPTIME-hard already for a given set $\mathcal{H}$, and so again, the complexity of the search is dominated by the complexity of the synthesis problem. Fixing the size of the secret, which is the source of the exponential complexity, does not not work either, as it also fixes the number of signals that we may need to hide. We address this challenge by moving to an even simpler setting for the problem, namely synthesis with privacy of a *closed* system. We are going to show that in this setting, the search for $\mathcal{H}$ is the only non-polynomial component in the algorithm.

In the closed setting, all signals are controlled by the system, namely $I = \varnothing$. Consequently, each transducer has a single computation, and realizability coincides with satisfiability. In particular, for $I = \varnothing$, we have that $\langle L_\varphi, L_\psi, \mathcal{H} \rangle$ is realizable with privacy iff there exists a word $w \in L_\varphi$, for which there exist two words $w^+, w^- \in \mathsf{noise}_{\mathcal{H}}(w)$ such that $w^+ \in L_\psi$ and $w^- \notin L_\psi$. We show that while synthesis with privacy in the closed setting can be solved in polynomial time for a given set $\mathcal{H}$ of hidden signals, it is NP-complete when $\mathcal{H}$ is not given, even when the function *cost* is uniform.

We start with the case $\mathcal{H}$ is given.

Theorem 4. *Given a finite set O of output signals, a set $\mathcal{H} \subseteq O$ of hidden signals, and DPWs $\mathcal{D}_\varphi$ and $\mathcal{D}_\psi$ over 2^O, deciding whether $\langle \mathcal{D}_\varphi, \mathcal{D}_\psi, \mathcal{H} \rangle$ is realizable with privacy can be done in polynomial time.*

Proof. First, we complement $\mathcal{D}_\psi$, which results in a DPW $\mathcal{D}_{\neg\psi}$ of the same size, and of index $k + 1$, where k is the index of $\mathcal{D}_\psi$. Then, we translate $\mathcal{D}_\varphi$, $\mathcal{D}_\psi$ and $\mathcal{D}_{\neg\psi}$ into equivalent NBWs $\mathcal{A}_\varphi$, $\mathcal{A}_\psi$ and $\mathcal{A}_{\neg\psi}$, respectively. All three NBWs can be defined in size that is polynomial in their deterministic DPW counterpart. Let $\mathcal{A}_\psi^{\mathcal{H}}$ and $\mathcal{A}_{\neg\psi}^{\mathcal{H}}$ be NBWs obtained by applying the construction in Lemma 1 on $\mathcal{A}_\psi$ and $\mathcal{A}_{\neg\psi}$, respectively. By Lemma 1, the NBWs $\mathcal{A}_\psi^{\mathcal{H}}$ and $\mathcal{A}_{\neg\psi}^{\mathcal{H}}$ have the same number of states as $\mathcal{A}_\psi$ and $\mathcal{A}_{\neg\psi}$ respectively. Let $\mathcal{N}$ be an NBW for the intersection $L(\mathcal{A}_\varphi) \cap L(\mathcal{A}_\psi^{\mathcal{H}}) \cap L(\mathcal{A}_{\neg\psi}^{\mathcal{H}})$. Note that $\mathcal{N}$ can be defined with size that is polynomial in $\mathcal{A}_\varphi$, $\mathcal{A}_\psi^{\mathcal{H}}$ and $\mathcal{A}_{\neg\psi}^{\mathcal{H}}$. Moreover, $\mathcal{N}$ accepts a word w iff $w \in L(\mathcal{A}_\varphi)$, and there exist two words $w^+, w^- \in \mathsf{noise}_{\mathcal{H}}(w)$ such that $w^+ \in L(\mathcal{A}_\psi)$ and $w^- \notin L(\mathcal{A}_\psi)$. Thus, realizability with privacy of $\langle \mathcal{A}_\varphi, \mathcal{A}_\psi, \mathcal{H} \rangle$ can be reduced to the nonemptiness of $\mathcal{N}$, which can be decided in polynomial time.   □

We continue to the case $\mathcal{H}$ should be searched.

Theorem 5. *Given a finite set of output signals O, DPWs $\mathcal{A}_\varphi$ and $\mathcal{A}_\psi$ over 2^O, a hiding cost function $\mathsf{cost} : O \to \mathbb{N}$, and a budget $b \in \mathbb{N}$, deciding whether $\langle \mathcal{A}_\varphi, \mathcal{A}_\psi, \mathsf{cost}, b \rangle$ is realizable with privacy is NP-complete. Moreover, hardness holds already when the specification and secret are given by DBWs.*

Proof. For the upper bound, a nondeterministic Turing machine can guess a set $\mathcal{H} \subseteq O$, check whether $\mathsf{cost}(\mathcal{H}) \leq b$, and, by Theorem 4, check in polynomial time whether $\langle \mathcal{A}_\varphi, \mathcal{A}_\psi, \mathcal{H} \rangle$ is realizable with privacy.

For the lower bound, we describe a polynomial-time reduction from the *vertex-cover* problem. In this problem, we are given an undirected graph $G = \langle V, E \rangle$ and $k \geq 1$, and have to decide whether there is a set $S \subseteq V$ such that $|S| \leq k$ and for every edge $\{v, u\} \in E$, we have that $E \cap S \neq \varnothing$. Given an undirected graph $G = \langle V, E \rangle$, with $E = \{e_1, e_2, \ldots, e_m\}$, we consider a closed setting with $O = V$ and construct DBWs $\mathcal{A}_\varphi$ and $\mathcal{A}_\psi$ over the alphabet 2^V such that for all $\mathcal{H} \subseteq V$, it holds that $\langle \mathcal{A}_\varphi, \mathcal{A}_\psi, \mathcal{H} \rangle$ is realizable with privacy iff $\mathcal{H}$ is a vertex cover of G. Accordingly, there is a vertex cover of size k in G iff $\langle \mathcal{A}_\varphi, \mathcal{A}_\psi, \mathsf{cost}, k \rangle$ is realizable with privacy for the uniform cost function that assigns 1 to all signals in O.

We define $\mathcal{A}_\varphi$ and $\mathcal{A}_\psi$ over the alphabet 2^V as follows. The DBW $\mathcal{A}_\varphi$ is a 2-state DBW that accepts the single word $\varnothing^\omega$. The DBW $\mathcal{A}_\psi = \langle 2^V, Q, q_1, \delta, \alpha \rangle$ for the secret is defined as follows. The set of states is $Q = \{q_1, q_2, \ldots, q_{m+1}\}$, the set of accepting states is $\alpha = \{q_{m+1}\}$ and the transition function δ is defined for all $S \subseteq O$ and $i \leq m$ by, $\delta(q_i, S) = q_{i+1}$ if $S \cap e_i \neq \varnothing$, and $\delta(q_i, S) = \bot$ otherwise. Finally, $\delta(q_{m+1}, S) = q_{m+1}$ for all $S \subseteq V$. That is, words in $L(\mathcal{A}_\psi)$ encode vertex covers of G. Indeed, if $w = S_1 \cdot S_2 \cdot S_2 \cdot \ldots \in L(\mathcal{A}_\psi)$, then for all $i \leq m$ we have that $S_i \cap e_i \neq \varnothing$. Thus, if for all $i \leq m$ we set $v_i \in V$ to be some vertex in $S_i \cap e_i$, then we get that $\{v_i, \ldots, v_m\}$ is a vertex cover of G.

In the full version, we prove that $\langle \mathcal{A}_\varphi, \mathcal{A}_\psi, \mathcal{H} \rangle$ is realizable with privacy iff $\mathcal{H}$ is a vertex cover of G.   □

4   Bounded Synthesis with Privacy

In the general synthesis problem, there is no bound on the size of the generated system. It is not hard to see that if a system that realizes the specification exists, then there is also one whose size is bounded by the size of a deterministic automaton for the specification. For the case of LTL specifications, this gives a doubly-exponential bound on the size of the generated transducer, which is known to be tight [28]. In [30], the authors suggested to study *bounded synthesis*, where the input to the problem includes also a bound on the size of the system. The bound not only guarantees the generation of a small system, if it exists, but also reduces the complexity of the synthesis problem and gives rise to a symbolic implementation and further extensions [12,19]. In particular, for LTL, it is easy to see that bounded synthesis can be solved in PSPACE, as one can go over and model-check all candidate systems. For specifications in DPW, the bound actually increases the complexity, as going over all candidates results in an algorithm in NP.

In this section we study bounded synthesis with privacy. As in traditional synthesis, the hope is to both reduce the complexity of the problem and to end up with smaller systems. In addition to a specification L_φ, a secret L_ψ, and a set $\mathcal{H} \subseteq I \cup O$ of hidden signals, we are given a bound $n \in \mathbb{N}$, represented in unary, and we are asked to construct an (I/O)-transducer with at most n states that realizes $\langle L_\varphi, L_\psi, \mathcal{H}\rangle$ with privacy, or to determine that no such transducer exists. As in the unbounded case, we can define the problem also with respect to a hiding cost function and a budget.

4.1 Hiding secrets by a bounded system is hard

We first show that hiding secrets in a bounded setting is hard. In fact, the complexity of hiding goes beyond the complexity of bounded synthesis with no privacy already in the case the specification and secrets are given by LTL formulas. In the case of DBWs and DPWs, hiding is also more complex than bounded synthesis without privacy, but the difference is not significant.

The key idea in both results is similar to the one in the proof of Theorem 2. There, we reduce realizability of NBWs to hiding of secrets given by DBWs. Essentially, we use the hidden signals to imitate nondeterminism. Here, with a bound on the size of the system, we cannot reduce from realizability, as the problem has the flavor of model checking many candidates. Accordingly, we reduce from universality, either in the form of LTL formulas with universally-quantified atomic propositions, or in the form of language-universality for NBWs.

Theorem 6. *Given two disjoint finite sets I and O, an LTL formula ψ over $2^{I \cup O}$, a set $\mathcal{H} \subseteq I \cup O$ of hidden signals, and a bound $n \geq 1$, given in unary, deciding whether there exists an (I/O)-transducer of size at most n that $\mathcal{H}$-hides ψ is EXPSPACE-hard. The problem is EXPSPACE-hard already when $n = 1$.*

Theorem 7. *Given two disjoint finite sets I and O, a DBW $\mathcal{D}_\psi$ over $2^{I \cup O}$, a set $\mathcal{H} \subseteq I \cup O$ of hidden signals, and a bound $n \geq 1$, represented in unary, deciding whether there exists an (I/O)-transducer of size at most n that $\mathcal{H}$-hides $\mathcal{D}_\psi$ is PSPACE-hard. The problem is PSPACE-hard already when $n = 1$.*

4.2 Solving bounded synthesis with privacy

We can now present the tight complexity for bounded synthesis with privacy for both types of specification formalisms. For the upper bounds, we construct an NGBW $\mathcal{N}^{\mathcal{H}}_{\varphi,\psi}$ that accepts exactly all words that satisfy φ and hide ψ, and search for an (I/O)-transducer of size n whose language is contained in that of $\mathcal{N}^{\mathcal{H}}_{\varphi,\psi}$.

Theorem 8. [Bounded synthesis with privacy] *Given two disjoint finite sets I and O, specification L_φ and secret L_ψ over $I \cup O$, a set $\mathcal{H} \subseteq I \cup O$ of hidden signals, and a bound $n \in \mathbb{N}$, represented in unary, deciding whether there is an (I/O)-transducer with at most n states that realizes $\langle L_\varphi, L_\psi, \mathcal{H}\rangle$ with privacy is PSPACE-complete for L_φ and L_ψ given by DPWs, and is EXPSPACE-complete for L_φ and L_ψ given by LTL formulas. Hardness in PSPACE holds already for DBWs.*

5   When the Observer Knows the Specification

In this section we study a setting in which the observer knows the specification φ of the system. Technically, it means that when the observer tries to evaluate the secret, she knows that only computations that satisfy φ should be taken into account. If, for example, $\varphi \to \psi$, then ψ cannot be kept private in a setting in which the observer knows φ. Indeed, the fact φ is realized by the system reveals that ψ is satisfied. Formally, we say that $\mathcal{T}$ *realizes* $\langle \varphi, \psi, \mathcal{H} \rangle$ *with privacy under the knowledge of the specification* if $\mathcal{T}$ realizes φ, and for every $w_I \in (2^I)^\omega$, there exist $w^+, w^- \in \mathsf{noise}_{\mathcal{H}}(\mathcal{T}(w_I)) \cap L_\varphi$ such that $w^+ \models \psi$ and $w^- \not\models \psi$. Thus, the satisfaction of the secret ψ in a computation $\mathcal{T}(w_I)$ cannot be deduced from the observable computation $\mathsf{hide}_{\mathcal{H}}(\mathcal{T}(w_I))$ even when the observer knows that φ is satisfied in $\mathcal{T}(w_I)$. The adjustment for the definition of the problem with respect to a hiding cost function and a budget is similar.

We start by showing the analogue of Lemma 2 for the setting in which the observer knows the specification. The construction is similar to that of Lemma 2, except that now, the construction of the DPW $\mathcal{D}^{\mathcal{H}}_{\psi|\varphi}$ involves an existential projection on $\mathcal{H}$ also in the automaton for the specification. Accordingly, the size of the DPW is exponential in both the specification and the secret even in the case they are given by DBWs.

Lemma 3. *Consider two disjoint finite sets I and O, a subset $\mathcal{H} \subseteq I \cup O$, and regular languages L_φ and L_ψ over the alphabet $2^{I \cup O}$. There exists a DPW $\mathcal{D}^{\mathcal{H}}_{\psi|\varphi}$ with alphabet $2^{I \cup O}$ that accepts a computation $w \in (2^{I \cup O})^\omega$ iff $w \in L_\varphi$ and there exist $w^+, w^- \in \mathsf{noise}_{\mathcal{H}}(w)$ such that $w^+ \in L_\varphi \cap L_\psi$ and $w^- \in L_\varphi \setminus L_\psi$.*

1. *If L_φ and L_ψ are given by LTL formulas φ and ψ, then $\mathcal{D}^{\mathcal{H}}_{\psi|\varphi}$ has $2^{2^{O(|\varphi|+|\psi|)}}$ states and index $2^{O(|\varphi|+|\psi|)}$.*
2. *If L_φ and L_ψ are given by DPWs $\mathcal{D}_\varphi$ and $\mathcal{D}_\psi$ with n_φ and n_ψ states, and of indices k_φ and k_ψ, then $\mathcal{D}^{\mathcal{H}}_{\psi|\varphi}$ has $2^{O((n_\varphi \cdot k_\varphi)^3 \cdot (n_\psi \cdot k_\psi)^2 \log(n_\varphi \cdot k_\varphi \cdot n_\psi \cdot k_\psi))}$ states and index $O((n_\varphi \cdot k_\varphi)^3 \cdot (n_\psi \cdot k_\psi)^2)$.*

Lemma 3 implies that all the asymptotic upper bounds described in Section 3 are valid also in a setting with an observer that knows the specification. Also, as the lower bounds in Theorems 1 and 3 involve secrets that are independent of the specification, they are valid for this setting too. Two issues require a consideration:

1. The need to search for $\mathcal{H}$: the NP-hardness proof in Theorem 5 is no longer valid, as there, $\varphi \to \neg\psi$, and so the satisfaction value of the secret is revealed in a setting with an observer that knows the specification.
2. The construction in Lemma 3 results in an algorithm that is exponential also in the specification, even when given by a DBW. On the other hand, the EXPTIME-hardness proof in Theorem 2 does not imply an exponential lower bound in the specification.

Below we address the two issues, providing lower bounds for a setting in which the observer knows the specification. Matching upper bounds follow the same considerations in Theorems 3 and 5, where $\mathcal{D}^{\mathcal{H}}_{\psi|\varphi}$ replaces $\mathcal{D}^{\mathcal{H}}_{\varphi,\psi}$. We start with a variant of Theorem 5, showing NP-hardness also in the setting of a knowledgeable observer. As mentioned above, the lower bound in the proof of Theorem 5 does not work when the observer knows the specification, yet, can easily be modified to work for the case of a knowledgeable observer.

Theorem 9. *Given a set O of output signals, a cost function* cost $: O \to \mathbb{N}$, *a hiding budget $b \in \mathbb{N}$, and DBWs $\mathcal{A}_\varphi$ and $\mathcal{A}_\psi$ over 2^O, deciding whether there is $\mathcal{H} \subseteq O$, with* cost$(\mathcal{H}) \leq b$, *such that $\langle \mathcal{A}_\varphi, \mathcal{A}_\psi, \mathcal{H} \rangle$ is realizable with privacy under knowledge of the specification is NP-hard. Moreover, hardness holds already when* cost *is uniform.*

We continue to the second issue, proving that synthesis with privacy under knowledge of the specification is EXPTIME-hard even for specifications in DBWs and secrets of a fixed size. Note that synthesis with privacy (without knowledge of the specification) can be solved in PTIME in this case (see Remark 1). The proof is similar to that of Theorem 2, except that here the lower bound needs the secret to be a of a fixed size, making the specification more complex. It follows that the exponential blow-up in $\mathcal{D}_\varphi$, which exists in Lemma 3 cannot be avoided even when it is a DBW and $\mathcal{D}_\psi$ is of a fixed size.

Theorem 10. *Given two disjoint finite sets I and O, DBWs $\mathcal{D}_\varphi$ and $\mathcal{D}_\psi$ over $2^{I \cup O}$, and a set $\mathcal{H} \subseteq I \cup O$ of hidden signals, deciding whether $\langle \mathcal{D}_\varphi, \mathcal{D}_\psi, \mathcal{H} \rangle$ is realizable with privacy under knowledge of the specification is EXPTIME-hard already when $\mathcal{D}_\psi$ is of fixed size.*

Remark 3. Recall that an observer that knows the specification can restrict the search for computations on which she evaluates the secret to ones that satisfy the specification. In fact, the observer can do better, and restricts the search to computations that are generated by an (I/O)-transducer that realizes the specification.

In order to see the difference between the two definitions, consider the case where $I = \mathcal{H} = \{p_1, p_2\}$, $O = \{q\}$, $\varphi = (q \leftrightarrow p_1) \vee Gp_2$, and $\psi = p_1$. An observer that knows φ does not know which of its two disjuncts is satisfied, and thus, even though she observes q, the value of p_1 stays secret. Formally, a transducer that realizes $q \leftrightarrow p_1$ $\mathcal{H}$-hides ψ from the observer, even if the observer knows that φ is satisfied. Indeed, for every observable computation $\kappa \in 2^{\{q\}}$, there is a computation $w^+ \in$ noise$_\mathcal{H}(\kappa)$ that satisfies $p_1 \wedge Gp_2$ and a computation $w^- \in$ noise$_\mathcal{H}(\kappa)$ that satisfies $(\neg p_1) \wedge Gp_2$. Hence, $\langle \varphi, \psi, \mathcal{H} \rangle$ is realizable with privacy even when the observer knows the specification.

On the other hand, a clever observer, especially one that has read [16,14], knows that a transducer $\mathcal{T}$ realizes φ iff $\mathcal{T}$ realizes $q \leftrightarrow p_1$. Indeed, if $\mathcal{T}$ does not satisfy $q \leftrightarrow p_1$, then φ is not satisfied in computations that do not satisfy Gp_2, which is the case for almost all the computations of $\mathcal{T}$. Accordingly, a clever observer that knows φ can learn the secret p_1 by observing the value of q.

Using the terminology of [16,14], the specification φ and $q \leftrightarrow p_1$ are *open equivalent*: for every transducer $\mathcal{T}$, we have that $\mathcal{T}$ realizes φ iff $\mathcal{T}$ realizes $q \leftrightarrow p_1$. Note that open equivalence is weaker than equivalence. Once we can simplify a specification to an open-equivalent specification that does not include *inherent vacuity*, the two definitions coincide. Such a simplification, however, requires further study. Also, as an unrealizable specification is open-equivalent to F, such a simplification is at least as complex as the realizability problem (which is also good news, as it means that an observer needs to solve a 2EXPTIME problem in order to benefit from the difference between the definitions).   □

6   Directions for Future Research

We suggested a framework for the synthesis of systems that satisfy their specifications while keeping some behaviors secret. Behaviors are kept secret from an observer by hiding the truth value of some input and output signals, subject to budget restrictions: each signal has a hiding cost, and there is a bound on the total hiding cost. Our framework captures settings in which the choice and cost of hiding are fixed throughout the computation. For example, settings with signals that cannot be hidden (e.g., alarm sound, or the temperature outside), signals that can be hidden throughout the computation with some effort (e.g., hand movement of a robot), or signals that are anyway hidden (e.g., values of internal control variables). Our main technical contribution are lower bounds for the complexity of the different aspects of privacy: the need to choose the hidden signals, and the need to hide the secret behaviors. We show that both aspects involve an exponential blow up in the complexity of synthesis without privacy.

The exponential lower bounds apply already in the relatively simple cost mechanism we study. Below we discuss possible extensions of this mechanism. In settings with a *dynamic hiding of signals*, we do not fix a set $\mathcal{H} \subseteq I \cup O$ of hidden signals. Instead, the output of the synthesis algorithm contains a transducer that describes not only the assignments to the output signals but also the choice of input and output signals that are hidden in the next cycle of the interaction. Thus, signals may be hidden only in segments of the interaction – segments that depend on the history of the interaction so far. For example, we may hide information about a string that is being typed only after a request for a password. In addition, the cost function need not be fixed and may depend on the history of the interaction too. For example, hiding the location of a robot may be cheap in certain sections of the warehouse and expensive in others. Solving synthesis with privacy in a setting with such dynamic hiding and pricing of signals involves automata over the alphabet $3^{I \cup O}$, reflecting the ability of signals to get an "unknown" truth value in parts of the computation. Moreover, as the cost is not known in advance (even when the cost of hiding signals is fixed), several mechanisms for bounding the budget are possible (energy, mean-payoff, etc. [3,6]).

References

1. B. Barak, O. Goldreich, R. Impagliazzo, S. Rudich, A. Sahai, S.P. Vadhan, and K. Yang. On the (im)possibility of obfuscating programs. *J. ACM*, 59(2):6:1–6:48, 2012.
2. R. Bloem, K. Chatterjee, and B. Jobstmann. Graph games and reactive synthesis. In *Handbook of Model Checking.*, pages 921–962. Springer, 2018.
3. A. Bohy, V. Bruyère, E. Filiot, and J-F. Raskin. Synthesis from LTL specifications with mean-payoff objectives. In *Proc. 19th Int. Conf. on Tools and Algorithms for the Construction and Analysis of Systems*, volume 7795 of *Lecture Notes in Computer Science*, pages 169–184. Springer, 2013.
4. U. Boker. Why these automata types? In *Proc. 22nd Int. Conf. on Logic for Programming Artificial Intelligence and Reasoning*, volume 57 of *EPiC Series in Computing*, pages 143–163, 2018.
5. C.S. Calude, S. Jain, B. Khoussainov, W. Li, and F. Stephan. Deciding parity games in quasipolynomial time. In *Proc. 49th ACM Symp. on Theory of Computing*, pages 252–263, 2017.
6. K. Chatterjee and L. Doyen. Energy parity games. In *Proc. 37th Int. Colloq. on Automata, Languages, and Programming*, pages 599–610, 2010.
7. K. Chatterjee, L. Doyen, T. A. Henzinger, and J-F. Raskin. Algorithms for ω-regular games with imperfect information. In *Proc. 15th Annual Conf. of the European Association for Computer Science Logic*, volume 4207 of *Lecture Notes in Computer Science*, pages 287–302, 2006.
8. M.R. Clarkson, B. Finkbeiner, M. Koleini, K.K. Micinski, M.N. Rabe, and C. Sánchez. Temporal logics for hyperproperties. In *3rd International Conference on Principles of Security and Trust*, volume 8414 of *Lecture Notes in Computer Science*, pages 265–284. Springer, 2014.
9. I. Dinur and K. Nissim. Revealing information while preserving privacy. In *Proceedings of the 22nd ACM Symposium on Principles of Database Systems*, pages 202–210. ACM, 2003.
10. J. Dubreil, Ph. Darondeau, and H. Marchand. Supervisory control for opacity. *IEEE Transactions on Automatic Control*, 55(5):1089–1100, 2010.
11. C. Dwork, F. McSherry, K. Nissim, and A.D. Smith. Calibrating noise to sensitivity in private data analysis. *J. Priv. Confidentiality*, 7(3):17–51, 2016.
12. R. Ehlers. Symbolic bounded synthesis. In *Proc. 22nd Int. Conf. on Computer Aided Verification*, volume 6174 of *Lecture Notes in Computer Science*, pages 365–379. Springer, 2010.
13. B. Finkbeiner, C. Hahn, P. Lukert, M. Stenger, and L. Tentrup. Synthesis from hyperproperties. *Acta Informatica*, 57(1-2):137–163, 2020.
14. D. Fisman, O. Kupferman, S. Sheinvald, and M.Y. Vardi. A framework for inherent vacuity. In *4th International Haifa Verification Conference*, volume 5394 of *Lecture Notes in Computer Science*, pages 7–22. Springer, 2008.
15. S. Garg, C. Gentry, S. Halevi, M. Raykova, A. Sahai, and B. Waters. Candidate indistinguishability obfuscation and functional encryption for all circuits. *SIAM J. Comput.*, 45(3):882–929, 2016.
16. K. Greimel, R. Bloem, B. Jobstmann, and M. Vardi. Open implication. In *Proc. 35th Int. Colloq. on Automata, Languages, and Programming*, volume 5126 of *Lecture Notes in Computer Science*, pages 361–372. Springer, 2008.
17. T.A. Henzinger, S.C. Krishnan, O. Kupferman, and F.Y.C. Mang. Synthesis of uninitialized systems. In *Proc. 29th Int. Colloq. on Automata, Languages, and*

Programming, volume 2380 of *Lecture Notes in Computer Science*, pages 644–656. Springer, 2002.

18. O. Kupferman and O. Leshkowitz. Synthesis of privacy-preserving systems. In *Proc. 42nd Conf. on Foundations of Software Technology and Theoretical Computer Science*, volume 250 of *Leibniz International Proceedings in Informatics (LIPIcs)*, pages 42:1–42:23, 2022.

19. O. Kupferman, Y. Lustig, M.Y. Vardi, and M. Yannakakis. Temporal synthesis for bounded systems and environments. In *Proc. 28th Symp. on Theoretical Aspects of Computer Science*, pages 615–626, 2011.

20. O. Kupferman and A. Rosenberg. The blow-up in translating LTL to deterministic automata. In *Proc. 6th Workshop on Model Checking and Artificial Intelligence*, volume 6572 of *Lecture Notes in Artificial Intelligence*, pages 85–94. Springer, 2010.

21. O. Kupferman and M.Y. Vardi. Synthesis with incomplete information. In *Advances in Temporal Logic*, pages 109–127. Kluwer Academic Publishers, 2000.

22. O. Kupferman and M.Y. Vardi. From linear time to branching time. *ACM Transactions on Computational Logic*, 6(2):273–294, 2005.

23. N. Piterman. From nondeterministic Büchi and Streett automata to deterministic parity automata. In *Proc. 21st IEEE Symp. on Logic in Computer Science*, pages 255–264. IEEE press, 2006.

24. A. Pnueli. The temporal semantics of concurrent programs. *Theoretical Computer Science*, 13:45–60, 1981.

25. A. Pnueli and R. Rosner. On the synthesis of a reactive module. In *Proc. 16th ACM Symp. on Principles of Programming Languages*, pages 179–190, 1989.

26. M.O. Rabin. Automata on infinite objects and Church's problem. *Amer. Mathematical Society*, 1972.

27. J.H. Reif. The complexity of two-player games of incomplete information. *Journal of Computer and Systems Science*, 29:274–301, 1984.

28. R. Rosner. *Modular Synthesis of Reactive Systems*. PhD thesis, Weizmann Institute of Science, 1992.

29. S. Safra. On the complexity of ω-automata. In *Proc. 29th IEEE Symp. on Foundations of Computer Science*, pages 319–327, 1988.

30. S. Schewe and B. Finkbeiner. Bounded synthesis. In *5th Int. Symp. on Automated Technology for Verification and Analysis*, volume 4762 of *Lecture Notes in Computer Science*, pages 474–488. Springer, 2007.

31. M.Y. Vardi and P. Wolper. Reasoning about infinite computations. *Information and Computation*, 115(1):1–37, 1994.

32. Y. Wu, V. Raman, B.C. Rawlings, S. Lafortune, and S.A. Seshia. Synthesis of obfuscation policies to ensure privacy and utility. *Journal of Automated Reasoning*, 60(1):107–131, 2018.

□

Author Index

A

Accattoli, Beniamino   II-24
Almagor, Shaull   I-191, II-229
Austin, Pete   I-79
Avni, Guy   II-229

B

Baillot, Patrick   II-70
Basold, Henning   I-121
Birkmann, Fabian   I-144
Blot, Valentin   II-3
Bose, Sougata   I-79

C

Comer, Jesse   II-137
Czerner, Philipp   II-116

D

Dafni, Neta   I-191
Dal Lago, Ugo   II-70
Dowek, Gilles   II-3
Doyen, Laurent   I-34
Draghici, Andrei   II-185

E

Esparza, Javier   II-116

F

Frohn, Florian   II-206

G

Gaba, Pranshu   I-34
Geatti, Luca   II-95
Giesl, Jürgen   II-206
Goncharov, Sergey   II-47
Guha, Shibashis   I-34
Guillou, Lucie   II-250

H

Haase, Christoph   II-185
Hausmann, Daniel   I-13, I-55

J

Jacobs, Bart   I-101
Jankola, Marek   I-234

K

Kassing, Jan-Christoph   II-206
Kop, Cynthia   II-70
Krasotin, Valentin   II-116
Kupferman, Orna   I-256

L

Laarman, Alfons   I-121
Lancelot, Adrienne   II-24
Lehaut, Mathieu   I-55
Leroux, Jérôme   I-3
Leshkowitz, Ofer   I-256

M

Mansutti, Alessio   II-95
Mascle, Corto   II-250
Milius, Stefan   I-144
Montanari, Angelo   II-95

P

Piterman, Nir   I-13, I-55
Prakash, Aditya   I-212

R

Ryzhikov, Andrew   II-185

S

Sağlam, Irmak   I-13
Samuelson, Richard   I-166
Santamaria, Alessio   II-47
Saville, Philip   II-160
Schmuck, Anne-Kathrin   I-13
Schröder, Lutz   II-47
Shamash Halevy, Naama   I-256
Sinclair-Banks, Henry   II-229
Stein, Dario   I-166
Strejček, Jan   I-234

N. Kobayashi and J. Worrell (Eds.): FoSSaCS 2024, LNCS 14574, pp. 279–280, 2024.
https://doi.org/10.1007/978-3-031-57228-9

T
ten Cate, Balder   II-137
Totzke, Patrick   I-79
Traversié, Thomas   II-3
Tsampas, Stelios   II-47

U
Urbat, Henning   I-144, II-47

V
Vale, Deivid   II-70
Villoria, Alejandro   I-121

W
Waldburger, Nicolas   II-250
Winterhalter, Théo   II-3

Y
Yeshurun, Asaf   II-229

Printed in the United States
by Baker & Taylor Publisher Services